D0910106

LOST IN SPACE

LOST
IN
SPACE

The Fall of NASA and the Dream of a New Space Age

GREG KLERKX

Secker & Warburg
LONDON

Published by Secker & Warburg 2004

First published in the United States in 2004 by Pantheon Books, New York

2 4 6 8 10 9 7 5 3 1

First published in Great Britain in 2004 by
Secker & Warburg
Random House, 20 Vauxhall Bridge Road,
London SW1V 2SA

Random House Australia (Pty) Limited
20 Alfred Street, Milsons Point, Sydney,
New South Wales 2061, Australia

Random House New Zealand Limited
18 Poland Road, Glenfield,
Auckland 10, New Zealand

Random House (Pty) Limited
Endulini, 5A Jubilee Road, Parktown 2193, South Africa

The Random House Group Limited Reg. No. 954009
www.randomhouse.co.uk

A CIP catalogue record for this book is available from the British Library

ISBN 0 436 20602 1

Papers used by Random House are natural, recyclable products made from
wood grown in sustainable forests; the manufacturing processes conform
to the environmental regulations of the country of origin

Printed and bound in Great Britain by Mackays of Chatham

For Marina—
may the journey never end

Contents

Prologue: Lost in Space   5

The Price of "Peace"   20

One Giant Leap   59

Escape Velocity   91

Back to the Future   115

The Belly of the Beast   142

Advance Reservations Required   174

Business in a Vacuum   209

Welcome to the Revolution   238

The Emperor of Mars   270

Mars on Earth   301

Lighting Out for the Territory   336

Reference Notes   357

Acknowledgments   373

Index   375

LOST IN SPACE

"It's only fair to warn you, right at the start,
that this is a story with no ending."

—from *The Other Side of the Sky,*
Arthur C. Clarke, 1957

Prologue: Lost in Space

When I was a boy, I wanted to be an astronaut.

It was an uncomplicated aspiration. I would ride the rocket and wear the space suit. I would gaze out through the fishbowl helmet into a star-flecked black ocean as the Earth spun blue-and-white below me. I would discover things. I would be a hero.

These simple ideas formed the landscape of my ambition, one that seemed to hold limitless possibilities. It was a common dream, as mine was perhaps the first generation to have space incorporated into its earliest hopes with any real chance of achieving that. By the time I was old enough to have idols, astronauts were at the top of the heap.

Astronauts, of course, came out of NASA—to a young boy, it was almost as if they were bred there like exotic plants. NASA was where space began; it was a miracle factory, a glorious place somewhere in Texas or Florida that somehow translated science fiction into reality. It would be years before I understood that NASA was a government agency created to fight a somewhat esoteric battle in the Cold War. NASA's political complexities may have been unfathomable to me, but its products—rockets, spaceships, interplanetary voyages—were immediately resonant.

For me, space was a synonym for adventure. As a boy born in 1963, I had space imprinted on my child's mind through the Apollo missions and Neil Armstrong—incredible, impossible Neil Armstrong, leaving us wide-eyed and gasping as we watched a soft powder of gray-white dust lift from the moon's surface as he planted that famous step. I was six years old, yet my memory of that moment is indelible. Had I known the word then, I would have called it surreal. Was he really not on Earth anymore? How could that be?

When Huckleberry Finn announced his intent to "light out for the territory," he spoke for every boy or girl who understood that the appeal of the unknowable There is simply that it isn't Here. Space travel gave new and tantalizingly infinite possibility to such a primal desire for adventure and escape: not only could we light out for the territory, we could leave the planet altogether. Each of the many times I've re-read *Huckleberry Finn,* I've always imagined that in a different life Huck, too, would have wanted to be an astronaut.

In my youthful dreaming, it didn't occur to me that I knew very little about what astronauts actually did. But their appeal was not connected to the specifics of their job description; the space suit, the fishbowl helmet and the rocket were enough for me, much as the hat, the horse and the Colt were enough for boys of an earlier era. The tools of the astronaut's trade, like those of the cowboy's, embodied the dream—as did the enigmatic environment in which the dream became real. The *Star Trek* creator Gene Roddenberry couldn't have captured it more succinctly for an American boy: space was the final frontier.

By the time I was a teenager, the idea of becoming an astronaut was one of many imaginings that I had lifted skyward to see how it captured the light and then, for one reason or another, put away again. Talents began to shape dreams, rather than the other way around.

And yet I stayed close to the idea of space all my life. I read science fiction avidly. I subscribed to science magazines and journals. I thrilled at each space shuttle launch and could rattle off the names of astronauts as others could those of sports stars. Whenever a science-fiction film opened, I had to be among the first to see it. In time, I found myself working in the space community and even with NASA, albeit at a somewhat oblique angle, as a senior manager with the SETI Institute, a space research organization based in the heart of California's Silicon Valley.

SETI—in its long form, the search for extraterrestrial intelligence—employs a novel combination of sophisticated radio telescopes and high-end computing to scan nearby star systems for evidence of communications technology that might be the product of distant civilizations. For nearly two decades, SETI was a NASA program: during that time, the agency spent more than $60 million refining SETI instruments and search strategies. In the dog days between the end of the Apollo program and the all-consuming focus on the space shuttle and the International Space Station, SETI was a bright spot for the agency. An endeavor rooted in solid astrophysics, SETI was light on NASA's budget even as it served as the inspirational ne plus ultra of many space dreamers—one is hard-pressed to name a work of science fiction, either in print or on screen, that doesn't involve some sort of encounter with aliens.

Unfortunately, because of its pop-culture ramifications, SETI science was frequently misunderstood and often derided as a taxpayer-funded hunt for "little green men." The critics finally got their way in 1992 and NASA's SETI program was terminated. But SETI was soon revived with the support of private donors, and at the same time the SETI Institute struck an operational truce with NASA. The agency has never again funded the search for intelligent life, but NASA and SETI have coexisted for a decade (if sometimes uneasily) through a variety of related research and education projects deemed beneficial to both enterprises.

From the vantage point of the SETI Institute, I gradually came to see NASA in a different light: not as a streamlined purveyor of space-faring know-how, but as a fractious bureaucracy roiling with politics and infighting, thick with red tape and feral self-interest. Even as this new perspective took hold, though, NASA still seemed like the place where "space" began. Through SETI/NASA, I met astronauts, engineers and other Space Age legends who rekindled my boyhood dreams of space travel. Quietly, secretly, I found myself imagining with growing excitement that space might yet hold a place for me.

Not long ago, I visited friends who have three very bright children. At the time, the oldest was six, the same age I was when I watched the first human walk on another world. In the long dance of sending him to bed, I visited his room. It was a place to warm my heart, filled with posters and photographs evidencing a love of space that I could immediately appreciate: Technicolor cross sections of stars and planets, Hubble Telescope photographs of distant galaxies, the ubiquitous Eagle and Horsehead nebulae in all their goggling scale, and more.

My friend's boy knew far more about space and science than I had known at his age. He could clearly describe the difference between "rocky" planets like the Earth and gas giants like Jupiter and Saturn. He even knew how many planets had been discovered outside our solar system and could explain the complex process by which astronomers searched for and, ultimately, discovered them.

Looking around his room and hearing the articulate, informed excitement of his words, I wistfully recalled my own Space Age dreams. Here was a boy who had the makings of an astronaut, who clearly loved space. In my mind there was nothing to keep him from blasting straight from the first grade to the dazzling horizon of my own youthful aspirations. High on the possibilities I envisioned for him, I asked him what he wanted to do when he grew up.

"Maybe computers" was his reply.

Dismayed, I asked him about being an astronaut. He pursed his lips and shook his head. Unlike me at age six, he had a very clear concept of what astronauts do these days: "They fly up in the shuttle and fix stuff." His lack of further comment told me that this didn't seem like much of a career to aspire to when stacked up against the wild, wired world of computers.

In that moment, I saw his room anew. It was a place filled with "space" in the abstract. He was fascinated by the Voyager probes and by the 1997 Mars Pathfinder, which had provided the world with a virtual-reality taste of Mars—a precursor, many hoped, of the "real thing," eventual human exploration. But human space exploration was notably absent in this room. There were no splashy posters of the space shuttle, nothing of Apollo or Mercury or Gemini. And there was not a single photograph of an astronaut.

That was perhaps the starting point of this book. In the weeks to come, I found myself returning to that encounter with increasing incredulity. In one generation, it seemed, the thrill of human space travel had faded. Modern astronauts were no longer bold explorers on the edge of the unknown. They had become high-priced handymen, anonymous to the world as they launched into space in expensive repair vans to service orbiting appliances, secondary to the machines they built and operated. Humans had ceased to be adventurers in space. Human spaceflight, it seemed, had ceased to be adventurous.

Once, human spaceflight was the most exciting thing going; an edge-of-the-seat marvel accomplished with slide rules, riveted steel, fire, bravery and incredible ingenuity. Between 1961 and 1973, America alone managed ten manned missions beyond Earth orbit and landed men on the moon—this before cell phones, the Internet, personal computers, DVDs or any number of other inventions we point to as evidence of living in the most modern of times. And yet, as of this writing, no human has traveled beyond Earth orbit in three decades. None of the world's space programs—not America's, not Russia's, not Europe's, not even up-and-comers like those of the Chinese or the Indians—have concrete plans to change that fact anytime soon. Here at the beginning of the twenty-first century, we look back wistfully, sometimes with mild shock, at an era of human space exploration whose promise seems to have evaporated.

Seeing the result of such diminished expectations through the eyes of a child was more than disheartening to me. It was distressing. It is arguably the case that human spaceflight, even in its diminished form, survives only on the strength of those generations who lived through the height of its influence: those who engineered it, who paid for it, or, in the case of my generation, whose very worldviews were shaped by it. I was born during the fevered race to the moon; my parents were young adults who believed human achievement in space was vital to human survival on Earth. Implicit in such achievement was the unequivocal promise that all of us would one day get a chance to be part of it.

Such beliefs don't exist anymore, at least not in terms of a broad cultural consensus. I write this only a few months after the explosion of the space shuttle *Columbia,* lost in the course of a mission that few understood and few, to be honest, even knew was happening until its tragic end splashed across newspaper pages, television screens and web sites around the world. With *Columbia*'s wreckage still smoldering, the chorus of critics calling for the immediate end of human spaceflight began their increasingly familiar refrain. Their rationale is often compelling: human spaceflight is expensive and dangerous, and it long ago ceased to serve the obvious purpose it had during the Cold War, even if in hindsight that purpose now appears callow. Robotic explorers are more cost-effective than human ones, satellites far more useful than space stations. The conclusion of such an argument is unambiguous: we don't need humans in space anymore.

Despite such criticism, NASA will survive *Columbia* just as it survived *Challenger* seventeen years earlier. If nothing else, the shuttle program employs tens of thousands of people in critical congressional districts. Just as important, the International Space Station, the completion of which depends on the shuttle, employs tens of thousands more. Such pragmatic realties will ensure that NASA's human spaceflight program will continue.

But for how long?

Without obvious purpose or popular connection, few enterprises last forever. The Apollo generations will give way, soon, to generations for whom human spaceflight is at best mildly entertaining and at worst a burden. If human spaceflight does eventually cease, it is not hard to imagine history being rewritten altogether. Once humans stop going into space, how many years will pass—fifty? twenty?—before the popular consciousness becomes infused with the idea that human spaceflight, in any form, never actually happened at all?

That may sound impossible, yet even now an astonishing number of people believe that the moon landings—events that remain confined to the already yellowing pages of twentieth-century history—were mere special effects engineered by Hollywood. To future generations, any surviving relics of human spaceflight may be imbued with the same mystic intent many people associate with the ruins of Stonehenge. What they achieved, how they were built and, critically, what they promised may well be lost forever.

"There is always one moment in childhood," Graham Greene once wrote, "when the door opens and lets the future in." For me, and for millions of my generation, that moment was Neil Armstrong's first step onto the moon. The generations that arrived after that epic moment have not had the benefit of any human spaceflight milestone so profound and inspiring; instead, their defining events have been the loss of *Challenger* and now of *Columbia*. It is no wonder, then, that not only have the Apollo generations lost faith in the original promise of human spaceflight, but subsequent generations have never been given reason to have such faith in the first place.

I am a child of the Space Age. I grew up believing in the power of aspiration, will and technology to lift humans off the planet and to a fundamentally new experience. I saw myself journeying into Earth's orbit to attain a perspective on my life and my world that was, in the truest sense of the word, unique. I envisioned people of different nations and cultures working together to establish a meaningful presence—for research, adventure and even commerce—in Earth's orbit and on the moon, and then heading to Mars and one day, perhaps, even beyond. I believed in the potential of such activities to improve the human condition.

All of these reasons—historic and personal, intellectual and emotional—are why this book was conceived and written. Try as I might to ignore it, this future in space is still the one I want.

To get there from here, one must necessarily determine at what point the future, or at least one concept of the future, became a thing of the past. Therefore, there are two questions that propel this book. What happened to the Space Age? And how do we get it back?

Answering the second question depends largely on defining what the Space Age, at its core, was all about. Many would argue that we still live in the Space Age, and it cannot be denied that space ventures continue to produce tangible benefits for humanity. Man-made satellites allow us to communicate instantly across great distances and help monitor our planet's environment (and also help us spy on unfriendly nations). Sophisticated robots have explored many of the worlds of our solar system, and orbiting space telescopes have imaged distant galaxies, adding immeasurably to our store of scientific knowledge about the cosmos.

Yet such inventions only make it clear that we live in an era in which their presence in space informs our lives. They cannot change the fact that we no longer live in an age defined by space, which is how the term "Space Age" came to be coined in the first place. The Space Age was never really about probes or satellites. It was about humans in space.

The Space Age was a drive to push humans to the limits of their physical, intellectual and emotional ability in an environment that had taunted, challenged and inspired humanity—philosophers, scientists, artists—for millennia. It was about human beings traveling into space—physically, not by robotic or observational proxy. We thrilled

at the Apollo launches because people, like us, were going boldly
where no one had gone before. We mourned the loss of the *Columbia*
and *Challenger* crews because real people, like us, had died in pursuit
of something we might, in our braver moments, want to do ourselves.
Certainly, unmanned exploration is an essential component of any sen-
sibly organized space agenda. But whatever the merits of remotely controlled machinery, we are still a species for which direct experience,
even if interpreted by another member of our tribe, is the spark that
lights our imagination and spurs us to better ourselves.

My friend's boy was emblematic of a world that no longer sees a
human future in space: he does not, as I did, see himself living in a
Space Age. One can hardly blame him. In the sour years following
Apollo, vested interests solidified, vision took a back seat to convenience, and the drive for a cogent rationale for advancing a "human"
space future disappeared altogether. Human spaceflight has continued,
but it is unmoored to any larger cultural or even geopolitical agenda.
As such, it cannot survive.

Above all else, then, this book seeks to look unflinchingly at how
and why such diminished expectations have come to pass and, importantly, to point the way to rebuilding the dream of a more meaningful
and engaging human future in space. Rather than waiting for our children to rescue that dream, we must instead hand it back to them in a
way that inspires them to claim it for their own, even when we are gone.

The primary tools we have at hand—namely, the space shuttle
and the International Space Station—are insufficient for the task. Perhaps the most disheartening aspect of the explosion of *Columbia,* other
than the human tragedy, was that it changed very few opinions about
NASA or NASA's human spaceflight activities. Both should continue,
the polls unanimously concluded, but with no more or less vigor than
at present.

Any extreme response would have been preferable, had it been a
rousing public cry to scrap the shuttle, a national drive to pump billions into a better spacecraft, or a top-level political campaign for a
human mission to Mars. Any such consensus would have been a sign
that human spaceflight mattered. But no such indignation was forthcoming. As defined by NASA, human spaceflight remains lodged in the

room of our collective consciousness like a middling work of art; its occasional contemplation provides mild pleasure, but it isn't really anything to stir the soul.

Desire is the beginning of any endeavor, and so it was for human space travel; whatever the ideological drives of the era, the early space pioneers—Russian and American—burned with an urge to send humans into orbit, and then to the stars. Remarkable achievements occurred in astoundingly short periods of time; just prior to the beginning of the Apollo program, more Americans thought that it was far more likely that mankind would discover a cure for cancer before anyone set foot on the moon. After the moon was achieved, people assumed that Mars would come next. After that, anything seemed possible.

It would be naive to claim that what became known as the Space Race was predicated primarily on exploratory zeal—without the Cold War, the Space Age as we know it would not have happened. How it happened, though, is just as important. The Soviet space program, while technologically impressive, was yoked to a harsh and unappealing sociopolitical agenda. While certainly not short on ideology, NASA had a fundamentally democratic appeal. After all, Neil Armstrong's first words upon stepping onto the moon were not "That's one helluva triumph for America." Instead, it was one giant leap for mankind. And so it was to NASA that the world attached its space-faring ambitions.

In the end, NASA proved to have not only more money and better technology than the Soviets (at least for reaching the moon), but also a better long-range marketing strategy. As Apollo was ramping up, NASA sold a twofold promise to the world as the rationale for its long-term significance.

Part One of the promise was that NASA would build upon the Apollo missions to establish a permanent human presence on the moon, then quickly move on to sending humans to Mars and eventually even farther, as technology improved and experience informed science and engineering. Such efforts would be the beginning of an exploratory expansion not seen since European ships set out in search of better trade routes (and in space, there wouldn't be indigenous people to contend with, at least not in our solar system).

Instead, caught in the web of the nationalist politics it was a product of, NASA scrapped the Saturn moon rockets for the space shuttle, a decision that ensured that human space travel would be bound to Earth's orbit for three decades and counting. When the space station was announced in 1984, it was promoted loudly as the staging post for missions to Mars and, critically, the first outpost of an orbital commerce revolution that would create a busy hub of human enterprise around the planet. Those promises, too, were quickly unmasked as little more than hype: the station was built, first and foremost, to give the space shuttle something obvious to do.

In the face of such criticism, the typical default position of NASA and its supporters is to cast the agency as the victim of a shortsighted political system that hacked and slashed at NASA's grand plans for the future. It's not NASA's fault that the Space Age we have is so dull, goes the argument; it's a lack of public will, a dearth of political leadership and, of course, a lack of money. It is the fault of virtually every constituency you can name—except NASA.

This argument holds water to a point. Apollo may have been a technological marvel, but it was surely a political creation. Only two months after the triumphant landing of *Apollo 11,* NASA submitted to Congress and the White House an ambitious blueprint for space exploration that at its maximum tempo (meaning a lot of money spent in short order, as was done with Apollo) would have established a base on the moon by 1978 and sent the first humans to Mars by 1981. This is a mind-whirling timeline given the turgid pace that actually came to pass, but considering that America had gone from lobbing grapefruit-sized satellites into orbit to landing men on the moon in the course of about a decade, it seemed wholly realistic.

That such magnificent plans went nowhere is certainly due to a lack of political will at some level. But NASA leaders always have been painfully aware that Apollo was, in political terms, both a means and an end; that NASA wasn't simply dismantled altogether after *Apollo 11* is a testimony to how deeply its success took root in the popular consciousness. Unfortunately for NASA, though, missions to Mars served no broad political purpose, nor did moon bases. The retrospective face-saving position for NASA is that such plans and timelines were too ambitious, and perhaps that's the case. Then again, the early consensus about the Apollo missions was that sending humans to the moon before 1970 was an impossible goal. Could an Apollo-level commitment have put humans on Mars only a decade after achieving the moon? The truth is, we'll never know.

After being stung repeatedly by the political system that created it, NASA shifted its strategy from creating exciting agendas of exploration to squeezing itself into whatever political agendas already existed. In time, the agency became uncannily good at anticipating these agendas, tailoring NASA programs to order in advance.

Whatever sympathy one might have for NASA, what the agency has never done is to fundamentally reinvent itself; its political minders have never insisted on such reinvention. There has been no shortage of commissions, reports, analyses, congressional testimonies, speeches and back-room briefings offering suggestions for NASA's overhaul. Most of these have attached NASA's future to greater interaction with the private sector and, particularly, the idea of regularizing human space travel—Part Two of the original Space Age promise.

The idea of spaceflight as a defining shift in the human experience was predicated not on cheerleading, but on participation. The Space Age promised, in deed and in word, that everyone at some point would have a chance to fulfill the age-old dream of traveling into space. NASA boasted that the space shuttle not only would revolutionize space flight, it would alter terrestrial transport by routinizing point-to-point travel through space—New York to Paris in an hour, or less. Hilton was designing "spacehotels," much like the one in Stanley Kubrick's *2001: A Space Odyssey,* while Pan Am signed up "space tourists" for spacecraft it planned to build with NASA's help. Pan Am had tens of thousands of takers, among them a popular Hollywood actor named Ronald Reagan—who as president would green-light NASA's space station, perhaps in the secret hope that his children or grandchildren could make use of his ticket.

Even now, with the space shuttle in danger of becoming the Edsel of space transportation and the International Space Station little more than an orbital trailer home, there is a large segment of the population that would willingly risk money and life for space travel. In October 2002, an exhaustive study by Futron, the well-known space-industry analyst, found that if it were openly available today, human spaceflight would draw fifteen thousand customers willing and—critically—able to pay up to $100,000 for the experience. That price tag is not much more than the cost of a guided climb on Mount Everest.

This response refers to a suborbital spaceflight—one that offers a few minutes of weightlessness and some spectacular views, a flight that essentially goes up and down without achieving orbit—a feat roughly equivalent to Alan Shepard's first Mercury flight more than four decades ago. Orbital spaceflights are necessarily more challenging and more costly. Still, the Futron poll suggested that at least hundreds, possibly thousands, of qualified and interested customers would exist even though the price tag for orbital flight jumped into the millions.

The Futron poll reflects only North America's interest. Yet similar polls in the United Kingdom, Germany, Japan and several other countries have yielded much the same result. All in all, this kind of enthusiasm for a more personal approach to human spaceflight has been remarkably consistent through at least a dozen such polls conducted over the past decade, even with potentially off-putting tragedies like the *Challenger* explosion. The most extensive poll ever conducted on the subject of popular human spaceflight, commissioned and funded by NASA itself in 2001, was no different in its basic conclusions.

Needless to say, Part Two of the Space Age promise has failed to materialize. Human spaceflight has never moved past the status of a demonstration sport; none of the fifteen thousand people who might be willing and able to go into space right now have done so, because the vehicles to carry them have not been built. Such vehicles have not been built because the bulk of the technology, funding and political influence needed to build them has been wrapped up in the shuttle and space station. In this two-headed albatross, NASA and its chief contractors, Boeing and Lockheed Martin, have what amounts to a state-sponsored market monopoly—with the market in this case being near-Earth human spaceflight. The situation is akin to one in which Microsoft became not only the owner and maker of every computer software program in the world, but also the owner and maker of every computer.

The typical NASA spin on what is derisively dubbed space tourism is that it reduces the noble enterprise of human spaceflight to mere jun-keteering for the rich. However, that is only the most cynical perspec-tive. Properly managed and encouraged, popular human spaceflight could serve as a means of engaging popular interest in space in a way that no NASA project is able to do. Underlying NASA's reticence is its fundamental disdain for sullying the grand spaceflight enterprise with the brassy sheen of commerce, but this again is backward thinking. Was Charles Lindbergh any less an inspiration to millions because he was, to put it bluntly, an aerial privateer chasing a hefty cash prize?

In that light, it is instructive to compare the different trajectories of popular human spaceflight and the commercial aviation industry. In 1935, not even forty years after the Wright brothers pioneered pow-ered flight at Kitty Hawk, Pan Am sent the Martin M-130 *China Clipper* into mail service across the Pacific Ocean. Passenger service followed the next year, flying an island-hopping route to Hong Kong that took several days. In 1939 Pan Am turned to the Atlantic, and on June 28 the *Dixie Clipper* carried twenty-two people from Port Washington, New York, to Marseilles, France. The fare was $675 round-trip, or about $4,000 in today's currency.

As routes expanded and drew more customers, airports, resorts and entire communities rose up to meet the needs of what quickly became a thriving commercial industry. By the end of World War II, thousands of people traveled around the world regularly in machines that were, in their day, nearly as technologically complex and as risk-prone as mod-ern spacecraft. Today, millions of people travel by commercial airliner each day, a reality that would have been truly unbelievable for anyone living at the beginning of the twentieth century.

Human spaceflight has never melded government and free-market interests the way commercial aviation did, largely due to NASA's active disdain for such a fusion of ideas. The result is that human spaceflight now happens via expensive, overcomplicated, irreplaceable govern-ment-sponsored vehicles—or, in the case of Russia, in tiny metal pods that have changed very little in nearly half a century. It's as if aviation never moved past the barnstorming days of the early twentieth century that so thrilled the public and made people eager to go along for the ride. (Lindbergh himself was a barnstormer before he was a transatlantic hero.) Had aviation never progressed beyond handcrafted vehicles operated by a small elite of crack pilots, no doubt the barnstorming crowds would have thinned out quickly.

Much like the turn-of-the-century aviation industry, the space industry has spawned a legion of entrepreneurs who believe that competition would improve both the cost and reliability of spacecraft, much as competition improved the cost and reliability of aircraft, automobiles and most other technologies once considered too risky for broader consumption. They are believers in both parts of the Space Age promise; importantly, they believe that achieving the goal of opening near-Earth human spaceflight would help realize NASA's plans for sending humans once again on great voyages of cosmic exploration. They want NASA's help; in return, they want to help NASA.

But NASA has failed to grasp the potential of popular spaceflight as a means of revitalizing public interest in, and support for, the agency's work. Perhaps this is the case because NASA has never grasped that it has a responsibility for laying the groundwork for a broader private-sector space infrastructure that would allow popular human spaceflight to blossom. In fact, NASA has worked against such a paradigm change whenever possible—a charge that is less surprising considering that the dawning of a near-Earth space-travel industry would necessarily mean the end of NASA's virtual monopoly on human spaceflight.

For all the billions of dollars spent, for all the media hype, for all the NASA spin, it cannot be denied that we are no longer living in anything resembling what we thought would be the Space Age. There are no passenger spacecraft, no orbiting platforms for business or pleasure. There is no human spaceflight at all that anyone would call ordinary. No one has returned to the moon; no human has gone to Mars. The human spaceflight programs that do exist are marred by foggy goals, ideological baggage and Rube Goldberg machinery.

Yet books are still written, films are still made and energy (and money) is still spent on pursuing something bigger, something grander, than what has come to pass for humans in space. None of these efforts of the imagination see the future in terms of the present; in none of the hundreds of fictional accounts of the future I've read, watched or listened to does NASA rule the galaxy; nor do the Russians, Chinese or Europeans.

The vision of a human future in space has always been just that—a *human* future. If the spark of such a future weren't still alive, no one would continue to write about it, or give it much thought at all. But the clock is ticking: human spaceflight needs to turn a corner soon, or face being consigned to history or, worse, oblivion. Given that, the only way forward is to put away the past, tear down the paradigm we have saddled ourselves with, and start over. When that happens, who can say what might be possible?

The Price of "Peace"

Our twin-prop Brasilia leapt off the runway like a startled bird, the turbocharged engines groaning as they struggled to pull the plane upward through the steamy funk that passes for air in the late-summer tropics. We had only a few seconds to watch the Fijian capital city of Suva—a hodgepodge of tin-roofed jungle, architectural brutalism—melt from view like a hazy ghost town. Then the world outside went white, and we became part of the clouds.

In the low-pressure turbulence the Brasilia popped and bobbed like a bathtub toy, but we sat in contented silence, happy to be airborne. It had been raining all day, sometimes in scattered wisps of mist, more often in pounding sheets of practically solid liquid. Our group, about thirty in all, had traveled for nearly three hours by bus from our base outside of the town of Nadi, a creaky tourist trap on the rounded curve of Fiji's southwest side. The rain had made for slow, treacherous driving on the island's snaking roads, which seemed to be losing the battle against the jungle's tireless campaign to reclaim them. In the end, the roads and the jungle had cost us time, a commodity we had only in the sparest quantity.

Nevertheless, here we were. We had traveled from all over the world and spent an anxious week in Fiji for a chance at witnessing two hundred seconds of history. As our plane roared through the late afternoon sky, none of us were thinking about the hoops we'd jumped through to shoehorn ourselves into a small plane buzzing southeast toward what one member of our group had dubbed the loneliest place on the planet. Instead, we were thinking about Mir.

The Brasilia was on course to a position where we could observe the aging Russian space station as it plummeted from the sky. For more than two years, space engineers around the world had done computer modeling and other scientific soothsaying to anticipate what Mir would do when it plowed into the upper atmosphere at seven times the speed of sound. As we took off from Suva that afternoon, there was general agreement that the models and predictions amounted to a collective shrug: Mir was a 140-ton Tinkertoy the size of a jumbo jet, massive and unpredictable. It was more than twice as large as the second-largest man-made object ever to return to Earth, the American Skylab space station, which in 1979 had skipped down from orbit slightly askew and accidentally rained debris across the Australian outback, mercifully scarring nothing but the landscape. Still, even Skylab offered only a limited precedent to Mir's reentry. The only certainty was that once Mir reached an altitude of about 130 miles above the Earth there would be no turning back. Mir would come home.

We were to fly parallel to a rectangular swath of ocean that stretched southeast from Fiji roughly toward Argentina. This was to be Mir's "debris footprint," a mostly empty patch of water with no populated landmasses, no shipping lanes and only a smattering of fishing boats whose captains had been duly warned of Mir's pending return. It was here that the Russians would try to dump what was left of Mir after it burned its way through the Earth's heavy blanket of air, which captures an astonishing one hundred tons or so of space junk each year and incinerates it before a trace can touch terra firma. Again, Mir was a different story. Because of its size, what was left of the station after reentry could be considerable; up to forty tons of metal that would be traveling faster than a rifle bullet, even after being slowed by the atmosphere.

This is exactly what we hoped to see. Our flight plan had been recalculated dozens of times, and the very date of Mir's demise—March 23, 2001—was finalized only two days before we took off from Suva. With real-time updates from both NASA and Russian space centers, we would be able to position our planes (a smaller Bandeirante would trail our Brasilia to increase the coverage zone) for an unsurpassed view of one of the most spectacular cosmic events of the century. We would be six miles up and only, at most, a few hundred miles away when Mir exploded in the mother of all modern fireworks shows.

The timing of Mir's final descent was bittersweet. Just the previous month, the station had celebrated its fifteenth year in orbit, three times its intended operational lifetime; undeniably, Mir was humanity's first long-term outpost in space. Yet it was hard to argue with Russia's decision to retire Mir. Pummeled by micrometeorites and even rammed by other spacecraft, Mir had come to be perceived as something of an orbiting death trap; a rickety, patchwork jalopy held together by hope, popular will and, famously, chewing gum. With the cash-strapped Russians committed as key partners in the NASA-led International Space Station, the official statement from the Russian government was that now was the time to retire Mir with grace and dignity.

But all of us on the Brasilia knew that Russia had not willingly decided to end its space station's career. In the year prior to Mir's termination, Russia had nearly succeeded in revitalizing Mir through an extensive capitalist makeover. The result would have been the world's first commercial space station. The possibility for such a stunning revival was tantalizingly real. There were signed contracts on the table, customers at the door and millions of dollars in the bank with hints of more to follow.

To anyone interested in the future of human spaceflight, the effort to turn Mir into a privately run, market-driven human-spaceflight platform was something truly new and exciting. In time, hoped Mir's backers, a successfully commercialized Mir would spark other entrepreneurial ventures—passenger space travel, media events, perhaps even the establishment of orbital recreational facilities—that might lead to more such ventures, to better and cheaper access to orbit, and to yet more people in space.

In time, a thriving human marketplace in Earth orbit would create the foundation for more ambitious voyages—back to the moon and on to Mars, voyages that would almost certainly require public- and private-sector partnership and international cooperation. Gone were the days when interplanetary human spaceflight could be sustained on the back of a single nation. A different kind of popular buy-in was needed if something as grand as Apollo was ever to happen again.

But Mir would not survive to justify the optimism attached, perhaps too hastily, to the plans for its revitalization. Long before we headed to

Fiji, it was clear that even if there had been millions of dollars more invested in Mir's makeover, money alone was never enough to save the station. Whatever else it promised, the possibility of a commercialized Mir also threatened a comfortable status quo. That fact spelled the station's doom.

In the wake of Mir's demise, however, the status quo would suffer body blows in the form of a diminutive California millionaire determined to fulfill a boyhood dream, the diminishing relevance (and growing expense) of the International Space Station, and the explosion of the space shuttle *Columbia*. Through it all, NASA—chief guardian of the status quo, and the binding agent that held the frayed threads of a fading Space Age together—would increasingly look less like an innovator of new frontiers than a desperate bureaucracy concerned, above all else, with its own survival.

For several days after our arrival in Fiji, there was little to do but lounge in the tropical swelter, nursing cocktails and conversation while waiting for the Russians to confirm the details of Mir's final descent. We were a mixed bunch: scientists, photographers, journalists and entrepreneurs. All of us, to a person, were space junkies. We asked one another what had become of the Space Age—that epic turning point in history whose ultimate legacy would not be one of petty nationalist squabbles but one of common adventure and discovery as we reached out to the stars toward new adventures and horizons. Convened on a tiny patch of humid paradise, we waited for the sky to fall—and wondered if our own Space Age dreams could survive the impact.

It took a lot of effort—economic, technological and most of all political—to kill the Mir space station. And whether talking over cool beers in the sweaty heat of a Fijian evening or on quiet morning walks along storm-clouded stretches of beach, we voiced the same question repeatedly, albeit in different languages and through the prisms of different agendas: was Mir's demise the result of euthanasia or of murder?

Where the proverbial human space odyssey is concerned, the date the Soviet Union launched its tiny Sputnik satellite, October 4, 1957, was the day the Space Age began. The following year, a hastily crafted congressional mandate created the National Aeronautics and Space Agency, or NASA, which just over a decade later would engineer humankind's first footsteps on another world. That period, between 1957 and 1969, represents the most stunning rate of technological advancement the world has seen, before or since.

Few have dated the end of the Space Age. But it did end. The year was 1986.

On February 20 of that year, the Soviet Union launched the Mir space station. Beaten to the moon and Mars, the Soviets specialized in space stations, the single area of human spaceflight in which they had regularly trumped the Americans; the Soviets can claim the first space station, Salyut 1, launched in 1971, two years before the American Skylab (although the first Salyut mission ended in tragedy when the Soyuz capsule returning the crew depressurized during reentry, killing all three cosmonauts). Several more Salyuts would precede Mir as the Russians learned more and more about living in space.

Yet all of the Salyuts, and Skylab, were little more than stretched space capsules. Mir was something altogether different—an elaborate, multimodule orbital platform that could host sophisticated science experiments and serve as home to half a dozen people at a time, including some who had lived on the station for over a year. It might not have been the gigantic spinning wheel of Kubrick and Clarke's seminal *2001: A Space Odyssey,* but it was the most significant step for humans in space since Apollo.

Until the International Space Station hit its stride in 2001, Mir was the largest and most sophisticated object humans had ever put into space; it was probably the Soviet space program's greatest triumph in the post-Apollo era. But it was an empty achievement. Only a few years after its completion, the cosmonaut Sergei Krikalev would find himself stranded on Mir as an abortive coup attempt delayed the launch of his return spacecraft—an event symbolic of the decline of the Soviet Union and its human spaceflight prowess. Launched as an emblem of Soviet power, Mir would end its days as a desperate capitalist experiment.

A few weeks before Mir's launch, on January 28, 1986, the space shuttle *Challenger* exploded seventy-three seconds after liftoff, killing all seven crew members—the largest loss of life in the history of the American space program. The *Challenger* disaster lives in America's collective memory as the diabolical doppelgänger of the soul-thrilling Apollo launches two decades earlier.

In technical terms, *Challenger's* demise was traced to the now-infamous O-rings, malleable gaskets that separate sections of the two solid-fuel rockets used to boost shuttles into orbit. *Challenger* launched on an unusually cold winter day; the O-rings stiffened in the chill and became brittle. One O-ring cracked and burned through, causing the explosion that destroyed *Challenger* and killed its crew.

The larger causes of the disasters were arguably more painful to absorb. Various *Challenger* inquiries excoriated NASA for mismanagement, for shoddy engineering and for putting public image above human safety. Far from a competent band of techno-pioneers, the NASA portrayed by the inquiries was fragmented, secretive, clumsy and more concerned with saving face than saving lives.

Seventeen years after *Challenger's* destruction, when the oldest of the space shuttles, *Columbia,* disintegrated over the southwestern United States, there was an understandably similar sense of horror and mourning. But *Columbia's* destruction, while certainly not anticipated (at least not by the public), wasn't quite as shocking as that of *Challenger.* For most people, the first eye-opener of the *Columbia* disaster was that there had been a shuttle aloft at all. That's because *Challenger* had long ago blown apart any remaining public interest, not to say confidence, in the American human spaceflight venture—what's more, *Challenger* had robbed NASA of what little boldness and imagination the agency retained after the swaggering heyday of Apollo. By the time *Columbia* was lost, not only had the shuttle program been downgraded as a technological exemplar, but NASA itself had largely mortgaged its vision for that most basic of ambitions, survival.

What both *Challenger* and Mir signified, in different ways, was that the accepted paradigm for achieving the two-part promise of the Space Age—sending human explorers to other planets and paving the way for "ordinary" humans to spend time in space—was not going to deliver either. The shattered promise of *Challenger* and the failed hope of Mir, taken together, effectively ended what we called the Space Age. From 1986 on, human spaceflight has been in a state of suspended animation. We're not quite regressing, but we certainly aren't moving forward.

The space shuttle was meant to herald the next phase of human-

kind's venture into space and restore NASA to its place of glory.
Instead, the shuttle tied humans more tightly to Earth than the moon
rockets it was built to replace. It did so not only in and of itself—by
being almost nothing that its supporters promised and virtually everything its critics feared—but by prompting the birth of its evil cousin,
the International Space Station, a money-gobbling albatross whose
original intent and design have become so subverted by politics, spin
and greed that not even NASA can give a straight answer anymore as to
what, in fact, it's for.

Whatever else might be said of them, the shuttle and the station
have kept NASA in business, since by winning the race to the moon,
NASA inadvertently made itself obsolete as part of America's Cold War
machine. Once the moon had done its part for the American war effort,
the sprawling caravan of politicians and pundits that had made space
the most important national concern since defeating Japan quickly
packed up their influence machines, boxed up their encouragement
and barely glanced backward at what was still being called the Space
Age—although not so loudly anymore.

In the process of assembling the $100 billion Apollo juggernaut, jobs
had been created and companies had become fat with space dollars: the
contract for the Saturn V moon rocket still holds the record as the
largest civilian contract in American history. By 1972, the year the last
Apollo mission went to the moon—and the year the space shuttle program was approved—NASA provided bread and butter to nearly thirty
thousand civil servants and three times that many contract employees.
Alarmingly for NASA, this was less than half the workforce the agency
had commanded less than a decade earlier. Something new had to be
thought up, built and perhaps even launched, or else all of those good
things would go away.

The upshot of this stew of circumstances was the space shuttle. After
its first few flights, beginning with the launch of *Columbia* in 1981, it
began to dawn on the public (and on Congress) that the shuttle didn't
seem to have any particular purpose. Previous NASA human spaceflight efforts, from Mercury to Apollo, had the clear goal of ensuring
that America sent humans to the moon before the Communists. At
about half a billion dollars per mission, the shuttle blasted into orbit,

conducted a few esoteric science experiments or occasionally released a satellite, then dropped back to Earth with all the grace of a freight car with fins. As the public yawned, Congress began to hack away at NASA's ever-declining budget with increased vigor.

In the early 1980s, before *Challenger* took the remaining luster off the shuttle program, NASA and its powerful aerospace partners stepped up an aggressive campaign to give the shuttle a bit more to do. As president, Ronald Reagan was sympathetic to concerns that a lack of global-scale war (even a "cold" one, the end of which was, in the early 1980s, appearing on the horizon) would be bad for the American aerospace industry and consequently bad for America; the shuttle program was a gravy train that needed to keep chugging forward. However, to make that case to Congress and America, NASA would need once more to build something new, this time to keep the shuttle plausibly in business.

And so, in 1984, Reagan announced that America was going to build a new space station. The station would be bigger and better than America's only other space station, Skylab, which was so provisional that NASA didn't like to call it a space station at all—to this day, the official designation of Skylab is "orbital workshop." Not insignificantly, so the propaganda ran, a new American space station would also put one in the eye of the evil Soviets, scoring yet another victory for democracy.

What was not trumpeted so loudly was that exactly the opposite was about to happen. Reagan, and certainly NASA, knew that the Soviets were about to launch their next-generation space station (their seventh space station, incidentally), the massive high-tech platform called Mir, which would be virtually everything the new American station hoped to become.

The NASA project first known as Space Station Freedom was grand and ambitious, at least at the outset. The station would house six to eight people plus research labs and facilities for fixing satellites and even building other spacecraft (including spacecraft that might someday send Americans to Mars—so it was spun by NASA). The station would be designed for assembly in orbit with parts lofted spaceward in the shuttle. Assembling the space station would require a lot of shuttle flights, nearly fifty all told—more than enough to keep the shuttle in business for quite some time, since by the time Freedom was announced it was clear that fifty shuttle flights would span about a decade and not a single year, as originally promised.

When the project was announced in 1984, NASA vowed to have Freedom orbiting in eight years for a relatively paltry $8 billion—a bold proposition, given what little experience NASA had with space stations. While Skylab was generally considered a success, its intended operational lifetime was much shorter than that proposed for Freedom (seven years compared to up to two decades for Freedom). Further, only three crews spent time at Skylab before its premature reentry in 1979, whereas Freedom was envisioned as a bustling spaceport with astronauts coming and going at regular intervals.

But most critically, Skylab was the product of adaptation, not original design: it was cobbled together from spare Apollo components and launched as one package on a mighty Saturn V rocket left over when Congress cut short the Apollo program. Freedom, on the other hand, would be custom-designed, custom-built and then assembled piece by piece in orbit over the course of several years, something NASA had never tried. Reality set in quickly: by 1988, the project's total cost estimate had more than tripled and the first launch was bumped from 1992 to 1995. By 1988, Mir had already been orbiting for two years.

Early in his tenure, President Bill Clinton was faced with two seemingly disparate challenges. One was the rapid implosion of the former Soviet Union, which by the early 1990s was in economic and political free fall and threatening to become more dangerous than ever to U.S. interests as its technology was leaked to the highest bidders around the world. The other challenge was the obvious fact that Space Station Freedom was a nonstarter: billions of dollars over budget and years behind schedule, the station had become a large, decidedly stationary target for congressional budget cutters. Between 1991 and 1997 alone, the space station survived nineteen vigorous congressional attempts to terminate it. They were all destined to failure, largely due to one of the most controversial policy decisions the Clinton administration would ever make.

Sift through any one of the congressional briefings that have over the years hammered at the space station—its politics, its economics, its scientific validity—and you'll invariably find that unique brand of wonderfully wry bureaucratic discourse that at some level must have been a joy to write. In a 1996 report to Congress by its Science Policy Research Division (SPRD), the veteran space analyst Marcia Smith summarized with brutal simplicity the transition from Freedom to what has become known as the International Space Station:

> Following a major redesign in 1993, NASA announced that the Freedom program had ended, and a new [space station] program begun, though NASA asserts that 75% of the design of the new station is from Freedom.

The report noted, in a gloriously deadpan ode to pork-barrel spending, that "most of the funding [for Freedom] went for designing and redesigning the program over those years; little hardware was built and none was launched."

If it was ever to get a space station into orbit, NASA clearly needed some help; that the Russians would be tapped to provide it was a geopolitical eyebrow-raiser but, in other respects, obvious. The Russians definitely knew how to build hardware and get it into space. Between 1957 and 1998, humans achieved 3,969 successful space launches (the Congressional Science Policy Research Division, which compiled these figures, defines "successful" as attaining Earth orbit or beyond). Of these launches, 65 percent were Soviet or Russian; less than 30 percent were American. Even between 1990 and 1994, with the Soviet empire coming apart at the seams, successful Russian launches outpaced U.S. launches by more than two to one.

A number of these launches had been assembly flights for Mir, which by 1992—the year the Clinton administration came knocking at the Kremlin gates—had already been spinning around the Earth for nearly six years, playing host to numerous human endurance records and hundreds of scientific experiments. Cosmonauts came and went with regularity, typically staying for five or six months, far longer than any American at that point had ever remained in space.

The Clinton proposal was straightforward: in exchange for a whopping amount of cash (eventually around a billion dollars), the United

States wanted to buy Russia's space station expertise and get Freedom off the ground. Being dead broke, the Russians accepted; the kernel of the idea was theirs in the first place. On September 2, 1993, Russia and America announced that they would pool their space station efforts, and under a new name—the International Space Station (ISS).

No one in the Clinton administration seemed concerned that Freedom actually *had* been an international space station for some time. In 1988, Canada, Japan and the European Space Agency (ESA) signed on as partners in Freedom to the tune of about $8 billion in hardware commitments, in addition to NASA's own funding. Each of these partners was putting up its own resources for a stake in the station; Russia would be the only partner actually paid to participate. The further insult to these early partners was that by being reconfigured as essentially a two-nation project with a cast of supporting players, the ISS had become much less international.

Mir was always part of the Clinton administration's thoughts in considering Russia as a full partner in the ISS. Under the agreement, Mir would be used as a template for building much of the ISS and as a training facility for ISS astronauts. But there was always the nagging claim that the ISS would do little more than ape what Mir had already accomplished. If the U.S. government was ready to spend big on humans in space, the critics were saying, why not establish a base on the moon or, dare it be said, move on to human exploration of Mars?

But as would be demonstrated time and again, the ISS was never really about space at the macro level. It was about nationalist politics. Mir was a reminder (passing overhead every hour and a half, just to rub it in) of an undeniable fact: the Cold War may have ended, but the Russians still cornered the market on human presence in space. NASA figured that the ISS would be successful only if it was a monopoly. The way to achieve that was by co-opting Mir, and then eliminating it.

Jeffrey Manber is a quick-talking, bright-eyed New Yorker who will tell you whenever he gets a chance that he is a businessman and that his business is space. Manber began his professional life in the media: while still a university student in the late 1970s, he started a film review magazine that he marketed well enough to have it sold on Manhattan newsstands.

It is one of the great ironies of Manber's life that what drew him into the business of space—and, eventually, head to head with NASA over the commercialization of Mir—was the launch of *Columbia* in 1981, the public beginning of the space shuttle program. His response to the destruction of the vehicle that inspired him was pointed and clear: "It is time to face reality," he wrote in an e-mail to me only a few weeks after the *Columbia* accident. "The space shuttle was a path we should never have taken."

Manber's path from shuttle enthusiast to detractor is as unexpected as how he got into the space business in the first place. He certainly didn't plan his life that way. He had received a degree in psychology at Northwestern University and by the early 1980s was on a trajectory toward a career in research or medicine. He was working toward a graduate degree in neurochemistry when "one day at six in the morning I was shocking white mice and I thought, 'You know, this isn't my future,' " he recalled with a short, sharp laugh.

When Manber was a boy, his older brother had gotten him hooked on space through such early science-fiction classics as *Tom Corbett* and *Space Cadet,* and since then space had never been far from his thoughts. With his nascent research career ended, Manber sold the film magazine and embarked on a career as a business journalist focused on space, which in turn led to a career as a space business consultant. At the time, to Manber and many others, the shuttle program seemed to signal a fresh start for the Space Age. "There was NASA talking about a meeting with the private sector in space, there was NASA talking about operations and profits," Manber remembered, the bitter edge still in his voice. "There was an entire generation of people, of young entrepreneurs like me, who believed NASA. And it was the last time I believed NASA."

Manber and others working at the entrepreneurial edges of the space industry feel a raw sense of betrayal that the space agency which inspired their careers possesses such a long and evident record of aggressively discouraging any attempt to expand human spaceflight beyond NASA's control. NASA's transition from innovative pioneer of new frontiers to truncheon-swinging government bully was long and tortuous, no doubt, for everyone involved. But that's what happened: scientists bemoan it, politicians rail against it—and even NASA, if you talk to the right people, admits it. For all intents and purposes, NASA owns space—but only in America.

In this critical exception lay Manber's future. Manber's road to becoming America's most savvy Russian space entrepreneur was short if not entirely smooth. In 1987, after a stint with the Wall Street investment firm Shearson-Lehman, where he helped develop the world's first space investment fund, Manber was summoned by the Reagan administration to help set up an office in the Department of Commerce devoted to exploring commercial opportunities in space. Manber's stay at the Department of Commerce ended in 1988 in frustration at the various political and bureaucratic roadblocks facing new commercial space projects.

Still, Manber was in a small but growing vanguard who believed that in order for there to be a human future in space, space activities needed—as much as possible—to be wrested away from the government domain. This meant not only the privatization or commercialization of government-developed "assets" like the shuttle, but also the creation of new businesses infused with private capital and tested, without government monopolies to hinder them, on the open market.

In the years preceding and immediately following the *Challenger* disaster, "space" was still very much associated with big government. However, private interest in space did exist. As Manber was leaving the Department of Commerce job, he was asked privately to help secure an export license for an American company that had negotiated a deal to "fly" a pharmaceutical experiment into space—on a Soviet rocket.

These were the days of Mikhail Gorbachev's perestroika, and as part of their nervous experiments in capitalism the Soviets were offering an alternative to the space shuttle: fly with us, for less. Believing in the potential for space commerce wherever it presented itself, Manber accepted the challenge but ran straight up against a complete lack of precedent for getting the pharmaceutical deal approved.

That the pharmaceutical mission's destination would be the then-new Mir space station complicated things even further. At the time, not only was Reagan's space station still on the drawing board, but NASA's entire space shuttle fleet remained grounded after the *Challenger* accident. In such a touchy environment, Manber's approach was to de-emphasize the location of the research, arguing that it was simply a business transaction between two countries—after all, even for all their Cold War saber rattling, America and the Soviets still did some business.

For weeks, Manber hawked this message to his comrades in the Department of Commerce and elsewhere in Washington. Eventually it worked. On February 21, 1988, the *New York Times* ran a front-page story that spelled out Manber's triumph. A framed copy of the story, titled "American Company and Soviets Agree on Space Venture: First Commercial Pact," still hangs on the wall of Manber's Washington office. "I was in Switzerland when the story broke and the shit really hit the fan," Manber recalled. "Congress was up in arms about doing deals with the Communists. NASA held a press conference and said this would never work. They were bent out of shape because an American company had the audacity to do a deal in space without NASA."

Adding insult to injury for NASA, the *Times* article predicted that more such deals would follow: "Now and for the immediate future, the Soviet Union has an advantage over the West in conducting some long-duration experiments in space because of its orbiting Mir station."

The pharmaceutical deal drew Manber into the orbit of Yuri Semenov (pronounced sem-YA-noff), a bearish engineer/manager who had been with the Soviet space program since 1959 and had survived to become head of the powerful space company RSC Energia. In the history of space exploration, only NASA can lay claim to as many space achievements as Energia, and even NASA can't claim as many firsts. Founded by the Soviet rocket genius Sergei Korolev (that mysterious Chief Designer who kept besting NASA in Tom Wolfe's Space Race epic *The Right Stuff*), Energia was effectively NASA's competition in the Space Race, since there was no civilian space agency in the Soviet era.

Under Korolev's leadership, Energia notched both Sputnik and Yuri Gagarin among its triumphs: the first man-made object to orbit the Earth and the first human in space. Until the early 1990s, Energia was run by the Soviet military, but as the communist Soviet Union made the transition to capitalist Russia its space program changed as well. Of the country's major space operations, Energia took the most commercial path; it has shareholders, stock, benefits, incentives and most of the other standard components of any Western company.

In 1992, Manber agreed to head up Energia's first American office, in Washington, D.C. That same year, he brought from Russia to the United States the first contract between NASA and Russia: an agreement to use the Russian Soyuz spacecraft as a rescue vehicle for the new space station. Other deals followed, including Manber's work on innovative space ventures like SeaLaunch, a partnership between Energia, Boeing and several other companies that envision using a converted deep-sea oil-drilling platform to launch rockets from the middle of the Pacific Ocean.

Even with such successes, it was rough going for Manber. Russia's economic and political situation deteriorated rather than improved, and to Manber's frustration he was unable to pull off many deals that were outside the NASA contractor orbit. This was critical, since Manber and Semenov believed that Energia's future lay in establishing it as NASA's competitive peer, not its vassal. In 1999, discouraged and burned out, Manber suggested to Energia that the Washington office be closed, or at least that they find a replacement for him. The time was right, Manber recalled, to try something new.

"And one day, in the door walks Rick Tumlinson," he said with gentle exasperation in his voice. "My lucky day."

"You've got a guy like Dan Goldin, a complete ass, who has spent all his years misleading Congress, lobbying on behalf of his Lockheed Martin and Boeing people to spend upwards of forty billion on a space station that will house between three and six people. How does that compare to Mir? God forbid, what if that were the place where they discovered the cure for AIDS? After such a sales job, how would that look?"

Rick Tumlinson unleashed this diatribe with barely a pause between bites of his enchilada as we sat in a loud, clattering Mexican restaurant squeezed into a busy corner of Burbank, California. Upon arriving,

Tumlinson swept an arm across the room and informed me that crowded into the worn red-leather booths around us were all the people who make Hollywood work: the sound technicians, the lighting wizards, the set designers, the editing gurus—the architects of cinematic spectacle, quaffing margaritas, crunching tortilla chips and otherwise taking a break from the business of creating the Next Big Thing.

Tumlinson himself is a filmmaker by training, having for many years produced both commercial and industrial films, and he clearly still feels comfortable in this crowd. With his ponytail-under-a-baseball-cap, blue-jeans-and-cowboy-boots uniform (often complemented by a long black cowboy duster, a hint of his Texas roots), there's a loose hipness to Tumlinson, a sense that he pursues his life and his work with passion and drive but also knows when it's time for a cocktail. Tumlinson has that easy command of schemes, ideas, philosophies, feelings and bottom lines that comes with the need to make your pitch and to make it quickly. Despite all this, though, Tumlinson's movie-biz days have long been behind him. For some time now, the Next Big Thing for Rick Tumlinson has been space.

Tumlinson's lunchtime harangue was aimed squarely (and with characteristic lack of subtlety) at Daniel Saul Goldin, who as NASA administrator for nearly a decade served as the bête noire of the community of investors, inventors and entrepreneurs who have been laboring, and lobbying, for a new paradigm where human spaceflight is concerned. They wanted NASA as a partner, not a competitor; they wanted inspiration, not spin—significant investment, not cheap handouts.

It had happened before. In the early twentieth century, an elite government research agency led advanced technology development that played a critical role in setting the stage for global conquest by the airplane. At the same time, the American government had provided the equivalent of billions of dollars in raw cash to lift new commercial airline ventures off the ground. This combination—government-supported R&D and government-provided cash infused into an embryonic industry—led to the creation and marketing of landmark passenger and cargo aircraft like the DC-3. Ultimately, this strategy produced the commercial airline industry we have today.

The agency that drove the aeronautical technology development that made the government dollars work was the National Advisory Committee for Aeronautics. In 1958, that agency, NACA, would become the National Aeronautics and Space Administration—NASA.

Obviously, NASA didn't pursue such an agenda, and the specific rap on Goldin was that he had done more to block entrepreneurial space interests than perhaps any other NASA administrator. That Tumlinson would ignite a chain of events that would help undo Goldin's tenure and further diminish NASA's public credibility would not bring Tumlinson any particular joy; it was just another battle in a long, long war whose greatest campaign to date began, but by no means ended, with Mir.

By the time he walked through Jeff Manber's door in 1999, Tumlinson had become a well-known rabble-rouser in the space community. His main vehicle for change was the Space Frontier Foundation, which he cofounded in the 1980s to advocate for innovative projects that would open space—for public travel and popular enterprise—to larger numbers of people. In the process, Tumlinson had built a reputation on Capitol Hill as an alternative voice to NASA's spin on space.

Tumlinson had also made waves with his provocative statements about the International Space Station. Testifying before the House Space and Aeronautics Subcommittee in April 1997, Tumlinson proposed turning the station into a purely commercial enterprise and getting NASA out of the picture altogether. "So why do we have a crack team of explorers acting as landlords?" Tumlinson asked. "After all, when stripped of the space mystique, the station is merely a combination port, hotel and research lab in orbit. NASA is an exploration and advanced research organization, not a construction, trucking and building management firm. Those are jobs that our private sector does, and does very well."

Tumlinson's primary agenda was to get NASA out of the trucking and building-management business, and thus help get it back on the long-range human exploration track. The way to achieve the former was through a gradual transition to the private sector; eventually, Tumlinson and others believed, that private sector could serve as a major catalyst for helping NASA pursue journeys of exploration even more significant than Apollo.

Along with opinions and ideas, Rick Tumlinson also had money, although not his own. Through his day job as a filmmaker, Tumlinson had become friends with Walter Anderson, founder of Esprit Telecom and a millionaire many times over before the age of forty. A bona fide space fanatic since childhood, Anderson was also frustrated by what he perceived as NASA's sluggish approach to space exploration, and in late 1997 he created the Foundation for the International Non-governmental Development of Space, or FINDS, to support entrepreneurial research in space technologies. Anderson launched FINDS with about $5 million, but with shrewd investment in the go-go late 1990s stock market, that initial funding grew to $25 million. Anderson put Tumlinson in charge of FINDS, and the two of them began thinking about the future.

Around 1997, word began to leak into the space community that the Russians were soon going to de-orbit Mir, a plan which Tumlinson and Anderson, among others, thought was absurd. "It was purely political, everyone could see that," Tumlinson told me. "Mir had its problems, but they could be fixed. It was going to be a waste, a total and absolute waste, of millions of dollars of very high-end space hardware. Ridiculous."

It didn't seem all that ridiculous at the time, considering that 1997 had not exactly been a banner year for Mir, which was by that point over a decade old. Doing ISS training duty on Mir in June 1997, the NASA astronaut Michael Foale recalled hearing "this big kind of ker-thump" as he was floating in the station's main module. Foale didn't know it at the time, but he was hearing the impact of a seven-ton auto-mated resupply ship hitting Mir after a botched remote docking attempt by Foale's Russian crewmate, Vasily Tsibilev. The impact sheared off a section of Mir's solar-power array and punched a hole in the station's Spektr science module. Foale recalled that his ears popped as air hissed through the jagged hole in the Spektr module and Mir started to depressurize. The station's power winked off and on, then failed altogether. If not for the seat-of-the-pants ingenuity of Foale and his Russian crew-mates, Mir might well have spun back through the atmosphere right then and there.

The crash may have been the worst of Mir's travails, but it wasn't the first. By 1997, the station had logged more than fifteen hundred mal-functions in more than sixty-five thousand orbits of Earth. Earlier in

1997, a backup oxygen generator sparked a fire that nearly forced the
crew to abandon the station. For all that, though, Mir was still in business; the damage caused by the fire and crash was ultimately repaired,
and Mir operated more or less without incident afterward. But the fire
and crash secured Mir's reputation as a leaky antique, one it would
never live down. In the larger context of Western geopolitical media
spin, Mir became one more example of the failure of the Soviet system.

With its ISS training nearly completed and the 1997 disasters lodged
firmly in the public mind, NASA begin pressuring the Russians to
bring Mir down. Anderson and Tumlinson wanted to do something
to help stave off Mir's demise, believing that, as the only permanent
human base in space, Mir had singular value. Shortly after the Spektr
crash, the Space Frontier Foundation mounted a public relations campaign called Keep Mir Alive, hoping to counter the mounting pressure
from NASA and Congress to terminate the station. Later that year, the
foundation staged its first conference and invited as a keynote speaker
Ivan Bekey, formerly an engineer in NASA's advanced projects office.
In his banquet speech, Bekey talked about research on a device called
an electrodynamic tether. "That was the spark, right there," Tumlinson
recalled, the real beginning of the campaign to remake Mir.

Electrodynamic tethers are one of those concepts that sound more
sci-fi than Space Age and yet have been the subject of considerable testing since the 1960s. At their most basic, tethers are analogous to the wire
that runs from a wall socket to a lamp—they are conduits of energy,
channeling it from a power source to a device. Expanded to a grand
scale, tethers are supertensile "wires" hundreds of feet or even miles
long that can be extended from an orbiting spacecraft and dragged
through the Earth's electromagnetic field—itself a power source—to
generate power.

NASA, industry and military organizations have experimented with
tethers as a way to keep satellites and space vehicles in their proper
orbits, which is a constant challenge. The Earth's atmosphere, while
growing progressively thinner with altitude, drags on objects orbiting
even hundreds of miles up and slows them down. Once they are low
enough and slow enough, gravity beats the vacuum and the objects fall
back to Earth or, more often, disintegrate.

A large object like Mir, with a relatively low orbit of about two hundred and fifty miles (known as low-Earth orbit), needed frequent blasts of expensive rocket fuel supplied by even more expensive spacecraft to keep it safely aloft. Even before the disasters of 1997, Russia was struggling to keep Mir "on orbit," since its ISS deal with NASA didn't cover the full cost of Mir's operation. Once ISS training was completed, if the Russians were to keep Mir from tumbling into the upper atmosphere and burning up, they would need help from another source.

In theory, electrodynamic tethers could be attached to Mir and then used to boost its orbit and buy the station a few extra years of life. If such a thing was to happen, Rick Tumlinson and his colleagues knew they had no time to lose: by the time the tether idea had surfaced, NASA and the Russians (under considerable pressure from NASA) had begun a PR campaign to pave the way for Mir's demise. In a February 1999 Associated Press article, the cosmonaut Viktor Afanasyev, on the eve of becoming Mir's next commander, said his crew might well be the last ever to work on Mir: "We hope that we will have a replacement," Afanasyev said sanguinely. The article further noted that if no additional funding was found to keep Mir aloft after Afanasyev's crew returned to Earth, "[Russian] Mission Control will fire the engines on Mir's cargo ship to send the station plummeting into the ocean."

Six days after the AP article appeared, Tumlinson sent a letter to Energia's Yuri Semenov offering $100,000 from FINDS to fund a more thorough analysis of how tether technology could help save Mir. Semenov quickly agreed, and as a March 1999 fax underscored, he was already thinking ahead: "Upon the completion of the first stage, we should meet to discuss a plan for further joint activities."

Semenov's enthusiastic reply to the FINDS proposition was based on nothing less than desperation. It was also a last grasp at national pride. Some months after Mir's eventual destruction, I would find myself in a Moscow café having lunch with Yuri Karash, a space journalist and adviser to the Russian space program who became a friend while we were both in Fiji waiting for our front-row seats to view Mir's fiery end. On that subject, Karash was pragmatic: "Mir was old, it had many problems," he said evenly. "Everything must end somehow." But on the subject of what Mir symbolized for Russia, Karash's measured tone became passionate, even angry. "Space to Russia is a source of immense pride, of power and great meaning," he said, leaning toward me across the table, the better to be understood clearly in the bustle of the busy café. "We have many problems in Russia, but even with so much difficulty, we can still say we did space first. For Russian pride, space means goddamn everything."

So eager was Russia to keep Mir aloft that at the same time Semenov was inking the tether study agreement with FINDS, he was unwittingly playing Energia into a con game involving the Welsh businessman Peter Llewellyn, who promised to pay $100 million for a weeklong trip to the station. In May 1999, the Russians announced Llewellyn's pending ride with great fanfare, but the deal fell apart only a few weeks later when media investigations revealed that Llewellyn had been forced to pay restitution for swindling an American businessman—and that, at various times in his colorful history, he had claimed (falsely) to be an admiral, a spymaster and Queen Elizabeth's bodyguard. When confronted with such shenanigans, Llewellyn announced he'd never intended to pay a penny for going to Mir, claiming instead that his flight would help raise funds to build a children's hospital. However, before the tale ended Llewellyn had managed to get in a few free rounds of cosmonaut training; for their pains, the Russians got nothing but more ridicule from the international press, and more pressure from NASA to bring Mir down.

FINDS immediately distinguished itself from Llewellyn by actually paying Energia the promised $100,000 for the tether study, which ultimately concluded that tethers could keep Mir in orbit and should be employed to that end. Testing the tether on Mir would involve actual spaceflight; now was the time to talk major money, which meant face-to-face negotiations—all of which prompted Tumlinson to appear on Manber's doorstep.

"Jeff and I were old, old friends from when Jeff was doing space business roundtable discussions," Tumlinson recalled, "and his credentials as far as free enterprise goes were great. Plus, he knew the Russians like nobody else." Regarding that last point, Tumlinson also knew that Manber would undoubtedly be sitting in on his upcoming discussions with Energia. He needed Manber as an ally, because what Tumlinson and Gus Gardellini (one of Walter Anderson's lieutenants from his Esprit Telecom days) were going to lay on the table in Moscow was a proposition far grander than just a single experiment.

"At the time, we were feeling pretty high and mighty," Tumlinson told me. "We were taking on bigger and bigger things, and I said 'Walt, what do you think about a space station?' See, Walt's plan had always been to become a billionaire and then based on that come back in and build his own space station. Here was a different opportunity."

"Rick calls me up and says, 'Man, we want to buy the Mir,' " recalled Manber. "And I said, 'You can't buy the Mir.' But the idea behind it was good."

The idea was to form a company, held jointly by Americans and Russians, for the sole purpose of privatizing the Mir space station. The way to privatize it, Western-style, was to own it: Tumlinson and Anderson didn't know at the time what that would take, but they were sufficiently encouraged by the tether studies to explore larger possibilities. "We didn't really know what we were going to do, who was going to be in charge of what," Tumlinson said. "It was 'Save the station,' then figure out what to do with it."

With that loose agenda, on September 29, 1999, Tumlinson, Gardellini and a Space Frontier Foundation board member, David Anderman, filed into a nondescript conference room in Moscow to meet with a group of Energia representatives. Also at the table, firmly on the Energia side, was Manber. The first day of discussions went well: everyone agreed that Mir was worth saving and that the tether concept was promising. The meeting went so well that on the second day, Semenov himself joined the discussion along with Valeri Ryumin, one of Russia's most famous cosmonauts and a major force in the Russian space program.

The Russians wasted no time getting down to business. Despite assurances that the FINDS money was good, Semenov rattled off reasons why the project was bound to fail: because the tether technology was untested on such a scale, Mir could be irretrievably damaged and cosmonauts could be injured or even killed. At that point, the Americans were convinced that the Russians would say *nyet* to the tether mission.

"Instead," Tumlinson said, "he [Semenov] grabs the agreement that's sitting on the table between us, signs it, and tosses it back to me. Then he stands up, shakes my hand and says, 'We're go.' "

For all his optimism, Semenov was not about to waste much time on FINDS. The agreement gave FINDS less than three weeks to arrange both funding and U.S.-Russia technology-transfer approvals to make the tether mission happen. Funding wouldn't be a problem; the tech transfer would.

Because tether technology originated in the space program, it automatically fell under export controls governed by the famously stringent International Traffic in Arms Regulations, or ITAR. FINDS marshaled a veritable army of attorneys to expedite a transfer of the tether plans (called the Mir Electrodynamic Tether System, or METS) from the United States to Russia, but ITAR was a process that could not be rushed. Worse, while no one seriously suggested that METS could be used as a weapon (in the letter eventually clearing the transfer, the Department of Defense described tethers as "1958 technology"), FINDS submitted the ITAR application at a touchy time for U.S.-Russian relations, since the countries were on opposite sides of the conflict in Bosnia.

In retrospect, Tumlinson admitted that he and his colleagues were woefully naive about the time and effort required to secure ITAR approval. "The way ITAR is set up is that you're guilty until proven innocent," Tumlinson said. As the end of October drew near, it became apparent that FINDS would not be able to get export clearance in time, and Semenov wouldn't extend his deadline. This was not wholly the result of skepticism; as 1999 came to a close, the time was approaching when Mir needed another orbital boost. Pressed to the wall by demands to produce ISS modules and the rockets to launch them, the Russians had only enough money to send one ship to Mir, which would either boost the station's orbit or drive it down to its doom. The greatest danger would arise if the Russians did nothing. Left to the laws of physics, Mir's orbit would decay irrevocably, and the station would tumble out of control and rain tons of flaming debris anywhere in the world. If no options to save Mir materialized by the end of January 2000, Semenov would have to bring Mir down.

Even without the ITAR clearance, both sides agreed to another meeting. In late October, Anderson, Tumlinson and Manber—who by this point was openly advising both sides—flew to Moscow on Anderson's private jet. To pass the time, all three played the world-conquest game Risk, perhaps in preparation for what they were going to propose to the Russians. Tether or no, Anderson was going to put up the money to save Mir.

"The original idea was you'd put the tether on Mir, you'd boost it to a higher orbit, and you'd shut it down," Tumlinson said. "That would give us time to put together the right elements of a business plan to make Mir pay for itself." But the ITAR delay changed their plans. "We knew we couldn't get the tether up in time to save the station, but we could save it for a year using conventional means," said Tumlinson.

"Conventional means" meant, in addition to the automated Progress spaceship already scheduled to send Mir up or down, a manned mission to supply the station and ready it for business—although what that business would be wasn't clear yet. That effort would cost about $20 million, which Anderson and some investor friends agreed to provide. This was far and away the most private funding ever committed to a manned space project.

In their second round of meetings with Energia, Anderson, Tumlinson and Manber proposed the creation of MirCorp, which would be owned 60 percent by Energia and 40 percent by Anderson's holding company Gold & Appel (pronounced "golden apple" and named for a shadowy organization in Robert Anton Wilson's sci-fi cult classic, the *Illuminatus! Trilogy*). The company would be based in a third country, the Netherlands, and have offices in Moscow and Washington; Manber would be its president. Although MirCorp would not own the Mir space station—it couldn't, since Mir was Russian state property—it would get an exclusive "lifetime" lease on the station and rights to market and conduct business there.

Even with such a staggering amount of money on the table, Semenov was nervous. Sending a manned mission to Mir, unrelated to (and as NASA would see it, in conflict with) ISS needs, would almost certainly put NASA on Energia's back. Knowing what was at stake, the night before the meeting Manber wrote a note to Semenov. "I said, 'Yuri

Pavelovich, this man is prepared to spend more money than any American has spent in the last two years on Energia.' And that got his attention," Manber said. In the end, not only did Semenov agree to the plan, he committed himself fully by signing on as chairman of MirCorp's board of directors. Semenov then gave the order for the January 2000 Progress mission to raise Mir's orbit.

On April 4, 2000, the veteran cosmonauts Sergei Zaletin and Alexander Kalery blasted off from the Baikonur Cosmodrome in Kazakhstan (Russia's manned-spaceflight launch facility). Two days later, they docked their Soyuz capsule with Mir and essentially dusted off the equipment and flipped on the lights, since the last crew had powered down the station and closed up shop seven months before. The mission established another Russian first in space: the world's first privately planned, privately funded human spaceflight.

Only a few months later, MirCorp announced that the California businessman and former NASA scientist Dennis Tito had signed up to become MirCorp's first privately funded visitor at a reported round-trip ticket cost of $20 million. A few weeks later, the company announced a deal to produce an American "reality" game show, for a cool $40 million. At this point, the Russians began to believe that they might actually be able to keep Mir running while also fulfilling their ISS commitments. In April 2000, Vladimir Putin, then Russia's president-elect, publicly hinted that he might dig up some government funding to help keep Mir aloft. "If there is a possibility to keep [Mir], then it should be done," Putin told the *Moscow Times*. "And, of course, there is such a possibility. It is not especially difficult."

With enough money Mir could ultimately be rebuilt entirely, constantly rejuvenating itself "like a baobab tree," said Manber, as modules were refurbished or replaced. But technology and funding were arguably the lesser factors in Mir's eventual demise. In trying to put together business for Mir, MirCorp ran smack into NASA and its Big Aerospace partners. For this conglomerate, the future of manned space was the ISS, not Mir.

Thus, what Manber and others at MirCorp quickly discovered was that anyone wanting to do business on a privatized Mir—whether that business was testing new materials, assembling and deploying small satellites or imaging the Earth from space—was being told by NASA to keep a distance from MirCorp. The implied "or else" was that the reach of NASA and particularly its two biggest contractors, Boeing and Lockheed Martin, was such that anyone who defied their interests might find himself effectively locked out of the shuttle, the ISS and, in theory, even the expendable rocket arena that Boeing and Lockheed—at least in America—owned lock and key.

In Washington, MirCorp made Russia an even more convenient scapegoat for the fact that the ISS was well behind schedule—indeed, it was often quipped that the only thing about the ISS that seemed on an upward trajectory was its ever-skyrocketing cost. The Russians had been beaten up on Capitol Hill and in the American press for keeping the ISS grounded, and while it is true that they lagged on some ISS delivery schedules, even Dan Goldin admitted in a February 2000 hearing before the House Science Subcommittee on Space and Aeronautics that "we had our own problems along with the Russians."

Goldin also singled out the ISS prime contractor, Boeing, calling it "a good company, but not an outstanding one" and cited delays in key ISS components, plus major cost hikes accompanying the delays. Despite Goldin's wrist slap, Boeing, as one of America's biggest corporations (and one of Washington's biggest corporate contributors), was never in much danger of bearing the brunt of the blame for the woes of the ISS. For the most part Goldin was quite happy to go along with and even encourage the incessant Russia-bashing if it helped deflect criticism of the ISS that might otherwise hit closer to home.

It was clearly in NASA's best interests that Mir come down and the ISS go up, whatever the cost or effort required. Well before MirCorp came along, NASA had kicked its public-relations machine into high gear to press home a popular portrait of Mir as a rickety obstacle on the very high road to progress. The machine was nothing if not inventive: case in point, the 1999 blockbuster film *Armageddon*.

NASA has long had tremendous clout in Hollywood where sci-fi is concerned, and the NASA seal of approval can, as George Steinbrenner once said, put fannies in seats. The agency takes its connection to the film industry very seriously. In 2002, for example, NASA hired the Academy Award producer Robert Shapiro as a "creative consultant" to better translate NASA's work for Hollywood. Not surprisingly, the agency's participation is based on calculated self-interest: it worked on the cheery *Mission to Mars* while declining to participate in the dour *Red Planet*, both of which were big-budget, big-star spectacles. The deciding factor was that the *Red Planet* producers refused to remove a scene in which one astronaut kills another, an unthinkable occurrence for NASA—even, apparently, in fiction.

NASA's involvement in *Armageddon* was designed to give the film a patina of plausibility. NASA provided technical advice and un-precedented access to its training and launch facilities, and helped to mount and operate nearly two dozen specially designed cameras to film unique perspectives of a real shuttle launch. *Armageddon* even features a NASA administrator named Dan Truman, played by the actor Billy Bob Thornton as no less than Dan Goldin with a Southern drawl and a perpetual five o'clock shadow.

In one of *Armageddon*'s key sequences, a souped-up space shuttle docks at the Russian space station (a specific name is carefully omitted) to tank up on fuel en route to saving the world from a killer asteroid. Manning the station is a lone cosmonaut straight out of Cold War Central Casting: during the entire film, he appears to be either drunk or insane. The space station itself is a leaky tin can with its equipment askew—as if the KGB had just raided the place. Inevitably, sparks fly, hoses snap and things begin to explode: the catastrophe begins with a lick of shooting flame, a nice touch to remind moviegoers of the highly publicized 1997 Mir fire. Everyone vacates the station for the seem-ingly indestructible shuttle, which gets pummeled with enough flying debris to pulverize a fleet of real shuttles. Our heroes take the hapless cosmonaut with them; as payback, he provides comic relief for the rest of the film.

The specific influence of such high-octane pop culture is hard to gauge, but few disagree that Hollywood has at least some impact on public perceptions where the space program is concerned. Lori Garver, a former NASA associate administrator under Goldin and now in the private sector, said, "When I go to schools and talk, it's shocking to hear what third-graders believe [NASA has] done, things like, 'Well, I saw this movie where NASA saved the world by shooting down that asteroid.' Of course, NASA doesn't do anything like that."

NASA also took its Mir campaign to the White House. On March 15, 2000, NASA submitted a briefing titled "Mir USG [United States Government] Observation Implementation Plan" which painted a grim picture of an uncontrolled reentry of the Russian space station. At no point did the briefing state that Mir was actually out of Russian control or was likely to be in the near future, but the briefing's insinuation was crystal clear: the Russians owed it to the world to bring Mir down as soon as possible.

There was a hint of panic in NASA's rush to terminate Mir. NASA knew that Mir, while old, was not unusable—a point made by certain members of NASA's own astronaut corps. The astronaut Norman Thagard, who spent nearly four months aboard Mir in 1995, said in an August 2000 chat session on SPACE.com that he would sign up for another tour of duty on the station if offered the opportunity. "There has been far too much concern about the condition of Mir and the safety implications of the station," Thagard wrote. "I won't downplay the significance of the fire and the significance of having a resupply module hit the station. The simple fact is that most of the risk is on the ride up to the station, not while you are on the station—that is a relatively low-risk part of the flight. I can state unequivocally that I was never concerned for my safety while on the Mir station."

What galled NASA most was how eagerly Energia devoted itself to the MirCorp effort. When Clinton mandated the ISS partnership, the American and Russian space programs had no formal mechanism by which to do business on such a scale; in fact, they were organized along fundamentally different lines. The Russian space program did not make the same kind of distinction between government and private organizations that the U.S. program did. In Soviet/Russian manned spaceflight, the major player had always been Energia, which for most of its history could best be described as a state-owned corporation.

Still, of Russia's space organizations, Energia was by far the most similar to NASA in terms of history and expertise; more to the point, Energia would bear almost sole responsibility for Russia's ISS partnership. Goldin, however, refused to view Energia as NASA's peer, instead directing NASA's ISS work through the Russian Space Agency, heretofore a small, relatively inconsequential administrative body that spent most of its time filing reports to the Russian government on what the nation's major space entities, like Energia, were doing. Goldin's decision created the uncomfortable situation in which this tiny agency was suddenly imbued with tremendous political clout while having no money, expertise, or even immediate responsibility where Russia's ISS participation was concerned—all of these were in Energia's court. The tension resulting from this arrangement seeped directly into the battle over Mir.

NASA was even more distressed that Energia appeared to be price-gouging the American government while giving sweetheart deals to MirCorp. In a February 2000 House subcommittee hearing, Goldin said that Energia had attached a price tag of $65 million to a single Soyuz capsule to be launched to the ISS as an emergency escape vehicle for station crews—which was needed because NASA's own ISS escape vehicle, the X-38, wasn't close to being ready (and was later canceled altogether). Energia had given MirCorp a four-for-one package—one Soyuz, two Progress supply flights and a forty-five-day stay on Mir—for about $20 million.

Rick Tumlinson chuckled unashamedly when recalling Goldin trying to explain to Congress why a group of private entrepreneurs got a better deal than the world's leading space agency. "I remember in one meeting with the Russians I said, 'Give us a price,' and they did that and then I said, 'What's the real price?' and they said another price and I said, 'No, give us the *family* price, the price you would charge your brother, because we are family.' I remember them groaning at that, but we kept talking." NASA, apparently, wasn't family. "NASA had insulted them, had abused them, had been rude to them, and here we were trying to be totally straight-ahead and talking about family," Tumlinson said. "And the fact was we were putting money on the table."

NASA, on the other hand, had given the Russians a take-it-or-leave-it deal for ISS participation in which, for a fixed price that would change very little over time despite massive changes in Russia's ability to perform its ISS duties, the Russians would have to provide as many Soyuz spacecraft and related services as NASA deemed necessary, virtually on demand. The pressures of this contract had stretched Energia to the point where, often, workers were building ISS items for no pay at all. "At one point, MirCorp money was actually helping to keep the

ISS supply line going," Manber noted dryly, a point privately conceded by many people I spoke with at NASA.

"Family" or not, by mid-2000 the Russians were beginning to bend under the U.S. pressure to kill Mir. In early October 2000, the Russian government, pressed by the Russian Space Agency, announced that it was going to de-orbit the station. The next day, MirCorp and Energia held a hastily arranged news conference to assure the world that the announcement had been "misinterpreted" and that Mir was not going to come hurtling down any time soon. The government responded a few days later by acknowledging that it was possible that "private options" could salvage Mir.

That same month, a representative of the Russian Space Agency announced that MirCorp's star attraction, Dennis Tito, "does not exist" as far as the agency's future flight manifest was concerned, despite the fact that Tito had already completed one round of training in Moscow with the agency's full knowledge. MirCorp's leadership was apoplectic, and their spokesman Jeff Lenorovitz summed up the whole debacle with barely restrained frustration: "This shows the gap between a space agency and a real commercial company."

Covering the first ISS crew launch in October 2000, the Associated Press returned to the fact that the name "International Space Station" was merely a placeholder since the partners were unable to come to consensus on something snappier. Would the launch of the first crew finally provide an occasion for the station's partners to agree on a sexier moniker? Members of Expedition One, NASA's can-do name for the first ISS crew, were asked for their thoughts.

Expedition One cosmonaut Sergei Krikalev was quoted as favoring one of the early suggestions, Alpha, because "it's not kind of specific [to] English or Russian." American astronaut William Shepherd, Expedition One's commander, replied that he was under orders not to voice an opinion on such a monumental topic, although he later convinced NASA controllers to use Alpha as the call sign for communications with the station.

The AP article, which ran on October 30, 2000, the day before the

Expedition One launch, is a gem not so much because it explores this thorny multinational branding issue in any detail, but for its end quote by Robert Cabana—retired U.S. Marine, former space shuttle pilot and, at the time, hard-driving deputy ISS project manager. Consciously or not, Cabana's words spoke volumes about the unavoidable fate of Mir.

"When we get rid of Mir," Cabana said, "then it [the ISS] will be THE space station. OK?" The quote's emphatic article, "THE," is verbatim and, one concludes, reflective of Cabana's exasperation with the whole Mir issue—an exasperation that, by late 2000, NASA's leadership no doubt shared wholeheartedly.

Cabana wouldn't have to wait long. By the time of the launch of Expedition One to the ISS, Mir's fate was sealed, although it would take some time for this to be clear to everyone involved. Despite its initial success, the walls of the space establishment were closing in around MirCorp. One after another, potential MirCorp business arrangements in Europe collapsed under pressure from the top brass in the European Space Agency (which, aside from Russia, is the most financially vested ISS partner), while the bullying from NASA and Congress had reached an almost unbearable level. "It was intense," Manber said quietly. "I received phone calls from NASA friends saying [Goldin] was calling people telling them not to work with us. There were people in America who told me I'd never work in the industry again. That hurt."

For all the mudslinging, Manber was quick to point out to me that MirCorp also had some allies within NASA. "Every time there was a meeting at the upper echelon in NASA on what to do about MirCorp, I was leaked the minutes," Manber said. Still, the NASA-led blockade was working. While MirCorp had put together some small deals, for the most part the company continued to run up against ISS contracts or hopes for contracts.

By the fall of 2000, the station was fast approaching the point where another mission, and more money, would be required to maintain Mir's orbit. Anderson and his partners had injected more cash into MirCorp earlier in the year and Tito had put several million into escrow for his trip, but that money couldn't be touched until he was cleared for launch, which wouldn't happen until there was more certainty that Mir would stay in space. Worse, by the summer of 2000, the dot-com bubble had begun to deflate, taking the net worth of Anderson and his MirCorp partners down with it. Pressed to the limit financially, they couldn't afford to invest more millions in MirCorp.

On October 23, 2000, just a week before the launch of Expedition One to the ISS, an unmanned Progress ship completed a refueling trip to Mir—funded by Russia, not MirCorp. That same day, Russian deputy prime minister Ilya Klebanov made the statement that would turn out to be definitive regarding Mir's fate: "We are planning to bring the Mir down into the ocean at the end of February." Manber responded a few days later with an open letter to Russian president Vladimir Putin in the Moscow newspaper *Kommersant,* reminding Putin of his pro-Mir statements that April and calling on him to commit government funding to keep Mir alive. Manber wrote that the destruction of Mir would "play into the hands of those who do not want Russia to be equal in space. I believe you understand that Russia won't be able to create such a masterpiece [as Mir] within the next 10 years."

This time, Putin was tight-lipped about Mir. The Russian president had in the early months of his presidency weathered a storm of criticism for his government's inept handling of the tragic loss of the nuclear submarine *Kursk* with more than a hundred crew members, and the Russian economy was continuing to struggle for its existence. In its ISS partnership, NASA offered Russia a steady paycheck. To sacrifice that for the sake of national pride was a luxury Putin could not afford.

Defying NASA and even its own space agency, Energia stepped up publicly, issuing a prescient statement that ISS funding cuts would reduce that station's effectiveness. ISS cutbacks, Energia argued, meant Mir was needed more than ever. This was not the first time Energia had attempted to directly link continued operation of Mir to the future of the ISS, proposing the use of existing Mir components as either substitutes or backups for the new station. In 1995, Russian space officials had proposed using Mir's core module as the ISS service module instead of building a new module, called Zvezda. Politically, this was no-go for NASA. The Russians tried again in November 1998 on the eve of launching the first ISS module, Zarya, by asking NASA if they could delay the launch by ten hours to allow Zarya to be placed closer to Mir. This would have allowed the transfer of equipment and crews between the ISS and Mir—thereby, the Russians claimed, making Mir a true backup to the ISS. NASA again said no, with its spokeswoman Lynn Cline saying simply that "we have all already agreed on another schedule."

Privately, some NASA managers were concerned that with Mir nearby, the Russians might persist in their efforts to make it competitive with the ISS. If for any reason the ISS never grew beyond Zarya, it would be relatively easy for the Russians to make that module compatible with Mir, thus further solidifying Russia's dominance of orbital space. Other, more paranoid whispers offered that the Russians might stage a cosmic coup by simply linking Mir to the completed or semi-completed ISS.

But Energia's requests were nonstarters for another reason, one that remains a point of bitter contention between the American and Russian space programs. In 1993, NASA grudgingly changed the proposed orbital inclination of the ISS when Russia was put in what NASA managers call the critical path of the station's development. "Orbital inclination" is an angular measurement taken from the Earth's equator, and measured in degrees; if something orbits around the poles, for instance, its orbital inclination is 90 degrees.

Since much of the ISS deal was predicated on keeping the Russian space program intact, using Russian spacecraft was considered a political necessity. Thus, when Russia was integrated into the project, the station's inclination was changed to 51.6 degrees, the same as Mir's. Prior to Russia's ISS involvement, the station's orbital inclination was going to be a shuttle-friendly 28.8 degrees—since Florida, where shuttles are launched, is much closer to the equator than Baikonur, the launch site of most Russian spacecraft. Changing the ISS's inclination to accommodate Russian spacecraft made it harder for shuttles to reach the station.

To this day, die-hard station advocates grumble about the time and resources wasted in positioning the ISS in favor of Russian, not American, access. In such a context, NASA was not about to do Russia any favors with regard to Mir. (As it turned out, the fact that the ISS is accessible to Russian spacecraft may be the only reason it is still aloft at all. After the *Columbia* disaster grounded the shuttle fleet, the only way to boost, supply and—importantly—evacuate the station was via Progress and Soyuz spacecraft.)

As 2000 rolled into 2001, MirCorp knew it had lost the Mir space station. But there was a final bit of insult added to that injury, at least for Tumlinson. Although it had shifted into the background, the plan to use a tether to boost Mir was still grinding its way through the ITAR process. Tumlinson said that by the fall of 2000, FINDS had spent nearly a million dollars on tether research and development, plus another $300,000 in legal fees to wrestle with ITAR.

"I'll never forget this," Tumlinson told me, closing his eyes and smiling. "End of November [2000], I get a call from one of the legal guys saying, 'You've been approved by the State Department to begin [tether] meetings with the Russians.' Twenty minutes later, I get a call from Jeff [Manber], 'Putin has just signed the formal de-orbit order.' I don't believe in conspiracies, but . . . well, I'll leave it at that."

At the scheduled moment of Mir's death in March 2001, flying smoothly over a South Pacific that sparkled turquoise and magenta in twilight, we eagerly looked up and saw the stars. We would see nothing else. Concerned that Mir might hit South America as pieces of the Salyut 7 space station did in 1991, the Russians at the last minute decided to lay down extra insurance against that possibility, firing the station's descent rockets sooner and burning them longer than planned.

As it turned out, Mir went down right off the coast of Fiji, several hundred miles northwest of our position. The cruel joke is that had we sat sipping mai tais on the beach of our comfortable hotel, we would have had the best seats on the planet. As our plane swung around and headed home, many in our group cursed our misfortune. The Russians in our company (which included several former Mir cosmonauts) were taciturn and polite, saying that they were just pleased the reentry had gone smoothly and adding, cryptically, that it was now a new day for the Russian space program. Later that night, some of them got drunk and cried.

An American tourist, standing on just that stretch of beach with a handheld video camera, caught what would turn out to be the definitive footage of Mir's destruction—a shaky, fifty-five-second-long video showing three or four streaks of light cutting a fast arc across the fading blue sky. Two years later, when I first saw the grainy footage of *Columbia*'s final moments, I was immediately reminded of Mir's demise. The recordings seemed eerily connected, and I felt, almost inexplicably, a similar sense of loss.

It is impossible to compare the destruction of Mir and *Columbia* on a human scale; only one involved loss of life. Yet their separate ends were, in the broader context, simply two more milestones in the slow, tortuous diminishment of a human spaceflight paradigm the world has too long bought into. Both Mir and *Columbia* had the chips stacked against them from the start. However hard their makers worked to keep them culturally relevant and technologically sound, they were doomed. They were part of a space future that couldn't be sustained.

Over time, the seven astronauts who perished later on *Columbia* will no doubt become etched in the public consciousness like those of *Challenger*—beacons of adventurousness and achievement, tragically heroic symbols of sacrifice. But the touchy question of whether or not their sacrifice was in service of anything really valuable was a distant second-order concern, even after the mourning had subsided. With the wreckage of *Columbia* still smoldering, its crew were instantly canonized as martyrs to the Cause of Space, just like the *Challenger* crew before them. In turn, the Cause of Space was quickly given the cheap cloak of homily: the shuttle missions were a quest for discovery, or a pursuit of knowledge, or critical to extending humanity's horizons. For all the speeches and commentary pieces, though, few pundits spent much time discussing, in any meaningful way, what the Cause of Space was anymore—or more importantly, what it should be.

Yet NASA, and America, responded to human spaceflight tragedy much more thoughtfully and specifically almost exactly thirty-six years before *Columbia*'s fiery end, when on January 27, 1967, Virgil "Gus" Grissom, Roger Chaffee and Edward White perished in a fire during a test of their Apollo capsule in the breathless run-up to the moon missions. It was, at the time, the worst disaster in American spaceflight history.

Like the shuttle astronauts, the *Apollo 1* crew died horribly (they were simultaneously asphyxiated and immolated) and, worse for NASA, their deaths occurred as the world hung on every test and advancement NASA made in its quest to beat the Soviets to the moon. While the *Challenger* and *Columbia* crews arguably had one celebrity apiece—the teacher Christa McAuliffe on the former and Ilan Ramon, the first Israeli astronaut, on the latter—all the *Apollo 1* astronauts were household names; Grissom, already the second American to travel into space, was a national hero.

As was ultimately revealed to be the case with *Challenger,* substandard engineering and rush-rush schedules were ultimately pegged as the cause of the *Apollo 1* disaster. Most likely, investigators concluded, a loose wire sparked what would normally have been a small and easily controllable fire; in the Apollo capsule, rich with nearly pure oxygen as was planned for in space, it became a raging inferno. Because of the extremely high atmospheric pressure inside the capsule, the crew hatch couldn't be opened from the inside; by the time the hatch was opened from the outside, it was too late.

Apollo 1 was an avoidable tragedy, a fact even NASA admitted. Public, media and political critics poured scorn on the Apollo enterprise, calling for its immediate end. Yet unlike *Challenger,* very quickly NASA emerged bloodied but emboldened by the *Apollo 1* disaster, and more determined than ever to pursue a daring, aggressive agenda that led to the moon. The NASA that emerged from *Challenger*—"emerged" in this case being in 1988, when the shuttle finally flew again—was more neurotic and inward-looking than ever.

The difference between *Challenger* and *Apollo 1*—separated by less than two decades—must be measured by more than the absence of a national mandate by a charismatic president. By the time of the *Apollo 1* disaster, John F. Kennedy—whose 1962 Rice University speech focused America on the moon—was four years dead. Kennedy's successor, Lyndon Johnson, was struggling to keep the Apollo program afloat in the face of an expensive war in Vietnam and boiling civil unrest at home. The great architect of the Soviet space program, Sergei Korolev, had died unexpectedly the year before the Apollo fire, throwing the Soviet manned space program—and its ability to reach the moon—into serious doubt.

No doubt Johnson, a clever politician, could have crafted a strategy for the nation to exit gradually and gracefully from Apollo, or to ease it to a more modest goal; even Kennedy had once proposed that America and the Soviet Union jointly pursue the moon (only to be soundly rebuffed by Nikita Khrushchev). If ever there was a convenient moment to end human spaceflight, *Apollo 1* provided it.

But by 1967 space had seeped into the popular consciousness, not only in America but also around the world. It had become the measure of modernity, in which "space" not only drove scientific discourse, but also influenced fashion, architecture, philosophy, literature and popular culture. Spacefaring really did feel like destiny. All things seemed possible for humans in space; in turn, space seemed possible for all people. Recognizing how deeply ingrained the human spaceflight endeavor had become gave Johnson, and NASA, the strength to push past the critics. Far from ending the Space Race, the death of the *Apollo 1* crew spurred America and its space program to pursue its lunar goal even more aggressively.

The eventual success of the Apollo program did nothing to diminish popular hopes for even greater things in space. But in Apollo's achievement were buried the seeds of its downfall as a driving force for further cultural interest and investment. And at the source of the disillusionment that now exists with regard to human spaceflight is NASA—the once pioneering space agency that decided, very early on, that not only would it keep the keys to the castle, it would shoot anyone who approached the drawbridge. In making such a decision, NASA changed from an organization that would risk its future for an outrageous goal— bold and daring, worthy of sacrifice—to one that wanted more than anything else to simply survive. *Challenger* solidified that posture; most likely, *Columbia* will further codify it.

Even before the first Apollo success, NASA had begun to distance itself from both its original Space Age promises: that human exploration would move beyond the moon, and that spaceflight opportunities would expand to encompass regular people, not just professional astronauts. Leaning hard on its Apollo triumph, NASA has convinced the world that it is through NASA, and only through NASA, that human spaceflight can move forward. This view is promoted by the mainstream media and purveyed through NASA's multimillion-dollar education and outreach programs, and through the hundreds of grants to schools, universities, small businesses and nonprofit organizations it makes each year. It is underscored in popular culture, bolstered by celebrities (Tom Cruise narrated the 2002 IMAX film *Space Station*, which is about the ISS) and, above all, rewarded by politicians whose enthusiasm for NASA is typically in direct proportion to the number of the agency's employees in their constituency.

The net effect of this strategy is that NASA's word became final on all things related to space. If NASA says human spaceflight is too difficult and dangerous for anyone but NASA to pursue, the public believes it. When NASA says the space shuttle program is safe, even though internal and external safety panels overwhelmingly conclude it isn't, people are convinced. If NASA says the best humankind can do is to spend billions each year to putter around aimlessly in Earth's orbit, most people accept that, too. If NASA insists that Mir is standing in the way of progress, then Mir must go, as it did. Thus, the pace of human spaceflight is whatever pace NASA says it should be. Anyone who dares suggest it might be different is, typically, dismissed as a dreamer or a heretic—either way, out of touch with NASA's "reality."

But over the years, as NASA hewed ever closer to this self-referential agenda, some of its most ardent fans—like Rick Tumlinson and Jeffrey Manber—became its harshest critics. As one NASA engineer confessed to me (albeit anonymously), the battle over Mir was a "warning shot" across the bow of a complacent space establishment that has lost its way and, consequently, lost its audience.

NASA and the alternative space community have distinct yet overlapping visions of what might be accomplished that have not been reconciled. The irony is that each side needs the other. Achieving a unified agenda for those who would create a near-Earth human spaceflight market and those who would once again undertake magnificent voyages of interplanetary exploration would further both causes. The fact that only the entrepreneurs seem to have realized this is not exactly startling; what the entrepreneurial community needs from NASA is not only technology and cash, but also the one commodity NASA refuses to relinquish—control.

Perhaps the strangest aspect of what might otherwise appear to be a classic David versus Goliath conflict—NASA's "space establishment" versus the alternative space community—is that everyone, at some level, has drawn inspiration from the same root sources. Like adherents to skirmishing faiths who somehow believe in the same god, with space (as with religion), the difference is all in the interpretation.

One Giant Leap

By its coldest definition, a frontier is what lies just on the other side of a boundary between the known and the unknown; beyond that boundary, in that unfathomable place, all bets are off as to what you know and what you don't. In its less austere and more popular usage, a frontier is an untrammeled geography of material potential, intellectual wonder and physical danger—the latter quality being typically re-labeled as adventure by the more romantic of our species.

Star Trek's bold "final frontier" pronouncement notwithstanding, space may be the most counterintuitive frontier humans have ever encountered. You cannot walk across it or swim through it, nor can you even survive in it except by enveloping yourself in a miniaturized replica of the singular environment from whence you came (space suit, spacecraft, space station). Perhaps because it is so unlike anything in natural human experience, the idea of traveling into space has intrigued humanity since the beginning of recorded history. Clay tablets dating from 2350 to 2180 B.C., discovered in that part of the world we now call Iraq, tell the epic story of Etana, who flew on the back of an eagle "far into the sky" toward the home of Ishtar, known to modernity as the planet Venus.

Unfortunately for Etana, the eagle got tired or nervous (the poem is unclear), and Etana ended up back on Earth without reaching Ishtar's home. Fictional astronauts of the future were to have more success than Etana, from Jules Verne's cannon-launched moon travelers to the swash-buckling escapades of Buck Rogers to the warp-driven voyages of Kirk, Spock and their many spin-off successors.

The motivations that drove these and other sci-fi heroes to extraor-dinary adventures in the cosmos were usually quite ordinary: they zipped around because of love or money, or for the sheer thrill of it. Other fictional space travelers have sought God, or the wellspring of time, or other such intangibles. The "space psychologist" Kris Kelvin, who served as the protagonist in the Polish novelist Stanislaw Lem's *Solaris,* discovered a sort of grim immortality out among the stars. Valentine Michael Smith, the hero of Robert Heinlein's hugely influential *Stranger in a Strange Land,* fell to Earth as an extraterrestrial Christ. (Much to Heinlein's horror, several "churches" were started with *Stranger* as their urtext.)

Real-life space travelers have also had some famously metaphysical moments. On Christmas Eve 1968, the crew of *Apollo 8*—the first humans to orbit another world—read from Genesis by way of describing what it felt like to be so close to the angels. The attachment of such spiritual import to space is nothing new. Before it was a frontier, for thousands of years "space" was merely the sky at night, and the sky was literally heaven—source of light, warmth and rain, and therefore home to God.

Whatever the motivation, when Stanley Kubrick and Arthur C. Clarke coined the term "space odyssey," they struck a note that was both immediately relevant and rich in historic resonance. For generations of philosophers and average folks, space has held a unique promise of enlightenment somehow lacking here on Earth.

The actual saga of space travel could certainly be described as an odyssey, although what it has revealed has been, for the most part, the rather limited impact of the odyssey itself. This truth is most keenly felt when it comes to human spaceflight, once the centerpiece of our Space Age future, but no longer. Is there any meaningful link between your life and the thundering engines of the space shuttle or the slow-motion balletics of space station astronauts? They're impressive, but what are they for?

Few people in the United States asked such questions at the height of the Space Race. The "for" was to stretch the limits of American ingenuity and therefore demonstrate the supremacy of the capitalist way in the face of the Communist menace. Yet even before the applause began to die down after Neil Armstrong's "one giant leap" for humankind, the sobering task of actually figuring out what to do next quickly dawned on all involved. After a few more missions, the powerful Saturn rockets used to send Armstrong and his colleagues to the moon would never leave orbit again. One Saturn would send Skylab on its three-year trip around the Earth, another would boost an Apollo spacecraft into service as an ambassadorial barge to dock with a Soviet Soyuz in an attempt to demonstrate that, really, the whole Space Race thing was just water under the bridge.

The rest of the Saturns ended up as theme-park attractions—if you want to see what $100 billion worth of 1960s technology looks like, just head for a NASA visitor center in Texas, Florida or Alabama and cast your eyes on one of these awesome metal dinosaurs, products of the best engineering of their day. When it came to sheer lifting muscle, the Saturns are still unmatched. The space shuttle can launch only about one-fifth the payload of a Saturn and can barely muster enough power to get into the most minimal orbit around the Earth—forget about reaching the moon. Nonetheless, the Saturns and the equipment to build them are gone. Even the paper plans for building the Saturns were gathered up and destroyed.

For their part, after "losing" the moon, the Soviets at first continued to build bigger and more powerful rockets, including one called Energiya that was at least as powerful as the Saturn, although it came along too late to matter. They would launch interesting things like Mir and even a version of the space shuttle that flew only once before being mothballed just like NASA's spare Saturns. But as communism proved to be an increasingly cruel mistress to the Soviet economy, Soviet and later Russian ambitions for spaceflight would wither. Ultimately, what ambitions were left would be sold to America—a cosmic coup de grâce whose defining moment was the forced destruction of Mir.

Before their rapid decline, the American and Soviet/Russian human spaceflight programs inspired a whole legion of entrepreneurs who figured that what would come next was what had always come next whenever military or political ambitions had opened a new frontier—be it on land, at sea or in the air. Wagon paths had become freeways, battleships had spawned cruise ships, and aeronautical advances financed from government coffers resulted in passenger and cargo planes that altered the perceived scale of our world. Since new technologies are inherently

expensive to develop, none of those things would have happened without government investment paving the way for private expansion,
which was what everyone figured space was poised for next.

In such a context, space in the post-Apollo world would have purpose and relevance in abundance. New air/space craft would shorten
planetary travel time by looping through space from one part of the
world to another, new drugs and materials would be created in the
singular perfection of microgravity, a whole new entertainment and
tourism industry would spring up around orbiting hotels and low-
gravity thrill rides. On the back of such close-to-Earth activities would,
in time, spring the realization of more distant ambitions: bases on the
moon, cities in free space, colonies on Mars. Far from believing that
such things were crazy, many thought they were inevitable; those who
did were convinced that private-sector ingenuity would provide the
spark.

And why not? As early as 1967, when ninety-five nations, including
the United States and the Soviet Union, signed the Outer Space Treaty,
prohibiting national ownership (although not private ownership) of
space or any real estate therein, the world had begun to acknowledge
that government control of space would soon be an obsolete idea, just
as government ownership of air travel had largely given way to the
private sector earlier in the century. So when the Space Race came to an
end, the entrepreneurs figured that, naturally, they would take it from
there.

It didn't happen that way. The effort to commercialize Mir was not
the first entrepreneurial space endeavor to flame out in the post-Apollo
era; if anything, its demise followed a depressingly consistent pattern.
Instead of opening the doors to competition and expansion in space—
and consequently opening space for the rest of us—NASA and its Big
Aerospace partners closed ranks. By freely wielding their considerable
political and economic influence, they soon created one of the most
complete and repressive monopolies in the history of American government.

Of the hundreds of entrepreneurial space enterprises like MirCorp
that have cropped up in the past two decades, only a handful have survived. Many of those that went over the cliff into oblivion had NASA

to thank for a healthy push. Those enterprises that survived did so largely at the pleasure of NASA or the military, usually because they could save some money for one organization or the other, not because anyone was interested in fanning the faint flames of a free-market space revolution.

Through the palliative lens of retrospection, Rick Tumlinson now views the Mir commercialization experience as a demonstration to the world that humans can do things differently in space—at least to some degree. Tumlinson is fond of opening speeches with a signature phrase: "Welcome to the revolution." That frames the idea that the best and most exciting time for humans in space hasn't happened yet.

In word and in deed, Tumlinson is perhaps the most catholic disciple of the man who, more than any other single person, is responsible for the fact that there is an alternative space movement at all. In 1985, when he met a mop-haired, soft-spoken Princeton physicist named Gerard K. O'Neill, Tumlinson's commitment to opening up space for the masses was confined to fervently wishing that it would happen and—when there was a job in it—working on short promotional or documentary space films. Meeting O'Neill changed Tumlinson's life forever, much the way that O'Neill himself altered the way the space community—alternative and other—saw itself, and its future.

Gerard O'Neill's is one of three visions of space that have come to embody both its value to the world and the manner in which humans might engage with its vastness. The first might be called von Braunian, after the man perhaps most responsible for shaping human spaceflight into a quintessentially military-industrial enterprise: Wernher von Braun, the technically brilliant, smooth-talking father of the legendary V-2, the Nazi missile that served as the basic model for every Space Age rocket to follow, including the Saturns that sent Americans to the moon.

Von Braun was among the first celebrities of the Space Age, although his initial fame was understandably of the notorious variety since his V-2 missiles killed thousands of people—mostly civilians—during World War II (von Braun was also an honorary member of the brutal

Nazi SS). However, von Braun's image got a makeover once he arrived in America. In the early 1950s, von Braun became indelibly linked with America's growing space prowess by authoring or coauthoring a series of articles in the popular magazine *Collier's*. The opening story in the series bore the provocative title "Man Will Conquer Space Soon."

Filled with fantastical propositions for orbiting space stations, powerful rocketships and missions to the moon and Mars, the *Collier's* articles were also full of faux-technical flourishes that, accompanied by the colorful artwork of the famous "space artist" Chesley Bonestell, performed critical public-relations spadework for America's ramp-up to the Space Race. The *Collier's* series also led to von Braun's collaboration with Walt Disney on a space-themed television series, in which the charismatic and telegenic engineer became even more famous. (Von Braun and Disney also collaborated to develop Disneyland's space-flavored Tomorrowland attraction.)

Wernher von Braun helped set a particular tone for how humanity's venture into the cosmos would be managed. According to the von Braunian doctrine, space is indeed a frontier full of new worlds to conquer. However, it is a frontier best left to the government to deal with (specifically the American government—but one suspects it would have been the Third Reich had World War II worked out differently). The expertise and, most crucially, the priorities for space exploration, goes the von Braunian thinking, are best controlled by those at the reins of the broader national enterprise, who by default surely must have the greater good firmly in mind.

Today as ever, the von Braunians find their most comfortable home within NASA and its Big Aerospace partners, who certainly accept on principle that the public should be engaged in space exploration, since, after all, the public is expected to foot the bill. As well, the von Braunians are deeply committed to human space travel, since that concept is the foundation of the salad days of the Space Age, for which they are ever hopeful of a reprise. For all this, actual public participation in human space exploration was never really on the von Braunians' agenda, whatever the NASA spin machine may have said to the contrary. Space is too technical, too complicated and just too dangerous for mere mortals to deal with directly.

On that last point, it is also a von Braunian tenet that space demands military imperatives of national scope, and NASA's work today links closely with many aspects of military research and operation, as it has from the beginning. In the March 22, 1952, issue of *Collier's,* von Braun wrote, "Small winged rocket missiles with atomic war heads could be launched from [a space station] in such a manner that they would strike their targets at supersonic speeds. . . . In view of the station's ability to pass over all inhabited regions on earth, such atom-bombing techniques would offer the [station's] builders the most important tactical and strategic advantage in military history." True to form, the first serious effort to develop an American space station was not Space Station Freedom or even Skylab, but the Air Force's 1960s-era Manned Orbital Laboratory program, which ultimately fizzled out when spy satellites and ICBMs proved to be a more effective combination of military space assets. The very term "aerospace" was coined by the Air Force in the 1950s as a way of asserting that military branch's desired primacy over both air and space.

"The Cosmos is all that is or ever was or ever will be," wrote Carl Sagan at the opening of his book *Cosmos,* the best-selling addendum to the television series of the same name which took the world by storm in 1980 and (coupled with strategic appearances on *The Tonight Show with Johnny Carson*) made Sagan a household name. Sagan probably did more for popular scientific literacy than anyone else in the latter half of the twentieth century, and the *Cosmos* series won a clutch of awards and was a huge commercial success, having been seen, it was estimated, by roughly 3 percent of the world's citizenry. Sagan's series-related book won a Pulitzer Prize and lived comfortably on the *New York Times'*s best-seller list for a year and a half.

While Sagan's interests ranged from nuclear disarmament to social evolution, he will forever be associated with his lyrical descriptions of the wonders of space, and his status as a media icon helped shape a second popular view of how space should be perceived. In the Sagan universe, space is not so much a frontier as it is a temple of the mind, a boundless source of brain-boggling mysteries being slowly unlocked by the patient hand of science. For the rest of us, space is best put to use as a source of constant contemplation on our own insignificance.

What could our worries and squabbles possibly amount to in a universe that is composed, as Sagan famously put it, of "billions and billions" of stars and galaxies?

As such, space is something akin to a virtual theme park; you can't actually go there yourself, but you can look at it, admire it and play out ideas and philosophies within its increasingly astounding theoretical confines, a view which was consistent with Sagan's occasional sideline as a sci-fi writer (*Contact* is perhaps his best, and best-known, work). With the major exception of military imperative (since Sagan was a vocal opponent of the global war machine), the Sagan view of space is remarkably compatible with the von Braunian one. Both advocate leaving space in the hands of the experts, with the public cheering from the sidelines and dutifully footing the bill.

Sagan had much in common with the man who would come to rival his position as a cosmic sage by fashioning a third modern framework for approaching space. Like Sagan, Gerard O'Neill was a native New Yorker whose stellar academic and scientific résumé earned him a high-profile teaching post at a major university (Sagan was at Cornell while O'Neill, seven years his senior, was on the physics faculty at Princeton). Both men were handsome, articulate and charming. Perhaps because they possessed that rare ability among scientists to communicate with a lay audience, both were often targeted for scorn by their more ascetic peers in the science community: in Sagan's case, he was infamously, and unjustly, rejected for membership in the prestigious National Academy of Sciences.

Sagan and O'Neill were also distinguished from their nose-to-the-grindstone colleagues by their high-profile positions on the signature issues of the late 1970s and early 1980s, at the height of their influence. Their ascendance to stardom followed closely on the heels of the Club of Rome's dire predictions about planetary sustainability as put forth in the 1974 book *Limits to Growth,* as well as the first real muscle flexing by the global environmental movement and the emergence of its own superstar advocates, like the Sierra Club's David Brower. Where O'Neill, Sagan, Brower and other seemingly disparate minds came together was through the overarching issue of planetary sustainability, which was itself comprised of four primary issues: overpopulation, the depletion of natural resources, the environmental degradation that accompanied that depletion and global conflict.

Sagan and O'Neill were profoundly concerned about such things, but where space was concerned, their visions of the future could not have been more different. For Sagan, human space activity merited little support in light of planetary sustainability issues, particularly the threat of nuclear war. "The cost of major ventures into space . . . ," Sagan wrote in *Cosmos,* "is so large that they will not, I think, be mustered in the very near future unless we make dramatic progress in nuclear and 'conventional' disarmament. Even then there are probably more pressing needs here on Earth."

Sagan was not opposed to the concept of human space exploration, even human expansion and settlement. He wrote frequently and with eloquence about future deep-space pioneering by human explorers, and one of his earliest brushes with fame came in 1961 when, as a twenty-six-year-old astronomer at the University of California–Berkeley, he made headlines with his proposal that the choking atmosphere of Venus might eventually be reconstituted to support human life. But Sagan believed that any real commitment to human space exploration, let alone paving the way for popular access to space, should come only after humankind had solved its most pressing problems. Until that happened, robotic probes—cheaper, more robust, more purely science-focused than human missions—should be the main components of space exploration.

Gerard O'Neill's ideas on such matters form a distinctly different vision of space exploration than that of Sagan or von Braun. Like Sagan and other intellectuals of his generation, O'Neill was anxious about the rate at which wealthy countries were getting wealthier and consequently consuming an increasing share of the Earth's dwindling resources, while at the same time the have-nots continued to outnumber the haves by an ever-growing factor. O'Neill had also concluded that such a volatile mix would inevitably lead to an escalating cycle of conflict, famine and environmental destruction. Halting this global death spiral, O'Neill believed, hinged on quieting that tireless handmaiden of overpopulation, poverty. Humanitarianism, while admirable, seemed to O'Neill to provide no real fix; to make a significant dent in the problems of the developing world (where according to current U.N. statistics, 85 percent of the world's population will live by 2050) and thus slow down the cycle of misery, the world's wealthiest nations would have to significantly downgrade their own standards of living.

O'Neill concluded that any such change was highly unlikely: if anything, the haves tended to hold more tightly to what was theirs as their wealth increased and outside threats to that wealth mounted. Yet even if such unprecedented first-world largesse were forthcoming, O'Neill didn't believe that reducing energy consumption in wealthy nations and distributing resources more evenly across the globe—the most popular conclusion of the day—was a viable, long-term solution to the problem. For O'Neill, the answer was simple: growth was good, but it had to be major, sustained growth and it had to be for everyone. The wealth of the Earth's underdeveloped regions needed to be increased "not just by a few percent per year, but massively, by factors of ten or a hundred," he wrote in *The High Frontier,* the 1976 book that set forth his vision of space and made him an icon. "We can't begin to do this by giveaway programs."

A future where everyone had more of everything would necessarily accelerate the depletion of the Earth's natural resources and the degradation of its environment, which was unacceptable: as a livable place, the Earth would be used up within a few decades, if not sooner. O'Neill had a radical yet appealing solution to this conundrum. Unlimited low-cost energy, endless new lands and an inexhaustible source of materials "available without stealing, or killing or polluting" were ready and waiting, in space.

In his 1977 book *Colonies in Space,* the veteran aerospace writer Thomas Heppenheimer described Gerard O'Neill as "tall, quiet and a modish dresser." O'Neill had been an astrophile all his life, devouring the science fiction that, in part, had motivated him to pursue a career in physics. In 1967, at the age of forty, he applied for one of the new "scientist-astronaut" positions coming open at NASA (all previous astronaut candidates had been military test pilots), making it to the final testing phases in Houston before being turned down.

O'Neill's interest in space was undeterred. In 1969, with the Apollo program at full steam and with the space shuttle and (so it was thought) a human mission to Mars on the horizon, O'Neill posed a deceptively simple question to a special section of his Physics 103 class at Princeton: is the surface of a planet really the right place for an expanding technological civilization?

In answer to his question, O'Neill might well have expected his students to propose the colonization of the moon or Mars, but they instead concluded that if you want to expand into space on any large scale, planetary surfaces were not your best bet. While it was dramatically convenient for the crew of the starship *Enterprise* to visit planets with breathable air and a nice summer breeze, no planet in our solar system has an environment that comes close to being naturally friendly to humans, save for the Earth. "Earth-like" planets might eventually be discovered elsewhere in the galaxy (although the bevy of recent planetary discoveries have not turned up anything remotely similar), but human technology is probably centuries away from being able to reach them physically for a firsthand look. The ability to bend local planetary environments to suit human needs—as Sagan once proposed for Venus—is probably about as distant a proposition.

Thus, O'Neill's students concluded that the best way to expand into space in the near term was to build artificial worlds from scratch. The concept that evolved from this unorthodox conclusion took the form of hollow, rotating space colonies designed to house thousands and eventually millions of people: undoubtedly a staggering proposition, but in theory more plausible than reengineering an entire planet.

O'Neill further refined this idea over time. The first of what he came to call islands in space would be built at Lagrangian Point Five, now known to would-be space travelers everywhere as L5: a gravitationally stable, perpetually sunny location (the better for constant solar-power generation) between Earth and the moon, and thus a good spot for free-vacuum homesteading. Such space cities would be built from materials mined on the moon or asteroids (thus avoiding expensive launch costs from Earth) that would either be hauled to L5 by robotic or piloted rockets or, even more efficiently, launched from the mining sites in large batches by an electromagnetic catapult called a mass driver.

From this relatively close proximity to Earth, more space cities would be built, each one self-supporting through on-site agriculture, manufacturing and the generation of solar power via large, purpose-built satellites that could also provide power to the Earth as a commodity, much as electricity and natural gas are sold on Earth today. Some materials, like hydrogen, would be imported from Earth, thereby setting up a basis for trade. The first colonists would come to work the manufacturing facilities and solar-power systems and eventually, like the intimate villages they were envisioned as replicating, everyone would have some role in maintaining the prosperity of their artificial world.

Gerard O'Neill landed on the public stage with these mind-stretching ideas in 1974 when he wrote back-to-back articles in *Nature* and *Physics Today*. Despite being published in such highly technical journals, the articles attracted a surprisingly mainstream audience: the *Physics Today* article, titled "The Colonization of Space," was probably "one of the most photocopied science articles in history," wrote the historian Michael Michaud. Spurred on by this enthusiastic response, O'Neill unified and detailed his vision of humankind's space future in *The High Frontier*—a book whose great triumph was to logically extrapolate, from existing technologies and well-known social exigencies, an urgent and thrilling future that might otherwise seem to belong wholly to the realm of fantasy.

Like many self-styled populists, O'Neill brazenly drew on other thinkers for his *High Frontier* vision. The proposal for self-sustaining human colonies in space had been sketched out more than half a century earlier by Konstantin Tsiolkovsky, the self-educated Russian peasant whose writings in the early twentieth century laid the theoretical groundwork for virtually every aspect of the Space Age, from the design of rockets to the various uses of artificial satellites. In Tsiolkovsky's 1900 novel *Beyond the Planet Earth* (published in the 1920s), his space travelers built what he called mansion-conservatories. These were free-floating space communities that grew their own food and drew their raw materials not from Earth but from the moon and the asteroids, much as O'Neill proposed.

Likewise, the basic design for O'Neill's habitats—a massive center tube set between stacks of doughnut-shaped hubs, with the visual effect being something like a giant spark plug—can be traced to the

British physicist J. D. Bernal, who described such structures in detail in his 1929 book *The World, the Flesh and the Devil*. Designs and uses for power satellites were introduced in the 1960s by the scientist Peter Glaser, who became an O'Neill ally. Detailed research on space mining and manufacturing had occurred within academia and NASA for at least a decade before O'Neill described their use in *The High Frontier*. O'Neill's major innovation was to assemble these ideas into a blueprint for humans in space that was coherent and compelling, and both comprehensible to a nontechnical audience and rigorous enough to hold up under scientific scrutiny.

Far from its being a science-fiction proposition, O'Neill claimed that 1970s technology was sufficient to accomplish the Herculean engineering his propositions required—all it would take was willpower, and money. Standard rockets and the then much-anticipated space shuttle could handle the human and cargo payloads, and by the time *The High Frontier* was published both Skylab and the Soviet Salyut stations had proved that humans could live in space for extended periods of time, although microgravity had taken a toll on their physiology—a problem O'Neillian worlds would mitigate through artificial gravity generated by rotating the space cities, a scientifically sound proposition then as now. Even O'Neill's "mass driver" was rooted in accepted concepts of the time: the U.S. Navy had tested the idea as a means of launching carrier aircraft, and the scientist-turned-novelist Arthur C. Clarke had adapted the concept for lunar work even earlier, in a 1950 paper that appeared in the *Journal of the British Interplanetary Society* under the title "Electromagnetic Launching as a Major Contributor to Space Flight" (O'Neill and his students later built working mass-driver prototypes).

O'Neill also indulged his creative streak by lacing *The High Frontier* with a series of "letters from space" ostensibly written by early residents of his space cities. A letter to Brian and Nancy, who are considering joining a space settlement, from Edward and Jenny, who are already there, makes it clear why O'Neill's propositions caught on so quickly, and with such a large nontechnical audience unnerved by predictions of a rapidly diminishing lifestyle on the home planet. Although O'Neill went to great pains to state throughout *The High Frontier* that his space habitation proposals were not utopian, the letters paint a picture of life at L5 that is at the very least idyllic.

Living in Island One, a habitat of ten thousand people, Edward and Jenny describe their world as having a Hawaiian climate in which they live in an apartment of comfortably Western proportions, complete with a garden. Edward and Jenny acknowledge that they get Earthsick from time to time, a situation alleviated somewhat by free telephone and video time to Earth and free round-trip shuttles home. While they are technically living on Island One at the pleasure of the Energy Satellites Corporation—the "multinational profit-making consortium under U.N. treaties" that built Island One to generate solar power for Earth from space—Edward and Jenny say that far from being a free-floating Matewan, ENSAT gives them considerable rein to pursue their own social system and form of government. The result: "Of the people who came with us, more than half intend to stay after their five-year contract is up." As for Edward and Jenny in particular, "I think we're more likely to move further out than go back."

The concept of moving forward was the key to Gerard O'Neill's ideas and their wide-ranging popularity. *The High Frontier* attracted advocates in political and scientific circles, as well as some environmentalists who saw in the O'Neillian future a way to relieve the Earth's population burden and thus its ever-growing hunger for natural resources. After the book's publication, O'Neill became a media darling, appearing on talk shows, writing for newspapers and magazines and speaking before congressional hearings on the merits of space manufacturing, space solar power and space settlement. His allies included the founder of the *Whole Earth Catalog,* Stewart Brand (who was inspired enough by *The High Frontier* to put together his own book on space colonization); California's governor and presidential candidate Jerry Brown; and the powerful congressman Morris Udall, who was sufficiently impressed by the energy ramifications of O'Neill's plans to lobby (albeit unsuccessfully) the Federal Energy Research and Development Administration to fund further design studies of O'Neillian concepts.

Even Carl Sagan became an O'Neill supporter. Sagan spoke favorably of O'Neill's ideas in several of his own congressional appearances, and according to Sagan's biographer, Keay Davidson, he planned to devote the final episode of his *Cosmos* television series to space colonization, with O'Neill's ideas front and center. He was reportedly talked out of the idea by one of his collaborators on *Cosmos*, who dismissed O'Neillian colonies as "orbital Levittowns." Even so, Sagan and O'Neill found common ground as leading members of what Davidson calls the pro-technology sub-counterculture, who believe that technology, in present and future forms, could solve many of humankind's problems. Of O'Neill's space cities, Sagan once wrote that they might "provide the social mutations that will permit the next evolutionary advance in human society."

A critical appeal of the O'Neillian vision was its emphasis on space as a *personal* frontier. In this, O'Neill had a ready-made audience in the fan base of Robert Heinlein, whose writing profoundly affected how the space community, alternative and other, saw possibilities for its work. Indeed, Rick Tumlinson was not the only entrepreneur who confessed to me that reading Heinlein changed his life (in Tumlinson's case, it was *Stranger in a Strange Land*).

While there are many fine authors of modern science fiction—David Brin, Stephen Baxter and Larry Niven come to mind—few have been as influential as the great triumvirate of Heinlein, Isaac Asimov and Arthur C. Clarke. Unlike Jules Verne and H. G. Wells, they had the advantage of writing at the dawn of the Space Age and were thus able to meld traditional aspirations for space travel with extrapolations from existing or emerging spacefaring technology. Arguably, these three writers are responsible for turning science fantasy—in which the future bore little resemblance to the present—into science fiction, which attempted to divine a reasonably realistic future from what was, or might soon be, possible.

Each writer brought a different sensibility and style to the sci-fi genre. Clarke was the lyricist, and his most famous works—particularly *Songs of Distant Earth, Childhood's End* and *Rendezvous with Rama* (Rama being an O'Neill-like "space city")—are laced with a kind of mystic humanism: a sense of humanity as part of a splendid cosmic design whose most distant edges we will only ever get a peek at. Asimov's seminal work, the sweeping *Foundation and Empire* series, conveys a more pragmatic futurism, one bristling with technology, ambition and conflict, a future in concert with our present, expanded to a galactic scale.

Heinlein's stories are wry and political and often focus on rugged individualists struggling for freedom in a universe that seeks mainly to control them. Stories like *Farnham's Freehold* and *The Moon Is a Harsh Mistress* pit scrappy pioneers against the sinister machinations of the government conglomerate. It comes as no surprise that Heinlein's tales, with their emphasis on frontiers and revolutions and self-made men (and the occasional self-made woman) had a particularly American appeal.

Importantly, Heinlein's stories reflect a zealous pursuit of individual freedom that, taken to its extreme, has long been identified with the political philosophy known as libertarianism. Libertarians take the American Constitution's promise of individual rights very, very seriously; the pursuit of happiness is elevated above all other things, and any rules governing the body politic must first and foremost not interfere with an individual's ability to do just about anything he or she wants, short of causing physical injury to other members of the community. Such a modus operandi makes for an awkward template for civic governance, since millions of citizens exercising their unbridled individual rights are bound to trample on one another. For this and other reasons, as a distinct political effort libertarianism has been only moderately successful.

As a social philosophy in America, however, libertarianism wields significant clout; in recent years, the Republican Party has found success by voicing increasingly libertarian positions on issues like health care and education, with the individual's "right to choose" trumping any broader communal needs. And so it has come to pass that many political libertarians with real aspirations to power have ended up as staunch Republicans. In turn, "libertarian Republicans" have been among the most passionate advocates for competitive enterprise in space, most notably the former congressional majority leader Newt Gingrich and the former Reagan speechwriter turned congressman Dana Rohrabacher, a long-standing member of the House Science Subcommittee (and one of NASA's most vocal critics).

Libertarianism has also reached into the space community through other avenues. Many in the "new technology" sector, particularly in Silicon Valley and Seattle, have rabidly libertarian leanings consistent with their belief that creativity unfettered by government meddling (for which, read "regulation") is the only acceptable path to individual success. Not surprisingly, the high-tech community is also home to a large contingent of the alternative space movement; indeed, many entrepreneurial space projects are seeded with new-technology money.

Gerard O'Neill himself was not a libertarian (if anything, he was apolitical), and rather than believing that the government should simply step out of the way where space was concerned, he believed government, meaning NASA, was essential to success. That said, it was clear that O'Neill's vision gave the average Earthling a far greater role in the space enterprise than anything NASA proposed. It was also clear that O'Neill's supporters took this idea—the individual, not the government, as the future of humans in space—very seriously. And thus in its earliest incarnation, the alternative space movement was very much a libertarian space movement—a philosophical and political foundation that was almost wholly orthogonal to that of NASA.

As odd as it may seem, the idea of space as an individual frontier was anathema to NASA's way of doing business. From its beginning, NASA has operated as a collective of collectives—centers, programs, contractors—in which individual ideas and actions are valued only if they contribute to the communal good. Such a description may, for *Star Trek* fans, bring to mind the Borg, a cheerless race of cyborgs whose unpleasant calling card is "You will be assimilated. Resistance is futile."

It would be unfair to take the Borg analogy too far, but it is true that the NASA collective is not only tight-knit (witness the immediate closing of ranks in both the *Challenger* and the *Columbia* investigations), but of an almost religious solidarity in the service of a governmental, and ostensibly public, agenda for space. The hostility faced by anyone opposing the collective—particularly its funding priorities—increases as one moves away from the agency's core. Mavericks within NASA have always had a hard time but are tolerated as long as they don't push the envelope too hard. On the other hand, anyone seeking to affect NASA's mission from the outside, in any way, is typically viewed as an enemy. Like the Borg, NASA views its enemies in a harsh light: they must either be assimilated or eliminated.

At the height of O'Neill's popularity, there were certainly those within NASA who supported an O'Neillian view of human space exploration, as there are today. The popular Apollo astronaut Russell "Rusty" Schweikart gushed that O'Neill was his "hero," and in the context of a space program in rapid contraction called him "a yes of fresh air in the no of a closed and stuffy room." O'Neill had a similar effect on some in Washington: in 1975 the House Committee on Science and Technology called for an incredible $750 million boost to NASA's budget "to lay the foundation for advanced projects, such as moon bases and orbital colonies."

Such funding was too rich for Congress, but the House proposal forced NASA to take O'Neill seriously. That same year, the agency gave O'Neill a small study grant and sponsored an O'Neill-led conference on space colonies. NASA then tasked O'Neill to lead a more extensive study that resulted in the 1979 report *Space Resources and Space Settlements,* a compendious technical opus that included treatises like "Extraterrestrial Fiberglass Production Using Solar Energy."

In funding O'Neill, NASA was engaging in one of its classic defensive tactics. By throwing O'Neill a study grant or two, NASA placated critics who claimed the agency has lost its vision and was too focused on mere self-perpetuation. This inexpensive and relatively threat-free exercise also bought time for the agency to deal with O'Neill by reeling him in for closer scrutiny. Such a ploy has been repeated, to great effect, throughout NASA's history—particularly when NASA's ultimate goal, as in O'Neill's case, was to neutralize a threat to its interests.

In his preface to the most recent edition of *The High Frontier,* published in 2000, the physicist Freeman Dyson writes of the "real NASA" and the "paper NASA," which are useful terms to help decipher NASA's motivation in funding O'Neill's work while at the same time laboring to neutralize his impact. The "paper NASA" has long been engaged in maintaining an image that is bold, daring and committed to a *Star Trek* kind of future—the image that kept the public happily spending its tax dollars on space during the Cold War. Task forces are formed and studies are conducted, from which emerge colorful viewgraphs, PowerPoint presentations or, lately, Web animations detailing fantastical futures that are always, so it is promised, right around the corner. Media releases almost always accompany "paper NASA" endeavors, which spark think pieces in scientific and popular media and thus perpetuate the desired image.

Impressive as such efforts may seem, nothing in the "paper NASA" process costs very much and few of the recommendations that emerge from these blue-sky activities ever see the light of day. Such remarkable and visionary things require money, which is controlled and administered solely at the discretion of the "real NASA," which is all about the convoluted bureaucratic mix of political back-scratching, government appropriations, industrial contracts and—at the core of it all—the one inarguably tangible result of their interplay: jobs. The "real NASA" is concerned with keeping the dollars flowing from its immediate and long-tested sources. Where they flow to within the agency, as long as they do, is of considerably less importance.

For the real NASA of the 1970s and early 1980s, the big dollars in human spaceflight were flowing toward one project only: the space shuttle. Keeping the fiscal floodgates open meant getting the shuttle built and flying. Anything that drew focus from this goal was a threat, and since Gerard O'Neill at the top of his game had the ear of the public, the media and Congress, he also had the potential to endanger the shuttle by persuading those constituencies to think even bigger than the shuttle—the result of which could be that the shuttle would be left at the congressional roadside, bypassed by more ambitious activities that NASA might not be able to entirely control.

Ironically, O'Neill was one of the shuttle's most eager cheerleaders, since the shuttle figured prominently in the earliest stages of his space settlement plans. It was needed to loft materials into space for the first of O'Neill's solar energy and manufacturing facilities, and it was to serve as the technology platform for building larger vehicles to transport more materials and people to more distant locations. So taken was O'Neill with NASA's stated plans to fly the shuttle dozens of times per year that he once proposed that its flight rate be boosted to dozens of times per *month* (which didn't seem out of the question, given NASA's shuttle sales pitch at the time), so that the first space outposts could happen even more quickly.

For every O'Neillian proposition, there was a polite but negative NASA response. When O'Neill's supporters once proposed that a bare-bones version of Island One could be built for roughly the same cost as the Apollo program, NASA's response was to say that it would actually cost about six times that amount. Probably, neither side really had any

idea what the cost would be, but even if the actual price tag was somewhere in between these two estimates, an O'Neillian mini-city housing
several hundred people would have been a bargain compared with
today's International Space Station, whose ultimate cost is pegged by
some at around $100 billion—about what it cost, in constant dollars, to
send men to the moon.

O'Neill made NASA so nervous that in 1975—the same year as the
founding of O'Neill's radical L5 Society—NASA launched the more
conservative National Space Institute, appropriately putting the aging
Wernher von Braun at its head to muster support for NASA's von Braunian approach to space exploration and, pointedly, to counter the growing enthusiasm for an O'Neillian vision of space the agency had no real
interest in pursuing. The two organizations would eventually merge
into the National Space Society, which even today walks an uncomfortable line between promoting O'Neillian dreams of space travel for
everyone and NASA's more tedious von Braunian reality.

As it turned out, NASA probably didn't need to worry about being
overwhelmed by legions of O'Neillians; the shuttle, simply by becoming operational, delivered the deathblow to an imminent O'Neillian
future. By the early 1980s, it was clear that the shuttle would be able to
fly only a fraction of the flights originally promised and at a price tag
that had private industry and the military—the proposed core of the
shuttle's customer base—scrambling to go back to the relatively cheap,
expendable, missile-based launch systems the shuttle was supposed to
make obsolete. Even O'Neill eventually had to admit that his space
future, when recalculated to incorporate the shuttle's actual cost and
performance, seemed implausible from even the most optimistic economic standpoint.

Amid the diminished expectations brought on by the shuttle,
O'Neill's most noticeable impact was felt as part of the Reagan-assigned
National Commission on Space, led by former NASA administrator
Thomas Paine, who was perhaps the most O'Neillian of any administra-
tor in the agency's history. The commission was established in 1984 by
Congress and the White House to define the nation's space goals for the
coming century.

The commission's report was bold and ambitious. It proposed the
establishment of an outpost on the moon by 2006, spaceports at several

Lagrangian points (although not the megacities proposed by O'Neill) shortly thereafter, a Mars base by 2015 and the active development of space-based resources for further exploration and settlement. Because the report was released only a few weeks after the explosion of *Challenger,* NASA wasted little time in burying it.

It was around the time the National Commission on Space was ramping up that Rick Tumlinson first met O'Neill. He and a colleague were hired to make a documentary on O'Neill that, Tumlinson now says with a laugh, "was going to evolve into Gerry's version of *Cosmos,*" and in the process they interviewed dozens of people, both inside and outside NASA and its Big Aerospace partners. Tumlinson quickly got to know many key players and was eventually invited to offer his thoughts to the commission—the first of what would become dozens of such politically charged testimonies, although it was the last in which Tumlinson expressed unbridled faith in NASA.

"When I testified, I said there was something wrong here, these people [at NASA] are heroes, they're risking their lives, they're doing great things," Tumlinson recalled in one of his conversations with me, slashing the air with his hand to punctuate each point. "I said, 'We've got to get the people involved, get the NASA astronauts on *Johnny Carson.*' " He hesitated, almost as if unable to believe he'd ever had such faith in the space establishment. "I was wrong," he quietly concluded. "I didn't know it yet, but the system itself was corrupt."

Initially, the alternative space movement that Tumlinson would eventually lead was characterized by outright defiance of the NASA-led status quo and of the belief that only private efforts, alone, could ensure a robust human future in space. It was also, to put it bluntly, a bit loopy; in its early days, members of the L5 Society were known to show up at congressional space hearings wearing *Star Trek* gear, complete with fake "phaser" pistols and rubber Vulcan ears. Over time, Tumlinson and others coaxed the movement onto the more sensible middle ground from which it operates today. Tumlinson himself seems, politically and philosophically, quite liberal. Yet I've watched him laugh easily and converse deeply with politicians of the most conservative stripe imaginable. All of this, it seems, is in service of the bigger picture: an expanded human future in space.

"The 'NASA will open space for us' crowd was wrong," Tumlinson said to me. "NASA was simply not going to build Hiltons and space liner fleets à la *2001*. The 'get the government out of it all the way' camp was also wrong. The private sector was neither able nor willing to simply jump in and throw open the frontier without some participation and support from government in the early stages."

Before O'Neill died in 1992, after a long battle with leukemia, space projects reflecting his legacy were beginning to emerge: projects with the stamp of practical purpose and individual intent. One in particular might have resulted in a very different outcome for Mir had it succeeded—as most, at the time, thought it would. Yet in the end, the fate of the project known as the Industrial Space Facility demonstrates that the vicious protectionism NASA employed to the detriment of MirCorp is nothing new. If anything, it shows that the agency's strategies have simply become more refined with time.

Gerard O'Neill and Ronald Reagan would, at first glance, seem to have little in common. Reagan was unflinching in his use of American economic, military and political might to encourage the spread of "the American way" across the planet. O'Neill was essentially an internationalist, although he was not above some pragmatic nationalism when it came to promoting his extraterrestrial agenda. "Should we not seek a role for this country that can be of benefit to humanity as a whole, and can at the same time benefit directly our own people and our own economy?" he asked in *The High Frontier*.

Such a proposition seemed to O'Neill to be wholly consonant with America as a technologically and socially advanced nation. Since America had emerged as the clear winner in the Space Race, O'Neill held that it was only natural that it should lead the rest of the world into space, as NASA had long promised.

On the other hand, for all his free-market enthusiasm for space, O'Neill was a patriot only in the broadest sense of the word. He believed the world would be a happier place if everyone had the opportunity to enjoy a Western standard of living, yet as he wrote in *The High Frontier*, "It should be clear to us that we have no special magic to export in regard to governmental systems. Most of us are passionately attached to a democratic form of government, but it doesn't travel well." While

O'Neill believed that universal wealth and leisure tended toward democratic forms of government, he suspected that "if the human race does
achieve general affluence, and with it an increase in real human free-
doms, it will do so within the outward forms of many different forms of
government, and with many of the old polemic catch-phrases still in
regular use."

In spite of these fundamental differences, where space was concerned, both Reagan and O'Neill believed that space activities, being
necessarily the result of significant investment, had to be paying propositions to be meaningful and, critically, to spawn more such activities.
O'Neill's "islands in space" were envisioned as multinational interests
with significant return on investment very much in mind (think of
Edward and Jenny's ENSAT-run colony). For Reagan, space payback
took one of two forms: economic or military. Science, the much-praised
but perpetually underfed child of human space exploration (and the
soul of the Sagan vision), played a relatively minor role for both Reagan
and O'Neill.

In terms of economics, one of the linchpins of both Reagan's and
O'Neill's plans for space was research and manufacturing, albeit for
different reasons: Reagan wanted to extend America's capitalist dominance into the vacuum, while O'Neill saw such activities as the building blocks for his space movement. The idea of doing space-based
research and manufacturing was particularly hot when O'Neill was at
the height of his influence. In addition to the usual university and
"pure" research experiments—the first of which were conducted during the Apollo missions—industry was showing real interest in going
into space, and on a much larger scale.

Among the early corporate pioneers in this area was McDonnell-
Douglas, which in 1975 began a space-focused program in electrophoresis, a process used by pharmaceutical companies to separate and
purify biological materials. The effectiveness of electrophoresis was
limited by gravity, but in space, gravity was not a problem, and new
drugs might be discovered and produced there that literally couldn't
be conceived of on Earth. McDonnell-Douglas's plans were to make the
equipment, operate it in space and sign up pharmaceutical companies
as customers.

McDonnell-Douglas found its first partner in 1978: the drug giant

Ortho Pharmaceuticals, who saw in the scheme a unique means of refining and eventually manufacturing a drug to help treat anemia. Both companies, of course, saw the potential for profit.

McDonnell and Ortho signed on with NASA in 1980, with the knowledge that a space-based research facility more robust than the shuttle would soon be built. When Reagan approved the project now known as the International Space Station, commerce—if not outright profit—was high on the list of his motivations for doing so; in fact, had NASA not convinced Reagan that industrial suitors interested in lucrative microgravity research and manufacturing could eventually turn the station into a for-profit enterprise, it probably wouldn't have been built. The idea may seem ludicrous today—the ISS is hardly a thriving hub of capitalism—but such was the power of NASA's persuasiveness and its desperation to find an obvious purpose for the faltering space shuttle that Reagan bought the concept hook, line and sinker.

It's safe to say that no one anticipated the near-decade delay and multibillion-dollar cost overruns that plagued the station's development, but everyone involved knew it would necessarily take some time to build. Meanwhile, other companies were lining up behind McDonnell-Douglas and Ortho to test their fortunes in the space research and manufacturing business. In this scenario, some saw a market opportunity.

In the early 1980s, a group of Houston businessmen approached one of NASA's living legends, Maxime Faget, about developing a private space station to address the growing demand for space-based business. Temperamental, intensely political and a brilliant engineer, Faget was one of the few personalities in NASA commanding the same reverence as Wernher von Braun. Faget had not only designed the Mercury spacecraft—America's first successful foray into human spaceflight—but was also widely credited as having created the two-stage design for the space shuttle.

In 1981, the year of the first shuttle flight, Faget had retired from NASA for a consulting career. But his discussions with the Houston businessmen eventually led him right back to the agency via the formation of a company called Space Industries and its signature concept, the Industrial Space Facility (ISF). While the ISF didn't involve Gerard

O'Neill, it did reflect a specific O'Neillian proposition: the establishment of the kind of small orbital manufacturing facility that was, in
O'Neill's scheme, a prior necessity upon which the first human outposts were predicated. Even the timeline for the ISF was in line with
O'Neill's agenda. In *The High Frontier,* O'Neill suggested that such a
facility could be operational by the mid-1990s. The ISF was targeted for
operation in 1991.

The ISF appealed directly to the Reagan administration's space-for-
profit ethos and drew early backing from companies like Boeing and
Westinghouse. Critically, the ISF was pitched as an *uncrewed* experimental facility to demonstrate space-based manufacturing in preparation for more such work on NASA's space station: a complement, not a
competitor. In addition to Faget, Space Industries was chockablock
with ex-NASA talent, including the much-liked ex-astronaut Joseph
Allen as its executive vice president. In short, it seemed the perfect
prototype for individual enterprise in near-Earth space.

The ISF didn't appear to threaten either the shuttle or the space station, at least at first. Its usable interior space was to have been a small
fraction of Space Station Freedom's, and it was designed to support
humans only long enough to swap out projects and perform maintenance, which would happen no more than a few times each year. It was
to be serviced by the shuttle, but such servicing could be done as
an adjunct to other shuttle missions. As for on-orbit assembly of satellites and the construction of massive Mars-faring spacecraft—both of
which were among the many outrageous promises that originally sold
Freedom—the ISF was hardly competition. Next to NASA's albatross,
the ISF would be a hummingbird.

Very quickly, NASA employed its signature tactic to get a closer look
at the ISF: on February 28, 1984, just over a month after Reagan announced Space Station Freedom, NASA signed a study agreement with
Space Industries to further flesh out the ISF concept. The NASA press
release announcing the initial ISF study agreement even said that the
ISF "could at a later time be attached to the Space Station or placed in
orbit in close proximity to the Station."

But as the ISF evolved and the development of Space Station Freedom began to lag, the ISF began to look less like a complement than a

threat. An internal NASA memo dated February 26, 1987, stated that "the marketing for ISF in Japan and Europe is reportedly going well" and that "there is reportedly great interest expressed in the ISF project by DOD [Department of Defense]." Servicing foreign and DOD payloads was also envisioned as a major part of Freedom's work.

At the time of the memo's writing, the first component launch for Freedom had been pushed from 1992 to 1994, while the first launch of the ISF had held firm for 1991. More to the point, the first ISF launch would be the only launch, since the ISF would be put into orbit by a single shuttle flight, much as Skylab had been sent up on a single Saturn rocket. NASA reasoned that if the ISF earned the early affections of military and foreign customers, those customers might be inclined to invest in its expansion, possibly to the detriment of Freedom.

As it turned out, the shuttle inadvertently shattered the ISF's carefully orchestrated ambitions much as it had doomed Gerard O'Neill's sprawling space agenda. The 1986 *Challenger* tragedy was the nail in the coffin of many would-be space manufacturing enterprises, including that of McDonnell-Douglas, which had lost Ortho as a partner the previous fall largely as a result of difficulties in pinning down NASA on pricing and launch timing, issues that plague NASA's attempts to do business on the ISS even today. After Ortho's withdrawal, McDonnell-Douglas began promising talks with 3M, which had conducted some previous shuttle-based research.

Instead of Space Station Freedom, however, this time McDonnell-Douglas pitched the ISF as its space platform of choice. Then came *Challenger,* and 3M walked away. McDonnell-Douglas—which had by this time about 150 employees committed to the space-based electrophoresis project—decided to cut its losses and abandoned the space manufacturing business altogether.

Its corporate customers in retreat and its only means of transport suspended indefinitely, Space Industries was left with only one option for keeping the ISF alive. The company proposed to NASA that the agency become the initial primary tenant for the ISF to prove out space-based manufacturing not only for the NASA space station but also for the private sector, which NASA was depending on to please the Reaganites. From that point forward, for Space Industries, NASA's warm embrace gradually became a death grip.

NASA began its formal assault on the ISF in a December 1987 report analyzing NASA's potential ISF tenancy—an analysis it was ordered to do by Congress over the agency's protests. The report began well enough, admitting that the ISF had "a complementary and adjunct capability to the Space Station" and citing a pile of projects queued up for the grounded shuttle and elusive Space Station Freedom that could work well on the ISF.

Things went downhill from there. Because it was to be "man-tended" rather than constantly occupied, the ISF would require more autonomous systems than Freedom, which NASA said would make it more expensive than Freedom. Considering that NASA's ever-changing station had yet to leave the drawing boards, another of the report's criticisms of ISF is almost funny. The ISF, the report sniffed, "is a concept—specifications and hard data [are] not available." The report also complained that the shuttle schedule, were such a thing ever to appear again, would need to be rejiggered to accommodate ISF flights, which meant planning and money. All of these concerns were true to some extent, but behind such language was a different worry alto-gether. With every ISF advantage, the more unnecessary NASA's mega-station seemed to be.

As is often the case, NASA itself was divided about the ISF. An inter-nal NASA white paper dated December 8, 1987, noted that the ISF enjoyed "strong support from the White House, OMB [Office of Man-agement and Budget], Congress, and the Departments of Commerce and Transportation." The message to NASA from all of these quarters was the same: find a way to lease the ISF and thus ensure its existence. Strangely enough, the white paper was issued by the then-extant NASA Office of Commercial Programs (an office that would come and go over the years), which also favored the ISF lease arrangement on the grounds that "NASA has an obligation to the private sector to provide a manu-facturing facility in space." The paper also pointed out the obvious, at least from a commercial perspective: "The [ISF] needs to be in place before commercial customers will seriously plan for its use."

Then as now, commercial interests within NASA paled in influence compared with the "real NASA" concerns of keeping tried-and-true federal funding channels clear of impediment. Unfortunately for Space Industries, NASA had a particularly formidable gatekeeper in place during its struggle to bring the ISF to life. Just after the *Challenger* accident, James Fletcher was brought back to NASA to become its only two-term leader. In Fletcher's first term, from 1971 to 1977, he had led the charge to bring the shuttle to life, and he was brought back to NASA principally, if not solely, to keep the shuttle program alive, a task that seemed increasingly dependent on the space station. It didn't take Fletcher long to conclude that the ISF was enough of a threat to Freedom that it could not be allowed to see the light of day.

In the pantheon of NASA administrators, Fletcher was probably second only to the Apollo-era administrator James Webb in terms of sheer political savvy, although Fletcher's NASA legacy is considerably more questionable. Among other dubious triumphs, during his first term Fletcher almost single-handedly convinced Congress to award the shuttle solid-booster contract to Thiokol even though by most accounts the company had submitted the least attractive bid, both economically and technologically. But Thiokol was based in Fletcher's home state of Utah; a devout Mormon, Fletcher was reportedly pressured by the state's Mormon leadership to make sure Thiokol got the job—a decision that would come back to haunt NASA, and the future of space travel. *Challenger* was blown apart when a rubber ring separating two sections of a Thiokol booster burned through—a problem which had occurred several times previously (although without the same catastrophic result) and which the company and NASA had tried, and failed, to fix.

Fletcher put all of his considerable political skill to work in killing the ISF. He fired off letter after letter to Washington's most powerful political leaders, urging them to give a thumbs-down to the proposition of a NASA lease on the ISF. To Jamie Whitten, then chairman of the influential House Appropriations Committee, Fletcher cautioned that the ISF shouldn't be funded because microgravity research techniques were immature and Freedom was the only facility that could properly prove them out—this despite the fact that NASA's own studies had concluded that the reverse was true, that the ISF could actually pioneer such techniques for the station.

To the influential treasury secretary James Baker, Fletcher wrote that he was "troubled that the [ISF proposal] foresees proceeding on a non-competitive, sole-source basis at a time when many other entrepreneurs, large and small, are seeking to establish a foothold in the commercial space area." This was disingenuous in the extreme, not only because companies were not exactly beating on NASA's door with leasing bids for private space stations (two concepts similar to ISF had withered and died even before the *Challenger* disaster, largely for want of NASA support), but because NASA itself was well known for its noncompetitive, sole-source contracts, as it is today: in August 2001, *Government Executive* magazine reported that NASA's head of public affairs and several other employees had resigned because of "intense pressure from senior staff to award outsourced projects in an anticompetitive manner."

By early 1988, the ISF battle had gone public. In a January 15 editorial titled "A Rival for the Space Palace," the *New York Times* anticipated future concerns about the merits of scientific research on the International Space Station, saying that "with no humans lumbering around, the [ISF] would be free of vibrations, an essential quality for many space experiments." It also noted that assembling NASA's space station in orbit required twenty shuttle trips (which would turn out to be a low estimate) as opposed to just one for the ISF. But most of all, the *Times* noted, "NASA seems petrified that Congress may next wonder why it needs a palace in space if a mobile home would do nearly as well."

Fletcher chose not to take on such criticisms directly. Instead, he and several station-friendly congressmen (supported by station-friendly companies in their political domain) led a flanking move around the ISF's supporters and convinced Congress to approve an open competition for a commercially developed space facility, in which Space Industries would be one among many bidders—and for which NASA would largely design the requirements. The resulting Request for Proposals required a facility with nearly twice the electrical power of the ISF and a host of other requirements that the ISF was, NASA knew, unable to meet. It should be no surprise that the final conclusion of the report was that no commercial space facility was needed at all and that NASA's space station would do anything a private space station could, better and even cheaper.

The ISF was dead by early 1989. Space Industries survived for several years afterward by marketing various components that would have been part of the ISF; but, robbed of its primary product, the company eventually folded. No serious company has since emerged to take on the task of privately developed space facilities for research and manufacturing.

As for the International Space Station, it is hard to find anyone who believes that any serious research, let alone anything of real value to industry or the creation of a stepping-stone to a more robust human presence in space, will take place there. And of course, after the demise of Mir a decade after the ISF failed to get off the drawing board, the "space palace" had no rival at all.

Gerard O'Neill continued to fan the flames of his space-faring vision until his death in 1992. Much of his energy during the final years of his life was focused on the Princeton-based Space Studies Institute, which O'Neill launched in 1977 to pursue independent research into space-based manufacturing. The institute remains a small nonprofit operation whose leadership reflects several generations of individuals inspired by O'Neill. The institute's president is O'Neill's longtime friend Freeman Dyson, and its various advisory boards include former ISF vice president Joseph Allen and MirCorp's Jeff Manber.

The institute's most notable success to date was initiating the concept for NASA's successful Lunar Prospector probe, whose mission to conduct geochemical mapping of the moon—the better to know where to dig for raw materials—ended in 1999 with a spectacular controlled crash into a lunar crater in an attempt to determine the presence of water ice (none was detected; a similar controlled impact by a later probe had inconclusive results). Lunar Prospector was a NASA mission in name only and was designed and built privately for $63 million, about a tenth of what most NASA-commissioned probes cost.

Lunar Prospector's lead developer, the former NASA scientist Alan Binder, acknowledged that it was the first step toward a very O'Neillian goal: "to get a commercial lunar base set up within a decade or so." Based on his own experience, Binder had a recommendation for success. "The way to get things done efficiently," Binder said, "is to keep NASA out of your hair."

The Lunar Prospector concept emerged from a more modest proposition for human development of space written by O'Neill only two years before his death. The "Alternative Plan for U.S. National Space Program" included many of the hallmarks of *The High Frontier:* manufacturing of materials mined on the moon, a drive to identify sources of lunar water ice (hence the Lunar Prospector concept), space solar power and mass drivers to lift materials off the moon cheaply and efficiently. Two aspects of O'Neill's original vision were conspicuous by their absence. O'Neill's islands in space had been reduced to a footnote—"safe, comfortable habitats for construction workers and their families" were one of many by-products of successful lunar mining under the wistful heading "Visions of Future Potential." And the space shuttle was nowhere to be found.

The Space Studies Institute and the Space Frontier Foundation have paid to keep *The High Frontier* in print since O'Neill's death, although its distribution and influence are now considerably less than they were in O'Neill's heyday, when the idea of rotating cities in space seemed within the world's grasp. "We made a promise before Gerry died that we'd keep it in print as long as we existed," Rick Tumlinson told me somberly.

To a world that is often more frightened of technology than inspired by it, O'Neill's blueprint for humans in space may seem simply naive. Yet O'Neill understood, clearly, that realizing a more robust and more relevant future for humans in space hinged on more than technology. When—indeed, if—his vision would come to pass was not determined by "science but a complicated, unpredictable interplay of current events, politics, individual personalities, technology and chance," O'Neill wrote in *The High Frontier.* As it turned out, technology has arguably been the least of these factors in blocking a future that contains at least something of the purpose that O'Neill envisioned.

Indeed, it is purpose, that strangely prosaic quality, that has been lacking in human space activities since the days of the Apollo program. Inspiration, the quality most often put forward by the von Braunians as the prime driver behind NASA's human spaceflight activities, has proved a very poor driver in and of itself. NASA says that space, much like Mallory's Everest, is important because it's there. There is nothing

wrong with such a rationale as a partial motivation for space exploration, but as NASA's post-Apollo fortunes have shown, it is a hard sell
in and of itself. O'Neill made a more resonant case, and one that com-
plements the value of exploration for its own sake: humans would
explore and eventually expand into space because they needed it and
wanted it.

For all of this, there are two overarching reasons why space has
remained a decidedly von Braunian game, at least so far. The first truth
is that despite its populist rhetoric, NASA is a government-sheltered,
corporate-supported monopoly that will fight competitive enterprises
with all of its considerable clout and resources. The second truth is that
what ultimately doomed Mir, the Industrial Space Facility and virtually every other independent space enterprise to date is the cost of
getting there.

Escape Velocity

Robert Heinlein famously remarked that once you've reached low-Earth orbit, you're halfway to anywhere in the solar system. Low-Earth orbit starts at about one hundred miles above the ground, although most satellites and spacecraft orbit much higher. Still, at one hundred miles up, you're above the thickest of the atmosphere and the strongest gravitational pull; thus, you have spent the lion's share of the fuel needed to travel to the moon, to Gerard O'Neill's L5, or to any number of other relatively nearby destinations. Mars, many millions of miles away, can be reached in as little as six months. From such a lofty perch above our home world, even the outer planets are not out of the question for human travelers.

Being first and foremost a novelist, Heinlein spends little time in his space dramas underscoring the fact that where space travel is concerned, those first hundred miles are the hardest part. Just getting the rocket to the launchpad is a triumph of hope over experience, since technical, design and funding glitches have delayed and budget-bled many a space project into oblivion well before it left the ground. And it's not just the entrepreneurs who have been plagued with what's commonly known in the space community as the problem of access— getting from the ground and into orbit reliably and cheaply. According to statistics compiled by the Space Consulting Resources Group, between 1988 and 2000 launches of rockets built by the established rocket companies Lockheed Martin, Boeing, SeaLaunch and Orbital Sciences failed twenty-three times, to the tune of hundreds of millions of dollars in lost hardware. The year 1986 was a dark one for space flight not only because of *Challenger*, but also for the less remarked-upon failures of the Atlas, Titan and Delta rockets, which were lost along with their expensive payloads.

At their most fundamental level, rockets are little more than controlled explosives, which makes it even stranger that the world takes them so much for granted as the obvious means of getting into space. The Chinese invented rockets around the twelfth century, and until the twentieth century they were used mainly for entertainment, only rarely for war—and as for the celebrated British "rocket troops" of the Napoleonic Wars, the rockets in question were as dangerous to the British as they were to the French.

It was only in the last half century, when America and the Soviet Union converted Nazi weapons into human vehicles, that rockets—the ballistic missile, Buck Rogers style—solidified in the popular consciousness as the obvious way of getting into space. But even at the beginning of the Space Age, it very nearly happened differently. In many ways, and at great expense, we've been trying to backtrack to that starting point ever since.

A cruise down the proverbial road not taken for spaceflight is most easily begun in southern California. Drive straight north through the clump of skyscrapers that gather awkwardly to form downtown Los Angeles—through the seething tangle of six- and eight-lane freeways with mysterious numeric designations (110, 101, 710, 605), past the anonymous warehouse towns that feed America's voracious consumer maw, and about one hundred miles northeast of Hollywood (a drive that, straight up, would put you in orbit) you will roll unmistakably into the scorched badlands of the Mojave Desert.

The Mojave is a bleak and lovely place that has served as the backdrop for countless movies, including the original 1923 silent version of Cecil B. DeMille's *Ten Commandments,* in which it stood in for the Promised Land. The Mojave plays a similar role in the history of human spaceflight: both the history we're familiar with and the one that might have happened but didn't.

It was here, on a hot autumn day in 1947, that a brash fighter ace named Chuck Yeager became the first human to travel faster than sound. Yeager's X-1 flight led to more X-vehicle flights (the X is for experimental), pushing ever higher, ever faster. Eventually, the X-program led to sophisticated rocket planes like the sleek, needle-nosed X-15 that in July 1963 took NASA pilot Joseph Walker to an altitude of sixty-seven miles. At that altitude, Walker was technically in space, although it was only suborbital space: high enough to experience some weightlessness and to see the curvature of the Earth, but not quite high enough to achieve orbit. Even so, sixty-seven miles was high enough for Walker to be awarded astronaut's wings only two years after Alan Shepard became the first American to achieve that feat (also with a suborbital flight) in a tiny pod blasted skyward on a retooled nuclear missile.

Unlike the rocket on which Shepard's Mercury capsule rode, the X-15 did not disappear in a blaze of glory after doing its job. The X-15 landed and flew again and again. In fact, before anyone concluded definitively that humans should go into space nuclear-warhead style, the X-pilots were building the logical bridge between humankind's amazingly fast conquest of the air and the limitless reaches of the vacuum beyond. All in all, eight pilots would earn astronaut's wings in X-15s: it is a little-known fact that Neil Armstrong first became an astronaut not on his Gemini or Apollo flights, but in the cockpit of an X-15.

In America, the parity between missiles and planes as would-be space vehicles was tipped when the Soviets launched Sputnik atop one of their converted ICBMs. Missiles were expendable, much like artillery shells—it was a German artillery officer, Karl Becker, who put together the Nazi missile program that eventually employed Wernher von Braun—and after Sputnik, the idea of space vehicles as fundamentally disposable objects permeated the Space Race. Choosing missiles over planes for human space travel had ramifications beyond whether astronauts went into space in a cockpit or a can: it fundamentally defined the economics of space travel for decades to come.

It's an enticing "What if?" to wonder how spaceflight might now be different had the Soviet rocket team been a little slower and the American X-team a little faster. Had reusable rocket planes, rather than expendable missiles, become the vehicle of choice for space travel, Gerard O'Neill's space cities might now be spinning about at L5, if only because it's likely that economies of scale, based on incremental improvement and reusability, would have taken hold much more quickly with the plane paradigm.

While that is certainly only speculation, it's easy enough to say where the missile route has led us. With the exception of the shuttle, none of the space-launch vehicles used today for civilian, commercial or military purposes are radically different from the SS-6 that launched Sputnik nearly half a century ago. Expendable rockets are built not even in small batches, but one at a time—they're practically hand-crafted, like fine furniture. They're expensive, problem-prone and technologically an easy lineal trace from the bamboo tubes stuffed with black powder first used by the Chinese nine centuries ago.

The missile standard has particularly solidified in the commercial space sector, which is built on the back of ICBMs first retooled for the Space Race and, later, privatized by the aerospace companies paid by the government to build them. As it stands, the world's commercial space industry generates about $100 billion in revenue annually, the bulk of which is from satellite communications. If you plan to launch something into orbit, the smallest bill you'll be presented with reflects a launch cost on the order of $3,000 per pound. Many satellites weigh several tons, meaning that just launching them—let alone building and operating them—costs millions of dollars.

It is doubtful that, say, FedEx would stay in business long if it charged $3,000 to ship a box of chocolates across the Atlantic. But where modern spaceflight is concerned, even that price is a relative bargain, and available mainly with Russian rockets: you'll pay about $6,000 per pound on most American rockets. To send something into orbit on the Cadillac of modern space vehicles, a space shuttle, costs roughly $10,000 per pound. Some sources put the shuttle's per-pound launch cost at double that amount, taking into consideration that NASA spreads shuttle costs across dozens of internal budgets. Since *Challenger,* the shuttle has been prohibited from launching primary commercial or military payloads, such as satellites, and many shuttle payloads are heavily subsidized by NASA or other government agencies.

Rather than improving over time, the space-launch status quo has become more entrenched, particularly in America, where the end of the Cold War effected far-reaching changes in the aerospace industry. Through mergers both friendly and hostile, an industry that in 1980 boasted seventy-five space companies (or in some cases, the space-focused divisions of larger companies) had shrunk, by 2000, to a mere five: Boeing, General Dynamics, Northrop Grumman, Lockheed Martin and Raytheon. Such brutal consolidation predictably cost many thousands of jobs. It also stifled competitiveness and curtailed the kind of creativity that diverse players typically inspire in any market. The survivors found themselves with more political and economic clout than ever, and fewer constraints on wielding it with abandon.

The contraction of the aerospace industry put by far the greatest share of the American space-launch market in the hands of two companies, Boeing and Lockheed Martin. Boeing gobbled up McDonnell-Douglas and thus inherited the Delta family of rockets, and Lockheed Martin (formed through the merger of Lockheed and Martin Marietta) makes and markets the Atlas series, having recently phased out its other dependable expendable, the Titan. These workhorses now compete head-to-head not only with European and Russian rockets but with those made and marketed by China and India.

International diversification has done little to make spaceflight cheaper. Expendable rockets have improved incrementally over time—marginally greater payload capacity, fuel efficiency and reliability—but no one believes the cost-per-pound paradigm can improve much more in the expendable arena. Thus the business of launching into space remains a closed loop in which small numbers of one-off launch vehicles are built for a handful of large companies and government agencies that are compelled to spend the tens, even hundreds, of millions of dollars to buy and fly them—and this cost in turn keeps prices up and the market small and effectively ensures that nothing changes.

This catch-22—a small market with high prices because it's a small market with high prices—has been the great barrier for the entrepreneurial space community, and the reason why those brave few who have embarked upon private launch-vehicle development have invariably failed. Boeing and Lockheed are multibillion-dollar corporations, and the expendable rockets they build are the product of billions of dollars of Space Age funding, most of it taxpayer-derived. Not only are space-launch entrepreneurs unable to get their hands on that kind of public funding, but most of the technology that resulted from the Space Race is now proprietary. Thus, the short and so far unhappy history of the entrepreneurial space launch is littered with the names of companies that, relying on private dollars, tried to beat government-funded aerospace conglomerates at their own game and failed.

When it comes to cracking the "access problem," the government has done no better, and as might be expected it has spent far more money doing so. In his 1986 state of the union address, only a week after the *Challenger* disaster, Ronald Reagan announced the National Aerospace Plane, which he dubbed the Orient Express, "that could, by the end of the next decade, take off from Dulles Airport, accelerate up to 25 times the speed of sound, attaining low earth orbit or flying to Tokyo within two hours." Tens of millions of dollars were farmed out to various aerospace contractors to get this space cruiser onto the drawing boards; billions were to be spent to get it from there to space.

The NASP, as it was alternatively called, burned through roughly $2 billion before being canceled in 1994. At least one government official, then–defense secretary Dick Cheney, saw the writing on the wall early on and suggested in 1989 that the NASP program be transferred entirely from the Air Force to NASA. As it was, both organizations and their industry partners lost money, and time, to a program that went nowhere.

Earlier still was the unfortunately named Dyna-Soar, a conceptual ancestor of the space shuttle that NASA canceled in the 1960s after its overoptimistic $800 million development budget turned out to be at least four factors short before it even left the drawing board (although this lesson was apparently lost on, or ignored by, those who pushed for the shuttle's birth).

The National Aerospace Plane, Dyna-Soar and the space shuttle all looked, to the untrained eye, like airplanes: the road not taken at the beginning of the Space Age. This is not to say that space vehicles must look like airplanes to change the current high-cost/low-volume paradigm for spaceflight. Indeed, the first and arguably greatest failure of the space shuttle was an obsession with the overliteral belief (on the part of NASA's spinmeisters and top managers, and certainly not of most of its engineers) that if the shuttle *looked* like a plane, it would work like a plane. Instead, what space travel needed was a vehicle with planelike functionality, the chief benefit of planes over rockets being that you can use them not just once but again and again. In that respect, what space travel has long needed, and what it needs now more than ever, is reusability.

Put simply, there are no fully reusable spacecraft in the world today.

That statement includes the space shuttle, which, at best, is semi-reusable: its enormous pod-shaped fuel tank incinerates in the atmosphere, and most of the shuttle's big-ticket components (including its main and strap-on engines) are either replaced, rebuilt or refurbished substantially after every flight, a process that can take from several months to several years.

The lack of true reusability means that the normal economies of scale realized in commercial air travel are virtually nonexistent in space-flight. In 1998, an Aerospace Corporation study found that a Boeing 777, by flying repeatedly, pays for its development cost in its first few years. After those first few years, it starts generating return on invest-ment. The Aerospace Corporation report concluded that, in general, commercial air companies end up charging about a penny per pound of payload, a situation that can be attributed almost solely to frequency of flight.

A 777 is not an uncomplicated machine. According to Boeing, a 777-200 has about 132,000 engineered, unique components; counting every rivet and bolt, the aircraft has more than 3 million parts—about the same number as a shuttle. Based on the sheer number of pieces that can break, wear out, fall off or suffer a lack of proper maintenance attention, the 777-200 and the shuttle can be said to have a comparable level of mechanical complexity.

Economically, however, the 777 and shuttle are different breeds of vehicle altogether. With eight or fewer flights per year and a cast of thousands (actually tens of thousands) needed to keep it running, fly-ing on the shuttle will never cost less than $10,000 per pound—in fact, as the shuttles get older and require more attention, that figure will go up. To produce a different economic outcome, the shuttle would need to fly dozens, if not hundreds, of times per year, and with a far smaller support team and much lower maintenance costs. The shuttle can never hope to produce anything resembling reasonable economic return or to serve as a platform for developing commercial vehicles that might do so. What "return" the shuttle does produce (in the form of payloads flown) is largely paid for in government subsidies or government pay-loads—meaning that the initial investor, the U.S. taxpayer, effectively pays twice.

It is grimmer to consider that eight flights a year would be a significant improvement over the shuttle's recent flight rate. The shuttle has
averaged about six flights per year and for the foreseeable future will be
limited to four or five flights annually (and even that number depends
on the long-term fallout from the *Columbia* disaster). The three remaining shuttle orbiters may look alike, but they're as different as three
siblings—another contributing factor in making the shuttle the most
expensive space transportation system in the world.

In the expendable launcher market, there are currently more than a
dozen rockets to choose from, varying in capability and cost. In the
human spaceflight arena, NASA's shuttle, Russia's Soyuz and China's
Shenzhou are the only vehicles available, with "available" used only in
the broadest sense. The Russians have sold spare seats on Soyuz flights,
but not everyone has the millions of dollars needed to secure a seat
even when it's available. The Shenzhou is likely to remain restricted to
the Chinese military for some time to come. As for the shuttle, you can
only get on board if you are a NASA astronaut, an astronaut from a
friendly government, or a politician (two congressmen have flown,
with little rationale other than their political status). The net result of
this situation is that as of the end of 2003, more than forty years since
the first human flew into space, just over four hundred people have
earned the mantle of space traveler—a fractional percentage of the
population of O'Neill's proposed Island One, spread across nearly half a
century.

Few disagree that reusability is the key to unlocking Part Two of the
Space Age promise—frequent, inexpensive and reliable popular access
to near-Earth space. The point where opinions diverge, and radically, is
whether the lack of a truly reusable spacecraft is due to insufficient
technology or insufficient motivation. The latter charge is usually leveled at NASA and its Big Aerospace partners by those in the entrepreneurial space sector: what real incentive do Boeing and Lockheed, and
by extension the shuttle's owner, NASA, have to change the way things
are?

You don't have to look far to see that the entrepreneurs have a point.
If you were driving through the Mojave Desert in the summer of 2001,
you needed only to take a turn toward the blowy demisuburb of Palm-
dale, where the Next Big Thing in reusable spaceflight—a prototype

rocket plane called the X-33—was being dismantled at Lockheed Martin's famous Skunk Works facility.

The X-33 may ultimately be viewed as the last gasp of the "real" X-vehicle program—the one that sent people into the lower reaches of space and promised, once upon a time, to send them even farther. Although the X-33 was initially unmanned, once it had proved itself flying to altitudes of up to sixty miles and reaching Mach thirteen (thirteen times faster than the speed of sound), Lockheed planned to build a bigger, more ambitious space vehicle that so tickled the company's instinct for exo-planetary capitalism that it had a name, VentureStar, long before the X-33 was ever built. The company even devoted a special web page to the VentureStar featuring lovingly computer-rendered images of a sleek space cruiser that looked astonishingly similar to the one in *2001: A Space Odyssey*. The web site's text ballyhooed the VentureStar as a commercial, fully reusable, bona fide space plane, and Lockheed wasn't shy about its ambitions: "We've been to orbit. We've been to the moon. In the new century, we'll be getting down to business in space as creative entrepreneurs join the next generation of brave explorers."

The VentureStar pages were quietly removed from Lockheed's web site in the middle of 2001, just before the X-33 program was shut down, victim of the fiscal vortex created by the International Space Station's cost overruns and, not insignificantly, Lockheed Martin's inability to actually make the X-33 work. Whatever its failings, the X-33 did not die for lack of ambition, since enough new stuff was invented for it to keep an office of patent attorneys busy for years: new engines, systems, materials—new everything. All that newness cost more than a billion dollars, most of them from U.S. taxpayers courtesy of NASA, which put up more than two-thirds of the funding. For all that, the X-33 never flew and it never will.

At first glance, the death of the X-33 may seem to be the sad result of NASA's and Lockheed Martin's reach exceeding their grasp. Yet a closer look reveals that, in fact, a truly reusable spacecraft is arguably the last thing either NASA or Lockheed wanted to materialize, since it would threaten not only Lockheed's lucrative expendable rocket market, but also NASA's golden goose, the space shuttle.

Lockheed also has a big stake in the shuttle program. Through a joint venture called the United Space Alliance, Lockheed and Boeing operate the shuttle for NASA, to the tune of $1.6 billion in sales of shuttle services in 2000 (the "sales" were primarily to NASA and other government agencies). USA, as it is known in space circles, was created in 1998 to answer criticisms that the shuttle was a government-held monopoly on human spaceflight. Far from silencing the critics, the creation of USA only succeeded in consolidating that monopoly.

Well before Boeing and Lockheed won the USA contract, the two companies had already locked up virtually every aspect of the shuttle's maintenance and upgrading work. Boeing has long handled most major shuttle improvements and has processed every shuttle payload since the fleet's inaugural flight in 1981, to the tune of billions of dollars (in 1996, Boeing also bought the space division of Rockwell International, which built the shuttle orbiters). For its part, Lockheed builds the shuttle's external tanks and has a hand in hundreds of other shuttle activities and operations.

By winning the contract to manage the shuttle program with their USA partnership, Boeing and Lockheed ensured that their separate shuttle contracts not only would remain impervious to competition but could actually expand over time; ergo in 2002, NASA's no-questions-asked extension through 2008 of Lockheed's long-standing contract to build the shuttle's external tanks, a deal worth about $1.5 billion.

In such a context, more than one skeptic has pointed out that Lockheed Martin, as the maker of pricey expendable rockets and cobeneficiary of most shuttle work, could do nothing more effective to keep those businesses intact than to co-opt any major effort to develop a reusable launch vehicle like the X-33. The easiest way to do that would be to win the contract to build one and then bleed away the contract money until NASA could no longer afford the program.

The X-33 also brings us back to NASA, whose zealous attachment to the shuttle has resulted in a deeply ambivalent position when it comes to considerations of the shuttle's successor. NASA has talked for years about the need to replace the shuttle, and for good reason, since the shuttle's design is more than three decades old. More critically, NASA knows that it is expected to lead the world toward a more robust space future, for which the shuttle long ago ceased to be an effective tool. But the NASA that calls the shuttle a stepping-stone to more ambitious, more O'Neillian things to come is, to repeat Freeman Dyson's term, the "paper NASA." Now as ever, the "real NASA" cannot envision a future in which the shuttle is anything but an end unto itself.

NASA, in fact, started the industry competition that resulted in the X-33: this competition was, ostensibly at least, a search for a better, cheaper and reusable launch system. Yet not only did NASA pay out a billion dollars for a vehicle that didn't come close to flying, but somewhere in the agency's collective consciousness it undertook to do so knowing that this was a likely outcome. The X-33 program effectively ensured that the shuttle would have no real rival for years to come, while at the same time answering the shuttle's critics by appearing to work hard toward its successor. This may sound too cynical by half, but there is no denying the following facts: the failed X-33 took the wind out of NASA's reusable launch-vehicle efforts, moneymaking Lockheed expendables like the Atlas are still flying and even after the *Columbia* tragedy the shuttle is nowhere near being superseded by anything else.

When the X-33 competition began, there existed a very real threat to the shuttle's hegemony—a far greater threat than the X-33 itself, which had many in the aerospace community shaking their heads at its over-complicated improbability from day one. The tactic employed to ensure the death of a stumpy, unassuming prototype spacecraft called the DC-X was much the same as the one used to sideline the Industrial Space Facility, and with the same result.

That the DC-X didn't begin as a NASA project is hardly a shock. That it ended up in NASA's domain, where it was quietly suffocated, was perhaps the cruelest blow to human spaceflight since the termination of the Saturn program.

Mitchell Burnside Clapp likes to joke that he's found the perfect setting for someone in his line of work, which for the past decade has been trying to find the right formula for opening up space travel to the masses. Early on a crisp winter morning, with the frost just beginning to sur-

render to sunlight, Burnside Clapp ("It's a family name so you have to use the whole thing") walked me out onto an spacious balcony just off the main conference room at Pioneer Rocketplane, the company he founded in 1995 and has nurtured ever since. We rested our steaming coffee mugs on a flat railing and from our third-story perch looked down at the sleeping town of Solvang, which has been Pioneer's headquarters for the past few years.

Solvang is one of the most incongruous places one is ever likely to visit, a picture-perfect replica of a Danish village seemingly lugged over from Denmark piece by piece then reassembled in a quiet California valley, some four or five hours northwest of the parched Mojave flats where the X-33 struggled vainly for birth. The two places couldn't be more of a contrast. Visiting Solvang—with its sober Tudor styling, giant wooden windmills and rows of pastry shops run by jolly pink-cheeked matrons—feels a bit like stumbling into Shangri-la via Hans Christian Andersen. For Burnside Clapp, a cheerful man who is never far from a humorous observation of one kind or another, the weird kitschiness of Solvang is a continuous source of mocking perspective. "If you want to spend your life tilting at windmills," he quipped, "this is the place."

Basing your rocket plane company in a faux-Danish village does make sense if you consider that Solvang is only about half an hour down the road from Vandenberg Air Force Base. Pioneer briefly made its headquarters on the base itself as part of a short-lived experiment in bringing some entrepreneurial verve to Vandenberg. Even afterward the Air Force didn't mind Pioneer Rocketplane's using some of the base's facilities for the development of its rocket plane, the Pathfinder.

The rocket plane business hasn't been easy going for Burnside Clapp. Since its founding in 1995, Pioneer has hung on from month to month with a financial jury-rigging of small government contracts, last-minute infusions of private capital and a lean staff working ridiculously long hours for not much money. Burnside Clapp has an insider's pedigree that seems tailor-made for space. He has degrees in aerospace engineering and physics from MIT and was a military test pilot who shook the skies above the Mojave Desert just like most of the Mercury, Gemini and Apollo astronauts did. Because of Burnside Clapp's military background, Pioneer Rocketplane had early access both to the military's considerable space expertise and to crucial test facilities, as well as to small R&D grants.

In the 1980s and early 1990s, Mitchell Burnside Clapp was part of the team that developed and flew a vehicle that was, in every way that counted, started from a clean slate. It wasn't a rocket plane, but the tiny spacecraft known as the Delta Clipper Experimental, or DC-X, warmed the heart of O'Neillian hopefuls throughout the world by promising to provide reliable, inexpensive, reusable access to space.

The DC-X emerged from Ronald Reagan's Strategic Defense Initiative (SDI), the proposed "shield" against intercontinental ballistic missiles dubbed Star Wars by its detractors. Billions were pumped into various SDI projects, among them the development of a reliable, reusable, inexpensive spacecraft to support a way-out concept called Brilliant Pebbles: tiny space-borne projectiles that would destroy incoming ICBMs. It was decided that a rapid-turnaround spacecraft—perhaps crewed, perhaps not—would be essential to servicing the Brilliant Pebbles concept. The shuttle, it was agreed, was definitely not that spacecraft.

The SDI specs on what became the Single Stage to Orbit (SSTO) program were straightforward and, where the economics of spaceflight were concerned, revolutionary. For $60 million, the SSTO contract winner was to design, test and fly a prototype of a fully reusable ship capable of carrying twenty thousand pounds into orbit at less than $5 million per flight and with better than 99 percent reliability (the shuttle has about 95 percent reliability, meaning that one out of twenty flights can be expected to have some kind of serious problem). At $60 million, the SSTO program was a bargain-basement proposition that left little room for new technology development. Thus, any vehicle developed would have to be conceived, built and flown with off-the-shelf hardware.

In 1989, the SDI tasked the Air Force with the SSTO program, which it put out to private industry for bids. McDonnell-Douglas came up with the winning concept: a ship that would take off and land vertically. McDonnell-Douglas called its would-be spacecraft the Delta Clipper Experimental, in honor of the company's successful Delta rocket and because the ship was, in performance, to be much like a nineteenth-century commercial clipper ship.

The DC-X looked like a white parking cone, four stories high. It had four holes in its base to vent rocket exhaust and retractable landing struts that receded into the ship after takeoff and popped out again just before landing. With pride, the DC-X team nicknamed their creation the junkyard rocket, since the DC-X didn't use a single piece of new technology, thus putting paid to previous development concepts (and future ones, like the X-33) predicated almost solely on the theory that newer is better. That said, DC-X engineers freely purloined technology that was once very new and developed at great expense: four Pratt & Whitney engines developed for the Centaur rocket, navigation equipment appropriated from F-15 and F/A-18 fighter jets, autopilot and avionics gear from Douglas's MD-11 airliner, an off-the-shelf global positioning system, and hinges, springs and other odds and ends from such fonts of high technology as Home Depot and Kmart.

On September 11, 1993, after twenty-one months of development by about a hundred people, the DC-X made its inaugural flight, at the White Sands Missile Range in New Mexico—the same place Wernher von Braun's V-2 rockets were tested after World War II. The ship was piloted by remote control by the Apollo astronaut Pete Conrad. Under Conrad's deft hand, the DC-X blasted straight up to three hundred feet above the ground, hovered in the sky for a few thrilling seconds, then began to move sideways on its swiveling, gimbal-mounted rocket engines. The DC-X then descended lazily and landed just a few feet short of the center of its designated target. The small crowd cheered wildly; another crowd cheered a few weeks later when the DC-X flew again. Only two people other than Conrad were qualified to fly the DC-X, and one of them was Mitchell Burnside Clapp. "It was an easy ship to fly, very smooth," he recalled.

The development of the DC-X was perhaps the first time in the short, tortured history of spaceflight in which "the right shoes were paired with the right choreography," as Burnside Clapp put it. "You could use [the DC-X] again and again without having a whole standing army to make it go," he said, speaking with hesitant pride. "That's a worthy goal and we largely achieved it."

Despite successful test flights and nearly universal acclaim, only a few months later the DC-X was sitting dormant in a shed at White Sands as a result of the complex, crazy-making dance between big government, big money and big ideas. In this case, it was a shift in military focus: the 1991 Gulf War had revealed new needs for America's military (more focus on "smart" offensive weapons, less focus on sweeping national defense strategies) and thus drew funding away from SDI. One casualty of this shift in priorities was the Brilliant Pebbles concept, which was scaled back so severely that the Pentagon found itself wondering whether or not it really needed a reusable spacecraft. By mid-1993, right around the time of the DC-X's inaugural flight, the answer was no: the Single Stage to Orbit program would not be part of SDI's future.

Still, the success of the DC-X had made an impression, and the Air Force and McDonnell-Douglas had friends on Capitol Hill. As a result, the Defense Appropriations Bill signed by President Bill Clinton on November 10, 1993, included funding to keep the SSTO program afloat. The money, however, went not to SDI but to the Defense Advanced Research Projects Agency (DARPA), with a directive to subcontract the work to the Air Force. DARPA had so little interest in stewarding a reusable launch-vehicle program that it actually tried to give the SSTO money back to the Clinton administration. When that didn't work, DARPA stalled the distribution of the money with the goal of cash-starving the DC-X to death and thus ridding itself of responsibility for the program.

NASA came to the rescue in early 1994 when Dan Goldin came up with $900,000 to keep the Delta Clipper program afloat, which bought enough time for DC-X supporters in Congress to browbeat DARPA into coughing up the rest of its money. Officially, NASA was excited about the DC-X. A NASA FAQ page on the DC-X glowed with predictions that human spaceflight on a DC-X derivative "could be less than the price for a round-the-world cruise on the QE2" and even predicted that future models "would be able to fly to the moon, land there, and then return to Earth."

After making such bold predictions, the NASA FAQ even dared to venture a bold bit of veiled self-criticism, putting the following question in the mouths of its readers: "Why should I believe all these claims for the Delta Clipper when similar ones were made for the shuttle twenty years ago?" The answer, partly, was that the shuttle's design was "frozen in the 1970s" and that "the Delta Clipper is being designed with supportability and operability as priority considerations"—which implied that the shuttle was not.

Unofficially, the DC-X scared the wits out of NASA, or at least rattled several key centers—not surprisingly, the centers with the biggest stake in the space shuttle. With its puny budget and do-it-yourself ethos, the DC-X hardly seemed like a threat to the shuttle monopoly. But then it flew and flew, again and again, to ever-greater acclaim. In July 1993, just as the DC-X was stretching its proverbial wings, NASA released the results of a study called Access to Space, which came to a conclusion that had long been reached by the alternative space community and by many in the space establishment: that for reasons of efficiency and economics, a single-stage-to-orbit vehicle was preferable to both the semi-reusable shuttle and expendable rockets.

A follow-up report released in January 1994 concluded that a reusable single-stage-to-orbit vehicle could capture "Delta, Atlas, and Shuttle missions at approximately 15% of the current combined annual operating costs of these systems." These reports seemed to signal, unequivocally, that the time had come for a real commitment to reusable launch vehicles. Instead, as so often happens at NASA, it scared certain factions within the agency into hastily drumming up the appearance of a commitment in order to buy enough time to figure out how best to protect existing projects.

On August 31, 1993, exactly one week after the first successful DC-X flight and a few weeks after the Access to Space study was released, a team from Marshall Space Flight Center—for all intents and purposes, the home of the space shuttle—unveiled a vehicle design called the X-2000, so designated because its first flight was to have been in the year 2000. The X-2000 was to have been a big beast of a vehicle, billed as being able to carry heavier payloads than the Delta Clipper. It would also be cheaper and—amazingly, since it was only a paper ship whereas the DC-X had already flown—ready for service at the same time as the final version of the Delta Clipper.

Most critically of all, the X-2000 called for joint NASA-Pentagon funding, a brazen attempt to cut the fiscal legs off the DC-X. Marshall tried to sell the X-2000 to Goldin as a response to the Access to Space study, but it was obvious that it was instead a knee-jerk reaction to the DC-X. Even NASA, in its official history of the X-33 program, admitted that "the X-2000 project officially had no connection with NASA's 'Access to Space' study."

Ultimately, the X-2000 withered on the vine because Goldin believed the DC-X held more promise. Speaking to the National Press Club on June 20, 1994, Goldin boasted that the "Delta Clipper is one of the most imaginative programs that this nation came up with for access to space. I will say that NASA put its money where its mouth is and this Administrator came up with money to save that program. We are enthusiastic about it." Although it didn't originate as part of Goldin's oft-mentioned "faster, better, cheaper" operational ethos, Goldin pitched it as such in his attempts to reinvent NASA along these lines. Still, the X-2000 would have its day. Cobbled together from previous designs left over from the early 1980s and slapped together in only two weeks, the X-2000 was the forerunner of the money-gobbling X-33.

In 1994, President Clinton signed into law the U.S. Space Transportation Policy, which split responsibility for space vehicle development between the Pentagon and NASA. The military would be responsible for improving expendable launch vehicles; NASA would be responsible for operating the shuttle while also working on "next generation reusable space transportation systems, such as the single-stage-to-orbit concept." Thus ended the military's ownership of the DC-X, which was then officially transferred to NASA.

On June 24, 1994, only four days after Goldin's enthusiastic speech on its behalf, the DC-X flew for the fifth time, achieving its longest (more than two minutes) and highest (half a mile) flight to date. Flying again a week later, the DC-X achieved an aerospace first when a ground equipment explosion caused a shock wave that ripped a four-by-fifteen-foot hole in the ship's graphite-epoxy hull. Immediately, Pete Conrad activated the command for "autoland" and the DC-X safely returned itself to the ground. No spacecraft had ever achieved such a feat.

Even though it survived the explosion, the DC-X had to be repaired, buying some time for those who would happily have seen the DC-X completely destroyed. For while Goldin's enthusiasm might have been

genuine, the DC-X continued to spook many at NASA, since it was
beginning to look like a legitimate threat to the shuttle. In October
1994, while the DC-X was undergoing repairs, Marshall once again
went on the attack by soliciting industry proposals for what was to
be called the X-33 Advanced Technology Demonstrator, spelling out
requirements for a vehicle that looked suspiciously like the X-2000
concept Marshall had failed to sell to Goldin only a year before.

It is not uncommon for NASA centers to issue "cooperative agreement notices" that spell out the parameters of a certain project and
invite public and/or private entities to bid for participation—such is
the primary artery through which taxpayer money for space flows to
private industry. However, it is less common to issue such notices in
defiance of known interests of NASA administrators. Almost certainly,
the only centers with sufficient political power to do that are Marshall
and Johnson, which together run the shuttle and station programs and
thus control a solid half of NASA's budget.

Learning after the fact about Marshall's preemptive strike, which
appeared under the mild-mannered bureaucratic moniker Cooperative
Agreement Notice 8-1, Goldin was reportedly furious. Yet Marshall was
following a tried-and-true tactic of NASA centers where NASA HQ is
concerned: "Act first, ask forgiveness later." Seemingly unable to retract
Marshall's request for proposals without embarrassment, Goldin agreed
that Marshall could go forward but admonished against influencing the
bidders with its voluminous report on the X-2000. That report was quietly sent around anyhow.

Three companies responded to Cooperative Agreement Notice 8-1:
Rockwell, Lockheed Martin and a joint Boeing–McDonnell-Douglas
team (McDonnell-Douglas was, at the time, in the process of being subsumed under the Boeing label). Not surprisingly, the Boeing-Douglas
team submitted a proposal for an evolved Delta Clipper. Rockwell, the
company that had built the shuttle orbiters, submitted a virtual carbon
copy of the X-2000 design (which it had helped Marshall develop), and
Lockheed Martin submitted a design that split the difference between
the X-2000 and the space shuttle.

On July 2, 1996, Vice President Al Gore announced to the world
a prototype for the next generation of space vehicle by pulling the

cover off a model of Lockheed Martin's design for the X-33. Lockheed won in no small part because it promised to spend more than either Boeing or Rockwell on the development of the X-33's successor, the VentureStar—money it would never have to spend.

The X-33 had robbed the Delta Clipper program of a clear future, but there was still money to keep flying after it was repaired, so fly it did. The rebuilt DC-XA (the A, for "advanced," was added by NASA, as was the agency's familiar logo to the ship's hull) had made its first post-repair flight in May 1996, only two months before Gore's X-33 announcement. The DC-XA flew again on June 8 and then, only twenty-six hours later, it flew a third time: a record turnaround for any rocket-powered vehicle. The third flight also set a new altitude and duration record for the program, as the DC-XA climbed nearly two miles into the sky and flew for nearly two and a half minutes.

The fourth flight of the DC-XA, on July 31, 1996, went beautifully until one of the ship's four landing struts failed to fully extend as the ship touched down. Still heavy with fuel, the DC-XA toppled to one side and exploded in a fireball that completely destroyed the vehicle—which was, because of the program's slim funding, the only one of its kind. A vigorous campaign, which counted Rick Tumlinson among its most vocal members—"We were making speeches, videos, anything to keep it going," he recalled—was mounted to find money to build a new Delta Clipper: an evolved vehicle to be called the DC-Y, which would cost about $120 million.

The Delta Clipper team intended to take the DC-Y into suborbital space. Building a space-capable vehicle for $120 million was unprecedented; then again, so was designing, building and successfully test-flying a reusable prototype for only $60 million. But the decision was NASA's, and with the X-33 program under way—a program that would burn through more than a billion dollars and fly nothing—NASA quietly ended the Delta Clipper program. Burnside Clapp was briefly operations manager for the X-33, which, as he put it, "sounds really impressive until you realize that nobody's working for you and there's nothing flying so it doesn't look that good on your vita." At that point, Burnside Clapp left the program to devote all his energies to Pioneer Rocketplane.

The shock waves from the death of the DC-X reverberate to this day. Opponents of NASA have made liberal use of the DC-X to make their case that it doesn't take big bucks and unknown technology to create an effective reusable launch vehicle. Burnside Clapp said that it was clearly in the minds of the DC-X team that in the not too distant future a DC-X successor would be piloted and carry passengers. The DC-X was perhaps the first real chink in the armor of the status quo; it proved that a combination of private- and public-sector agendas for space could, to use Dan Goldin's words, create a space vehicle faster, better and cheaper than either sector alone.

At least on paper, Goldin himself supported the effort that has to date been the DC-X program's most significant legacy, the X Prize. "Make no mistake," Goldin is quoted as saying on the X Prize web site, "it is the private sector that will finally build the machines and provide the access to space to make the dream a reality for all Americans. We encourage the participation by as many people and as many organizations as possible in this noble venture."

The X Prize was launched on May 18, 1996, the day before the repaired DC-XA first flew after surviving its near-death experience. The shadow of the DC-X looms large over the X Prize: the organization's advisory board includes General Simon "Pete" Worden, who oversaw the creation of the DC-X program under its military aegis, and William Gaubatz, who served as the program's manager at McDonnell-Douglas (a third DC-X denizen, NASA's Maxwell Hunter, was on the X Prize advisory board until his death in 2001).

The driving force behind the X Prize is Peter Diamandis, a St. Louis–based physician who read *The High Frontier* while still a university student and quickly became enamored of an O'Neillian future in space. Diamandis concluded that suborbital passenger spaceflight was the needed stepping-stone to such a future, and among his many entrepreneurial space efforts, Diamandis cofounded Space Adventures, the "space tourism" company that had a hand in arranging space tourist Dennis Tito's activities in Russia.

While the X Prize also has space tourism as its driving rationale, it has taken a different tack toward that end. The X Prize is modeled directly on the Orteig Prize, which spurred Charles Lindbergh's solo crossing from New York to Paris in 1927. Much as Lindbergh's backers did, Diamandis raised the X Prize funds by convening a group of St. Louis businesspersons to put up the prize money, and hopes for the X Prize are similar to those attached to Lindbergh's historic flight. "Through a smaller, faster, better approach to aviation," reads the X Prize web site, "Lindbergh and his financial supporters, The Spirit of St. Louis Organization, demonstrated that a small professional team could outperform a large, government-style effort."

The Orteig Prize was one of more than a hundred such prizes offered between 1905 and 1935 to those who would stretch the limits of aviation. Whereas the Orteig Prize awarded Lindbergh a relatively slim $25,000, the X Prize promises $10 million (backed by an insurance policy to guarantee payment) to the first team that can send three people to an altitude of sixty-two miles—the same achievement as Joseph Walker's 1963 X-15 flight—and return them safely . . . and then do it again, with the same vehicle, within two weeks' time.

To date, twenty-three teams—fourteen American, three British, two Canadian, and the rest from Argentina, Romania, Israel and Russia—are registered as X Prize competitors, with a wide variety of schemes in play. The Canadian Arrow looks like a V-2 with windows, and even uses the same fuel mixture and a similar graphite vane guidance system. (The Arrow's flight will end not by crashing into London or Antwerp, but by parachuting into water.) The other Canadian entrant, the Da Vinci Project, plans to cut down on fuel costs by launching from the world's largest helium balloon, which would lift the ship to about eighty thousand feet before firing its engines; Israel's Negev-5, conceived by IL Aerospace Technologies, uses a similar concept.

The X Prize has been gradually gaining not only entrants but also momentum and support. It boasts endorsements from a number of astronauts and celebrities and even has two of Lindbergh's grandchildren in advisory roles. The X Prize received a significant morale booster in late 2002 when the U.S. Department of Commerce's Department of Space Commercialization called the prize a "potent catalyst" for the suborbital transportation industry, citing, in addition to the initiation of a space tourism market (the prize's primary aim), the potential for "fast package" delivery, high-speed commercial transportation and Earth imaging.

What is most obvious about the X Prize is that no one, in seven years, has come close to claiming it. This is not necessarily a show-stopper, since eight years went by between the hotelier Raymond Orteig's announcement of his prize and Lindbergh's flight to claim it. As X Prize supporters are quick to note, the nine entrants in the Orteig Prize spent about $400,000—about $4 million today—in pursuit of the $25,000 victory, thereby underscoring a consistent theme of exploration prizes throughout history: that the monetary value of such prizes is secondary to the achievement they represent.

The most promising X Prize effort was unveiled in April 2003 by the veteran aircraft designer Burt Rutan, who is most famous for designing the *Voyager,* the first powered aircraft to fly around the world nonstop without refueling (the airplane now hangs in the National Air and Space Museum). Rutan's company, Scaled Composites, is not new to spacecraft design; it built the external frame, or "aeroshell," for the DC-X. Rutan's White Knight aircraft will climb to an altitude of about fifty thousand feet—above the heaviest layer of atmosphere—and launch from its belly a single-engine rocket plane called SpaceShipOne that will take three people (Rutan has not said whether he'll be among them) to the required suborbital altitude.

In designing the White Knight–SpaceShipOne configuration, Scaled Composites looked to the past: the airplane–rocket plane combination is strikingly similar to the method by which the X-15 reached suborbital space. True to his famously maverick reputation, Rutan claims that he isn't particularly interested in getting into the space tourism business: "I want to do something different and fun, and show it can be cheap," he has said. Even so, Rutan's effort is clearly intended to make a point, perhaps most of all to NASA—among those attending the Mojave Desert rollout of White Knight and SpaceShipOne were Maxime Faget and Dennis Tito.

Whether or not the Scaled Composites vehicle wins the X Prize, the leap Rutan and other X Prize contestants are being asked to make in their pursuit of a new paradigm for human spaceflight is far greater than that faced by Lindbergh and his counterparts. While Lindbergh and his team did build the *Spirit of St. Louis* more or less from scratch, they were not inventing a fundamentally new vehicle, nor was the best technology in existence owned by the government and a handful of corporations who weren't about to give up their proprietary information to help. (Lindbergh contracted an experienced builder of aircraft, Ryan Airlines, to build his prizewinning plane.) By the time the American government first thought to provide military funding for aircraft development, in 1912, there were more than two hundred types of planes in existence to work with as prototypes, along with more than seventy propulsion systems to power them. All of these, notably, were developed in the private sector. It was only later that government cash was provided to advance their work for broader commercial and military purposes.

It is unlikely that any of the X Prize contestants have approached Boeing or Lockheed for assistance with their X Prize spacecraft; for starters, they probably couldn't afford it. They might reasonably expect some help from NASA, but despite Goldin's fine words on the X Prize web site, the agency has kept its distance from the X Prize endeavor—calculating, perhaps correctly, that any significant or high-profile support of the X Prize, through either technology or funding, could underscore the expense and limited aspirations of its own human spaceflight programs, most particularly the shuttle. A more visionary government space program would be dangling more significant carrots in front of X Prize competitors (and their current and potential backers). If there is an X Prize winner, NASA should help ensure its survival through hands-off technology grants to improve whatever structural and propulsion systems succeed in attaining the prize in the first place: such systems would be, or should be, of tremendous interest to a space agency trying to lower the cost and improve the efficiency of its own spacecraft.

Yet nothing of this sort is forthcoming from NASA, which remains myopic in its obsession with the shuttle and the space station, even more so after the *Columbia* disaster. In this and other ways, NASA has not yet taken the lesson of the DC-X to heart. Far from opening the door to new ways of pursuing human spaceflight, the failure of the X-33 and, later, the destruction of *Columbia* provided ready excuses for NASA to block the way for partnerships that might pick up where the DC-X left off.

The effect of such closing of ranks has been stifling for the entrepreneurial space-launch sector. Outside of the X Prize contestants—who are only focused on suborbital space—fewer and fewer entrepreneurs are pursuing launch technology as their primary business venture. Those that have, like SpaceX (started by the Internet entrepreneur Elon Musk), have focused on the small satellite market, rather than challenging the large, expendable rockets, let alone the shuttle.

There is one entrepreneurial company, Kistler Aerospace, that has pressed forward with its plans to develop a large, reusable launch vehicle capable of achieving orbit. It may be the last hope of the current generation of space-launch entrepreneurs to spark any change in the way things are. Even so, most of the alternative space community eyes Kistler Aerospace with caution. After all, can any entrepreneurial company that is filled, top to bottom, with ex-NASA leaders really change anything at all?

Back to the Future

Walter Kistler's eponymous company makes its ambitions clear on its web site: "To be a leader in a revolution that will make commercial space activities as pervasive as world aviation has become today."

Kistler himself seems tickled at the thought that his company, the culmination of a lifelong dream, could make this happen. He recalls being a somewhat bashful ten-year-old boy in his hometown of Biel, a small Swiss village nestled alongside a lake north of Bern, when he first learned of the pioneering work of the German scientist Hermann Oberth and the American Robert Goddard, the men who, along with Russia's Konstantin Tsiolkovsky, provided much of the scientific and technical framework for modern spaceflight. Since the year in question was 1928, Kistler, who is now in his eighties, was learning about such work more or less as it was happening.

Before he was a teenager, Kistler was making not only small rockets but also the volatile fuel needed to launch them. For the latter requirement, Kistler combined nitric and sulfuric acids to make nitroglycerin, using a tattered chemistry book as his guide. "In those days, it was easier to buy extremely dangerous chemicals," Kistler said with a mischievous laugh. Even so, nitric and sulfuric acids weren't exactly convenience store items, particularly for a teenager; Kistler says that he convinced his father to get him a bottle of sulfuric acid for "an experiment," then dumped it into an empty wine bottle, fudged the label and went back to the pharmacy to get a refill, this time with nitric acid. "I was very determined," Kistler said proudly.

Such determination has characterized Kistler's pursuit of his own spaceflight legacy. The possibilities Kistler saw in the complex engineering detail of Oberth's and Goddard's work set the pattern not only of his boyhood, but of his entire life. Although the route from there to here would be circuitous, to Kistler the journey always had a single goal. "The main reason I came to the United States was to work on rockets," he told me with a chuckle. "Since 1928, I was hooked. I wanted to build a rocket to the moon and I've never let up."

Kistler Aerospace is not building a moon rocket, but it is in the final stages of testing a two-stage, reusable launch vehicle with an unassuming name that belies its history-making potential: the K-1. It is intended to be competitive with the costly expendable workhorses of today's launch industry, and investors from around the world have put up nearly half a billion dollars to develop the twelve-story K-1—more than the investment in every other entrepreneurial launch company, present and past, put together.

Most of that money would never have been invested had Walter Kistler himself continued to run Kistler Aerospace, a fact that Kistler offers without any bitterness, since he planned it that way. Like most successful businessmen, Kistler recognizes that ego must take a back seat to investment, and from the beginning Kistler Aerospace has needed a lot of cash. In addition to the half billion already in the bank, the company needs about that much again to get the K-1 off the ground. "It just all hangs on the money," Kistler said evenly, his careful English still bearing strong traces of a Swiss accent. Kistler added sanguinely that the K-1 is "practically all built. But it took too much time and meanwhile expenses grew and grew and now we need one billion in all." When Kistler and I first talked, in 2002, the company's leadership had been scouring the investment community worldwide for the better part of two years, with some encouraging results but with funding still incomplete, a by-product of the fallout from the bursting of the tech-industry bubble, terrorist fears and general planetary recession. "It's very tough," Kistler conceded.

Things got even tougher in July 2003, when Kistler Aerospace announced that it had filed for Chapter 11 bankruptcy protection. But unlike many Chapter 11 filings—often little more than formalities in advance of outright corporate liquidation—Kistler Aerospace seemed intent on using its filing to restructure, not disassemble, the company.

The company's leadership secured a $5 million loan to keep operating during the bankruptcy period and continued to work with its contractors and potential customers, including NASA. The company's goal, according to official statements released with the Chapter 11 announcement, remains the same—to see the K-1 fly.

While he expresses appropriate concern for his company's future, Kistler doesn't seem particularly worried about how things will turn out. It's not that he doesn't have something to lose: although he stepped down from his position as company chairman several years ago, Kistler remains a major investor, both financially and emotionally, in the company he started. Part of his calm-blue-ocean demeanor can be attributed to the fact that he founded two companies before Kistler Aerospace, both of which he sold for "many millions," and that his entrepreneurship in general has rarely missed. Perhaps it will be the same with Kistler Aerospace, perhaps not. For Kistler, it seems, just taking his shot is what has always mattered most to him.

"I'm a bit of a philosopher, and I like to think about the long-range future," said Kistler, who has kept personal diaries of his ideas and ventures for nearly half a century. "We have to have a larger horizon. The only way to go farther, the only place to go, is into space. In my view, humanity has two ways to go, either down the drain or into space. But that may be hundreds or thousands of years ahead, and someone has to keep pushing."

What is here and now is the future of Kistler Aerospace and, quite possibly, the near-term future of the entrepreneurial launch community. In the wake of the X-33 and *Columbia,* the death of Kistler Aerospace could make the entrepreneurial launch industry a no-man's-land for major investment, thus ensuring the current status quo for years, if not decades. If the K-1 succeeds—both technologically and in drawing paying customers to lower-priced launch opportunities—Kistler Aerospace will build investor confidence in alternatives to Big Aerospace launchers and, many believe, begin to drive launch costs down to the point where space might become a place where smaller players with big ideas can find a way to do business. Once that happens, the thinking goes, economies of scale will begin to take hold and the gates to Gerard O'Neill's high frontier will at last swing open.

For all his lifelong interest in space, Walter Kistler never expected to find himself in the position of cofounding not one but two entrepreneurial space companies. Kistler studied physics at university and was a meticulous amateur rocketeer on the side, graduating just as another dedicated amateur, Wernher von Braun, was ascending to fame in neighboring Germany. But Kistler didn't pursue rocketry as a career, at least not at first. He stayed in Switzerland during World War II and worked on diesel engine instrumentation. Kistler came to the United States in 1950 and was recruited to work for Bell Aircraft, builder of the X-1 that sent Chuck Yeager into the record books. Kistler's primary work was on instrumentation (gyroscopes and accelerometers) for a V-2 derivative called the Rascal, which as Kistler described it, "never quite matured and was canceled."

In 1957, Kistler left Bell with some of his colleagues to start Kistler Instruments, a company focused on developing and marketing pressure-measuring gauges for diesel engines and accelerometers for missiles. By 1968, Kistler said bluntly, "we were ready to see some money" and the company was sold to Sundstrand. Not ready to retire, Kistler stayed on with Sundstrand and transferred with the instrumentation group in 1970 to Seattle, where his entrepreneurial drive prompted him to spirit away another team of engineers to form yet another company, Kistler-Morse, this time for developing instrumentation to weigh extremely large objects like storage vessels and grain silos. By Kistler's recollection, the company began with about a dozen people, grew to several hundred, and eventually did tens of millions of dollars in business each year. More than a manager or engineer, Kistler was also an inventor, and over time he invented enough products or processes to earn more than fifty patents.

In his twenty years with Kistler-Morse, entrepreneurship became something of a hobby for Kistler as he began to invest in other companies and ventures. He was a so-called angel investor before that term became somewhat notoriously associated with the good-money-after-bad funding that launched a thousand dot-com start-ups in the 1990s. Fortunately for Kistler, most of his gambles turned to gold. "These investments where I helped friends and colleagues were even more profitable to me than the companies I started myself," he said, giggling with genuine mirth at the crazy thought of all that money and success. "By the time I met Bob Citron, I had quite a little fortune."

Bob Citron, the man who has been Kistler's partner in space entrepreneurship for nearly two decades, often speaks in casual O'Neillian terms of his own commitment to opening a broader human future beyond the Earth. "If you take a long-term perspective on human evolution and particularly on human cultural evolution, going out into space is a natural progression," Citron once told me. Such big thinking is echoed within the nonprofit organization Citron cofounded with Kistler, the Foundation for the Future, whose web site boasts of the foundation's devotion to "the increase and diffusion of knowledge concerning the future biological and cultural evolution of the human species." The foundation's offices occupy a tranquil corner of a tiny business complex in Bellevue, just outside Seattle. Its small library includes *The Road Ahead* by the local icon Bill Gates, Alvin Toffler's 1970 techno-cautionary masterpiece *Future Shock*, Aristotle's *Poetics*, any number of science-fiction books and, of course, *The High Frontier*.

Despite such broad-ranging and very big thinking, Citron is a pragmatist where the future of humans in space is concerned; he claims Gerard O'Neill was, too. "Gerry talked a lot about space colonies and stuff, but when I talked to him he didn't really believe that was going to happen in the near term," Citron said. "I never believed it." What Citron and O'Neill had in common was a guiding belief that expansion into space was humankind's next obvious step as a species; they also agreed that, to make that happen, near-Earth space had to become more accessible for business and adventure, the better to galvanize public interest and create a solid space infrastructure for further exploration.

Where Kistler is small and dapper in appearance and politely concise in conversation, the seventy-something Citron is tall, casual and enthusiastically voluble, particularly when the subject is spaceflight. Once, as we were sitting in the living room of his comfortable Seattle-area home, Citron practically bounded over to a photo album to extract a wonderful black-and-white snapshot of himself and his younger brother, Rick, as reedy teenagers proudly holding a slender homemade rocket, just one of many they built and launched in their early days, dreaming of space travel.

As a young man, Citron spent much of his time in the Mojave Desert as a member and later president of the amateur Pacific Rocket Society, whose participants developed multistage rockets and even the systems to track them in flight. Rockets and space have been part of Citron's life ever since: in 1957, the same year Sputnik changed the course of the twentieth century, Citron joined the Smithsonian Astrophysical Observatory, where he would spend the next eighteen years of his career developing satellite tracking systems to help monitor environmental events on Earth.

Citron has always combined his space dreaming with an equal knack for divining the bottom line. While at the Smithsonian, he founded Educational Expeditions International, later called Earthwatch, which allowed the lay public to help scientists conduct field research around the world for a fee (a model now common among large research universities seeking to augment their budgets). Only when Citron left the space business did he make enough money to pursue his interest in space on his own terms, selling his successful Adventure Travel Publications, which grew out of the Earthwatch project, to the publishing giant Ziff-Davis in 1980.

The Citron brothers (Rick, now a Los Angeles–based attorney, has provided legal guidance to dozens of space start-ups) were also the masterminds and chief organizers of the Fiji expedition to view Mir's demise. It would be a considerable understatement to say that the Citron brothers were disappointed at not seeing Mir's final blaze of glory: I will never forget the sight of Rick sobbing in his older brother's arms as our plane spun back around toward Fiji, both men realizing that a year's work of hard planning had come to naught.

Bob Citron had come to record the spectacle for posterity, personal interest and science, and out of respect for the Russian space program's achievements (the Russians on our trip were guests of the Citrons). He was not, pointedly, among the believers that a commercialized Mir stood a real chance of success. "I thought it was great that MirCorp took its shot, but I never thought they'd be able to do it," Citron said. "The timing was all wrong." By timing, Citron meant that Mir could never hope to compete with the International Space Station juggernaut.

On April 12, 1961, Soviet Army Colonel Yuri Gagarin became the first human to travel into space. Gagarin is seen here on the way to the launchpad on the morning of his flight, a 108-minute ride once around the Earth. He would never fly in space again: Gagarin died in 1968, at age thirty-four, while on a training flight that he hoped would help earn him a chance at flying to the moon. *(Photo courtesy of NASA)*

This iconic photograph of *Apollo 11* moonwalker Buzz Aldrin represents the apex of the American space program. *(Photo courtesy of NASA)*

NASA officials relax at the Johnson Space Center following *Apollo 11*'s successful launch on July 16, 1969. Second and third from left are Wernher von Braun and George Mueller, whose stewardship of the Apollo program helped keep John F. Kennedy's promise to land an American on the moon before 1970. Mueller and von Braun were also important champions of the space shuttle, although the shuttle would ultimately frustrate the space-faring dreams of both men. *(Photo courtesy of NASA)*

In this November 1984 photograph, astronaut Dale A. Gardner holds up a homemade sign referring to two malfunctioning satellites hauled into the cargo bay of the space shuttle *Discovery*. Although intended as a joke, the photo became symbolic of NASA's unwillingness to embrace entrepreneurship in human spaceflight. The astronaut taking the photograph, Joseph P. Allen, would later find himself battling NASA in his efforts to develop the Industrial Space Facility. *(Photo courtesy of NASA)*

The forty-foot-high Pizza Hut logo emblazoned on this Russian Proton rocket, which launched in July 2000 with the International Space Station Service Module on board, highlights the extent to which the Russians have turned to capitalism to fund their cash-strapped space program. Pizza Hut scored another publicity coup in April 2001 when the Soyuz delivering space tourist Dennis Tito to the ISS also delivered pizza to the station's crew.
(Photo courtesy of Pizza Hut)

The Delta Clipper-Experimental, or DC-X, launching on its maiden flight, on September 11, 1993. Developed for about $60 million as part of the Reagan-era Strategic Defense Initiative, the DC-X was the prototype for a new kind of inexpensive and fully reusable spacecraft. The DC-X was succeeded by NASA's X-33, which never flew—despite nearly a billion dollars in funding—and was terminated in 2001. *(Photo courtesy of White Sands Missile Range)*

In this January 1960 photograph, Neil Armstrong, the first human to walk on another world, stands next to the rocket-powered X-15, which in the 1960s became the first reusable vehicle to fly into space. *(Photo courtesy of NASA)*

The legacy of the X-15 extends to privately developed spacecraft like Scaled Composites' SpaceShipOne, pictured here during a test flight in 2003. Like the X-15, SpaceShipOne is launched from the belly of an aircraft and then fires a rocket engine to reach suborbital altitude. *(Photo courtesy of Scaled Composites)*

Space entrepreneurs Robert Citron *(left)* and his brother Rick were teenagers in this 1958 photograph taken in the Mojave Desert, where they were among many space enthusiasts who built and tested homemade rockets. *(Photo courtesy of Robert Citron)*

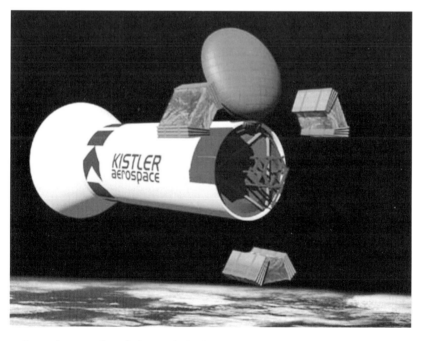

Robert Citron co-founded Spacehab, the most successful entrepreneurial space company to date, and Kistler Aerospace, whose K-1 rocket, pictured here, promises to dramatically lower the cost of space launches. *(Photo courtesy of Kistler Aerospace Corporation)*

Perpetually over-budget and behind schedule, the International Space Station is nonetheless the largest man-made object ever to orbit the Earth. This December 2002 photograph was taken by the crew of *Endeavour*, the last shuttle to visit the station prior to the February 2003 destruction of *Columbia*. *(Photo courtesy of NASA)*

Citron hopes the timing is better for Kistler Aerospace, and, like Kistler, he seems fairly unruffled by its financial problems. "As far as I can tell," Citron wrote to me in the wake of the company's Chapter 11 filing, "the K-1 will become the breakthrough launch vehicle that Walter and I had planned it to be in 1993." Their cool pragmatism may simply be the by-product of an unusually long-term perspective, since opening space to interplanetary adventuring is what Citron and Kistler had in mind when they created Kistler Aerospace in 1993, the year they crossed paths with Max Hunter, a NASA engineering legend and the driving force behind the design of the DC-X. "They built it, they flew, and again and again," Kistler said of the DC-X, still with enthusiasm. "That was our inspiration. That was the way to go."

Hunter convinced Kistler that the time was right to get into the reusable spacecraft business. The DC-X, Hunter enthused, would change the face of the space industry, and anyone on the early track to capitalize on it would make a fortune. The original estimate for building a working, orbit-capable vehicle based loosely on the DC-X concept was a not inconsiderable $2 billion, but Kistler—for whom every entrepreneurial venture he'd touched became gold—was undaunted. "I called Bob on the phone and said, 'Max Hunter says we should go for it and build a reusable rocket. Do you think you can raise two billion?' " Kistler recounted with a regret-free laugh. "And Bob said, 'Well, let me think about it,' and then a week later he said, 'Let's try it.' Bob is very good at organizing things and at raising money."

Kistler had intimate experience with Citron's investment prowess, since he was the first major investor in the company that made Citron something of a legend in the space community well before Kistler Aerospace was on the scene. In 1984, Citron founded a company called Spacehab, which builds modular cargo, personnel and science modules that fit into the payload bay of the space shuttle orbiter. Under a typical Spacehab arrangement, the company pays NASA to fly its modules and then sells space on the modules to universities, corporations or anyone else needing something done in orbit. Although its existence has depended entirely on NASA's willingness to fly the modules (which benefits NASA by maximizing the shuttle's usable space), Spacehab did not spring from the head of any Big Aerospace company nor has it based its viability on government subsidies. For these reasons, Space-hab is widely considered the most successful entrepreneurial space venture to date.

For Spacehab, the year 2003 began with a one-two gut punch. First, its double-wide experimental module was lost on *Columbia* to the tune of about $67 million, only $18 million of which was insured. A week after *Columbia* was destroyed, a Spacehab subsidiary lost a NASA contract for building and maintaining ISS training facilities that was worth another $30 to $80 million over a period of several years. (Wisely, Spacehab didn't bank its entire future on the shuttle; it also has a media and education division.) The combination of these events, which by all accounts were unrelated, sent Spacehab's share price dipping below a dollar, but the company gradually recovered. Such dependence on NASA makes for a rough ride indeed, yet over the course of its history Spacehab has generally prospered, with revenues holding steady at more than $100 million annually over the past few years.

Spacehab is a rare example of a new space company making it based on an idea that the entrepreneurial community has long promoted: NASA as customer and enabler. Citron practically had to box NASA into a corner to make that happen, and his upbeat demeanor about the future of Kistler Aerospace is all the more unexpected considering what it took for him to get Spacehab off the drawing board and into space: Citron says frankly that it cost him his first marriage. "He'd come in in the morning and disappear into his office all day, working the phones," recalled Donna Hines, Citron's longtime assistant, about the early Spacehab days. "Sometimes I didn't see him go in or out. It was like he was there all the time."

"It was a real struggle," Citron admitted. "All these companies in the eighties couldn't believe I was developing a module for the shuttle that would cost a million dollars instead of four or five million. They couldn't believe we could reduce the cost by a factor of that much. But that's what Spacehab did."

Because he started a company that actually succeeded, Citron is something of a guru to other space entrepreneurs, who want to know the secret of his success. Tenacity is obviously a key component: when Citron started the company, he figured there were about fifty entrepreneurial space ventures around at the time. Other than Spacehab, only one company survived, Orbital Sciences, which makes and markets small, plane-launched rockets for light payloads: one of Orbital's founders, David Rossi, later became Spacehab's president. (The company's woes in early 2003 were compounded by Rossi's death, of cancer, at the age of forty-six, about two weeks after *Columbia*'s destruction; Rossi had stepped down as Spacehab's president in late 2001.)

Getting NASA on its side was another linchpin of Spacehab's success, and that most certainly was the hard part. Citron may have benefited by the fact that he first approached NASA about a completely different proposition, albeit one that is in many ways closer to his heart. He started Spacehab with the idea of kitting out the shuttle payload bay to carry passengers—a concept not dissimilar to Gerard O'Neill's plan for passenger-carrying shuttles to service his space colonies.

The concept interested the shuttle's designer, Rockwell, sufficiently for some of its engineers to sketch out a configuration for how such a retrofit might be accomplished. Citron showed me a sketch of a Rockwell design that accommodated up to seventy-four passengers on two separate decks; the engineers and Citron even made a cost estimate of the work needed for the reconfiguration, about $200 million. "We designed seats where everybody had video monitors where people could see the launch while it was happening, because there are no windows in the payload bay," Citron said, the excitement still in his voice. "I even had a five-page scenario of what passengers would do on their flight. They would be busy minute to minute."

NASA eventually shot down the passenger idea for safety reasons, but discussions continued about alternative ways to utilize the shuttle's interior, with sketches and concepts shifting toward maximizing the shuttle's capacity as an orbital research facility. By the time of the shuttle's first flight, such capacity had become a serious issue for NASA due to its aggressive marketing of in-flight research opportunities on the shuttle (many of which were subsidized so heavily as to be virtually free). The result was a massive backlog of projects, most of them science experiments.

Citron proposed to build modular inserts for the shuttle's payload bay that would increase the usable space for payloads of many shapes and sizes. The Spacehab concept was not unique. Around the same time, the European Space Agency had finished spending a decade and about $2 billion developing something called Spacelab, which also fit into the shuttle and which, like Spacehab's modules, effectively turned the shuttle orbiter into a semi-reusable space station. But a single Spacelab flight cost about $400 million, whereas the flight of even the largest Spacehab module was about one-tenth that amount. "We put them out of business," Citron said gleefully of Spacelab, which was mothballed in 1998.

Despite its enthusiasm for his ideas, and their clear benefit to NASA, the agency initially said no to Citron. Lurking in the background was what one entrepreneur has dubbed "the brother-in-law problem." In the space business, the story goes, everyone has a brother-in-law who works for NASA; thus, any investors approached to fund an entrepreneurial space project will find themselves measuring their potential investment against the advice of someone who works for, or knows someone who works for, NASA. "And of course, that person will say, 'You can't do that,'" the entrepreneur sighed, "or, 'If you could do that, NASA would be doing it.'"

After turning Citron down, NASA tried to build its own version of Spacehab's cheap, flexible modules—only to find that Spacehab had secured patents on the key designs needed to make that happen. Meanwhile, Citron was reconsidering his approach with NASA. What would make NASA trust Spacehab sufficiently to give it a shot at proving itself? Rather than fight the brother-in-law problem, Citron embraced it. He seeded Spacehab with NASA insiders like the former NASA administrator James Beggs, who became the company's chairman, and worked his NASA connections to secure flight agreements with organizations who had been promised future time on the shuttle. Gradually, and not without more sweat and toil than Citron would have liked, NASA embraced Spacehab.

"It was a couple of years of very tough selling," Citron recalled. "We couldn't get NASA to give us a launch, and then I paid NASA a hundred thousand dollars just for a launch reservation, funds I had to raise from private sources. We almost went belly-up about half a dozen times between 1984 and 1987." Ultimately, Spacehab modules not only allowed hundreds of experiments to fly sooner and more cheaply on the shuttle than otherwise might have been possible, but Spacehab eventually got NASA as a customer, too: its modules played a critical resupply role during NASA's ISS training on Mir.

Ironically, both Kistler Aerospace and Spacehab are counting on the ISS to ensure their future viability. While Spacehab grew to prosperity building modular research facilities tailored for the shuttle, its focus turned to ISS resupply in the late 1990s when the shuttle's primary role changed from mobile research van to delivery truck. Although he tries not to play favorites with his two corporate creations and is no longer actively involved in running either company, Citron confesses he spends more mental energy on Kistler Aerospace these days. He believes that the success of the K-1 could be the key to change that the alternative space community has long awaited.

"If the K-1 vehicle succeeds . . . then we will reduce the cost, we'll make things efficient, and because of the reduction in cost we may have new companies emerge," Citron said. "After that, many things are possible."

The K-1 looks much like any other modern rocket, but it operates more like the shuttle. The K-1 would lift off with three Russian NK-33 engines, chosen because they are reliable, powerful and cheap. The engines were originally built to power Sergei Korolev's failed N-1 moon rocket, intended as the Soviet answer to the Saturn V. The N-1 never flew, and its last failure caused an explosion that killed dozens at the launch site—in terms of sheer loss of life, the worst disaster in the history of spaceflight. Although ordered destroyed when the N-1 was canceled, dozens of unused engines were hidden away by Russian engineers and later sold to Kistler Aerospace for the post–Cold War bargain price of about $4 million apiece.

The generic K-1 mission profile is elegantly simple. About two minutes after liftoff and at an altitude of about twenty miles, the first and second stages of the K-1 separate, with the second stage firing a single, smaller Russian NK-43 engine and continuing upward. The first stage drops back to Earth and deploys a bloom of parachutes to slow its descent. Shortly before touching down, it would sprout a cushion of bright orange airbags to provide a soft landing. Meanwhile, once the second stage—which looks something like a badminton shuttlecock—reached its target altitude (about one hundred miles up), a disk-shaped hatch at the rocket's nose would swing open to release one or more satellites, which would then fire their own small motors to boost themselves to wherever they needed to go.

After deploying its payload, the K-1 second stage would flip itself around using maneuvering engines, then head down into the atmosphere, nose first. After reentry, the second stage would also deploy parachutes and airbags and land back at the launch site. The airbags and parachutes have been tested successfully, as have the Russian engines and most other major components of the K-1.

Following each flight, the K-1 would be turned around for another flight in only a few days. Each K-1 is designed to be usable for up to a hundred times, the same lifetime advertised for each shuttle. However, the K-1 would be a lot cheaper to use than the shuttle, both because it will be a smaller, less complicated vehicle and because it will not be piloted and therefore will not have need of the extensive "human rating" requirements that NASA employs for the shuttle. The K-1 is designed to send up to ten thousand pounds into orbit, making it competitive with Boeing's Delta rocket.

Until the K-1 flies—an event yet to be scheduled—the sheer scale of Kistler Aerospace's fund-raising success will remain the company's most impressive feat. But what attracted the big capital to Kistler Aerospace was not technological innovation; it was organizational design.

Leaf through the Kistler Aerospace prospectus and you will see a veritable who's who of ex-NASA leadership staring back at you in bold, beautiful color. Listed as senior technical adviser and prominent in the company's promotional video is Aaron Cohen, the former czar of NASA's Johnson Space Center, arguably the most powerful position in NASA next to the administrator (and some don't bother with that qualifier). Another senior adviser is Dale Myers, former NASA associate administrator for manned space flight and former leading light at North American Aviation and Rockwell International. The K-1's chief engineer is Richard Kohrs, who once directed NASA's space station program, served as deputy director for the shuttle program and played a key role in systems engineering for Apollo. And at the top of the company heap, serving as CEO, is George Mueller, former head of the Apollo program, and one of the last living legends of NASA leadership from the days when the future of Western civilization seemed to hang on NASA's every flight.

In short, rather than trying to be as different from NASA as possible, Kistler Aerospace deliberately snapped up the most senior NASA people it could find in order to end-run the brother-in-law problem and convince NASA that the K-1 would be a complement, not a competitor, to its work.

In the alternative space community, the structure and operational style of Kistler Aerospace has generated as much controversy as the K-1 itself. Many feel that the company made the wrong choice in trying to emulate NASA. "You see Mueller and [Dale] Myers and those other ex-NASA guys at Kistler and, well, you wonder if they'll just end up spending a lot of money and not flying anything," said one entrepreneur, who didn't wish to be identified. Others aren't ready to write off the company or its approach: "It's hard to say what [business] model will eventually deliver the goods," Rick Tumlinson said. "Maybe Kistler's got it right and everybody else has got it wrong."

The company's fate may well validate one position or the other, but what can't be denied is that its NASA-friendly structure gave Kistler Aerospace a running head start. "When we hired George [Mueller] to run Kistler Aerospace," recalled Citron, "the first thing he did was pick up the phone and call Dan Goldin and say, 'I'd like to have lunch with you next week.' The following Tuesday they had lunch in Washington." That lunch eventually led, in 2001, to Kistler Aerospace's receiving a hefty $135 million from NASA, the second-largest award in the first round of a new technology-demonstration project called the Space Launch Initiative.

Depending on whom you ask, as originally proposed the Space Launch Initiative was either a breakthrough effort to finally build something better than the shuttle, or a cleverly disguised chunk of pork destined to produce something like the failed X-33, therefore ensuring that the shuttle would be around even longer. As advertised by NASA, the Space Launch Initiative (or SLI, as it is commonly referred to) was developed "to meet the goal of substantially reducing space transportation costs" by creating a reusable launch-vehicle template upon which vehicles for NASA, the military and the commercial sector could be based. NASA made the remarkable pronouncement that SLI would reduce launch costs by as much as 90 percent—bringing the cost-per-payload-pound down to about $1,000—and thereby would open up the near-Earth space market that had long been dreamt of.

When it was announced in early 2001, SLI generated considerable excitement in the entrepreneurial launch community because NASA went out of its way to invite entrepreneurs to bid for its first-phase funding, which totaled $800 million (all in all, SLI was initially slated to spend nearly $5 billion through 2006). While Kistler Aerospace received by far the biggest chunk of SLI entrepreneurial funding, two companies even smaller than Kistler received $6 million and $3 million each in the program's first round of funding—unusually large NASA awards for entrepreneurial launch ventures.

Oddly enough, serving NASA's needs was not really on Kistler Aerospace's radar screen when it was founded. In its formative days, the company drew not only topflight space managers like Mueller but also investor/board members like John McCaw, cofounder of the McCaw telecommunications empire, and the financier Robert Wang, who became the company's chairman. Such fiscal eminences weren't enticed simply by gee-whiz technology or, necessarily, even by Citron's and Kistler's very real hopes of opening the gateway for broad-ranging human space activity. Critically, the K-1 came to light as a result of what seemed to be the most auspicious opportunity in the brief history of the entrepreneurial space-launch community: a new, satellite-based communications boom whose advent would absolutely require better, cheaper access to space.

In the mid-1990s, the cellular telephone revolution was well under way, but its limits were already clear. Despite their ubiquity, cell phones work well only in areas where there are a lot of antennas planted into the Earth to provide the overlapping "cells" of electronic coverage which allow you to carry on conversations while driving for miles and miles, and which give the entire process its name. However, as most users know, cell phones don't work well in areas where gaps in antenna coverage cause calls to fade, crackle or cut out altogether. If you're someplace really remote, you may not find any cells at all.

In theory, satellite transmission offers a more elegant solution by ensuring that wherever you are on Earth there is always a satellite overhead to serve your communications needs. The barrier to using satellites in such a broad and ubiquitous fashion has been cost. While communication satellites have been around since the 1960s and currently transmit everything from voice data to television programs, the satellites used for such purposes are typically large, extraordinarily sophisticated machines that fly about twenty-two thousand miles above the planet in what's known as a geostationary orbit, an altitude that gives a satellite a speed matching that of the Earth's rotation and makes it appear stationary for transmission purposes.

Geostationary orbit is hard to reach, and each satellite launched to such an orbit costs hundreds of millions of dollars to build, launch and maintain. If one of these pricey birds conks out, ends up in the wrong orbit or gets hit by debris—a real possibility considering that NASA tracks about ten thousand pieces of space junk whirling around the planet—one's investment and possibly one's entire business become worthless.

By the 1990s it had become feasible to build satellites that were small, relatively cheap and able to operate in low-Earth orbit. A company that deployed a "constellation" of such satellites could, in theory, cover the communications needs of its customers more efficiently and more cheaply than could cellular systems. Not only would a constellation still function if a single satellite failed, but a replacement satellite could be built and shot up into orbit with relative ease. In other words, what geostationary satellites did with altitude and sophistication, constellation satellites would do with sheer numbers.

This combination of predicted lower cost and better service led many to believe that satellite-based phones could, in the space of a decade, make antenna-based cell phones obsolete. A thousand constellation satellites could cover the transmission needs of the entire planet, embracing the globe like a sparkling blanket. If a thousand satellites sound like a lot, they are. The Teal Group, an aerospace consulting firm, estimated that as of late 2001 there were a total of only six hundred active satellites in orbit, along with hundreds of inactive ones.

Serious money accompanied the satellite constellation boom, particularly in the three biggest constellation schemes: Iridium, Globalstar and Teledesic. The first two companies were focused on going head-to-head with the cell phone market while Teledesic hoped to capitalize on the world's growing hunger for broadband data communication. Iridium was backed by the electronics giant Motorola, while Globalstar's high-profile patrons included Loral, Qualcomm, China Telecom and Vodafone. Of the three, the most ambitious was Teledesic, whose original plan was to launch a whopping 840 satellites, later scaled back to a still-astounding 288. Founded by the telecom guru Craig McCaw, Teledesic was also backed by McCaw's Seattle neighbors Bill Gates and Boeing (which put up $50 million in cash and another $50 million in R&D help), along with a Saudi megabillionaire who handed over a reported $200 million for a stake in the project.

The exponential growth in numbers of satellites promised by the constellation boom also required a massive increase in launch rates and, importantly, launch vehicles. Since their orbits would be lower and thus easier to reach, constellation satellites could be placed in theory by simpler spacecraft. Certainly, the early constellation satellites could ride on existing converted ICBMs like the Delta. But Deltas and their ilk were pricey, and the constellation boom was all about profit. Whoever presented the constellation-satellite industry with a cheap, reliable alternative to Boeing's and Lockheed's luxury rockets would be looking at a wide-open market worth billions.

The period between 1993 and 1995, the height of the constellation development phase, saw the founding of Burnside Clapp's Pioneer Rocketplane, Rotary Rocket (whose founder, Gary Hudson, worked on an early version of the K-1), Universal Space Lines (started by Pete Conrad) and a host of other serious-minded companies intent on grabbing their share of the constellation business by providing an alternative to expendable launch vehicles. Much of the initial funding for the new entrepreneurial space ventures came from the nascent constellation companies themselves, which anticipated the launch availability problem and provided early funding for new approaches to getting their satellites into orbit.

By 1997, Kistler Aerospace had already secured a contingency contract from the would-be constellation giant Globalstar (contingent on Globalstar's need for launches and Kistler's ability to provide them) for ten launches in support of the Globalstar constellation. The pact was worth more than $100 million, meaning that each launch would cost about $10 million—a low price even for Kistler Aerospace, which advertised a per-launch cost of about $17 million. Yet even at $17 million, the K-1 "would shatter industry price structures," as a June 1997 *Aviation Week* article put it. More to the point where spaceflight standards are concerned, Kistler Aerospace initially boasted that the K-1 would also be launched up to fifty times a year, far outpacing the annual launch rate of any other space vehicle.

Other companies were similarly ramping up to address the new market. What was left was for the constellation business to take hold. Iridium was first out of the block with the launch of five 1,500-pound satellites on May 7, 1997, atop a Delta II. But even before the first Iridium satellite reached orbit, the project had already been scaled back from the originally planned seventy-seven satellites to sixty-six, which would cost the company $5 billion to develop, build and launch. This presented a minor branding problem, since "iridium" is the name of the element with an atomic number of seventy-seven. When the system was reduced to sixty-six satellites, a name change was briefly considered, then quickly abandoned when company leaders learned that the elemental designation of their scaled-back system was dysprosium, which sounded less like high technology than bad indigestion.

Unfortunately, the company's name was the least of its problems. Once its satellites were in orbit, Iridium fluffed its commercial rollout in two ways, first by selling its phones at an exorbitant price (about $3,000 each) then by offering an expensive service with calls averaging about seven dollars per minute. Not illogically, the company targeted high-end corporate executives, but most of those executives already had one or more overlapping cell-phone services, essentially giving them worldwide coverage that, even with occasional interruptions, was comparable to the Iridium system unless they needed to do business from high in the Himalayas. The Iridium phones were also clunky (about the size of a World War II walkie-talkie) and didn't work particularly well indoors, which is where most executives tend to do business.

In its first two quarters as a commercial service, Iridium attracted a mere ten thousand customers. By August 1999, the total Iridium user base was up to about fifty thousand, but that wasn't nearly enough to bail the company out from under several billion dollars worth of debt. That same month, Iridium filed for bankruptcy protection. It was one of the largest bankruptcy filings in U.S. history.

Globalstar launched the first batch of its forty-eight satellites, scaled back from sixty-four, in February 1998; its constellation would ultimately cost about the same as Iridium's. Although the company claimed more customers than Iridium—about sixty-six thousand, a much-disputed figure—in February 2002 Globalstar followed its competitor to bankruptcy court, also unable to find enough market to cover its investment. Iridium's satellite infrastructure was later bought for the fire-sale price of about $25 million; the new company, which somewhat ominously kept the name Iridium, is struggling to make its revamped service profitable. (As of this writing, Globalstar is continuing its restructuring attempts.)

Teledesic hung on until mid-2003 and then announced that it would close shop, never having launched a single satellite. While wisely staying out of the business of competing with cell phones, Teledesic was outpaced by an even more primitive technology: old-fashioned cable and fiber lines, which have slowly snaked their way around the planet to provide ever-greater broadband access, thus cutting the legs out from under Teledesic's market.

The constellation revolution was dead by the end of 2000. In the absence of the constellation-launch bonanza, the K-1's stated ability to handle fifty launches a year would at present cover most of the world's annual space-lift needs. Put another way, the K-1 would have to put the Delta, the Atlas, the Ariane and all the Russian launchers out of business to hope for such a flight rate. Following the constellation bust, the satellite industry also slumped. The Teal Group predicted only sixty-one commercial launches in 2003; even that figure is an improvement over 2001 and 2002.

The constellation crash was only one of several major challenges confronting Kistler Aerospace as it raced to get the K-1 flying before the end of the millennium. At the apex of the constellation boom, the investment market was stumbling over itself to pour millions into dot-coms of any shape or sort, and therefore had less money to funnel elsewhere. As Kistler himself acknowledged, Kistler Aerospace needed more of those millions than anyone had anticipated. For all its simplicity of concept, nothing like the K-1 had ever been built before, at least not past the prototype stage. And while Kistler Aerospace had significant financing, it had nowhere near the Space Age billions that fueled the creation of the Delta, Atlas and other current launch vehicles.

The company also ran into problems with many of its subcontractors, most of which were used to doing their work with an open funding tap from NASA or the military. Jobbing out big-ticket items like the K-1 hull structure, pressurized fuel tanks and engines (which, although Russian-built, had to be reconfigured for the K-1) was the only way Kistler Aerospace could reasonably hope to realize its goal of providing a reusable, low-cost launch alternative. To build such expertise and machinery in-house was out of the question.

Yet the cultural investment in Kistler by Lockheed, Aerojet, Northrop and other Big Aerospace players contracted to do work for Kistler was questionable, at best. Many of these companies had deep vested stakes in the space-launch status quo. There were delays, glitches and, through it all, the steady cost creep that government agencies like NASA simply absorb into their multibillion-dollar annual taxpayer stipends.

"I don't think any of the contractors bought into the Kistler concept," said Jason Andrews, a former Kistler engineer who left the company after the constellation boom went bust. "They saw it as a revenue source." Andrews said Kistler Aerospace also fought a constant battle to get its Big Aerospace contractors to work with the company on its own terms. "You can have all these great, innovative managers but when you have to turn around . . . and try to get them to do business in new ways, it just doesn't happen."

The cumulative effect of these problems ground Kistler Aerospace to a halt in 1998 and nearly caused it to fold: Andrews left the company when it went from more than a thousand employees to "a few hundred." Private investment was increasingly stop-start, and the NASA money Kistler Aerospace had already secured through the Space Launch Initiative was of relatively little assistance since it came with a giant caveat. Kistler Aerospace would get only $10 million of the $135 million promised by NASA until the K-1 actually flew, which amounted to a significant chicken-and-egg issue for the company.

The Space Launch Initiative itself ran aground in 2002, at least as it was initially conceived. In November 2002, NASA announced that it was no longer planning to pursue a fundamentally better space vehicle and, instead, was refocusing most SLI funding on something called the Orbital Space Plane, or OSP, which would serve as a crew taxi for the ISS. As currently envisioned, the OSP will be something akin to a smaller, dumbed-down version of the shuttle: it will blast into orbit atop an expendable Delta or Atlas and glide back to a runway landing.

With the advent of the Orbital Space Plane, gone was about $3 billion to be spent on conceiving a new space vehicle. The money didn't disappear from NASA's budget; it was rerouted, not only to the OSP but also to the ISS and shuttle programs. Some SLI funding remains in place for new technology research, but nothing like what is needed to produce a radically new approach to space transportation.

The OSP will certainly not be that radical new approach, since it will be less planelike in every sense of the word than the shuttle: the OSP will have no significant cargo capability, nor will it be able to conduct any noteworthy maneuvers other than docking with and undocking from the station. (Its chief selling point seems to be quick turnaround for crew rescue, although even that attribute is much-debated.) Unless it is supplemented by the Soyuz capsules now used as emergency crew-return vehicles, the OSP may also have the effect of permanently reducing the ISS's crew size from a hoped-for six to a maximum of four: the OSP design requirements call for the vehicle to be able to carry "no fewer than four ISS crew," therefore implying that designing a vehicle to carry more people is only a discretionary requirement.

While Kistler Aerospace will still get its SLI award if the K-1 flies, the OSP sucked the life (and funding) out of another program that offered arguably greater promise for the company. Almost exactly one year before the *Columbia* disaster, Jason Andrews offered a chillingly prescient rationale for the importance of NASA's having alternative means of reaching the ISS. "NASA is terrified that one of their international partners is either going to bail on them or the space shuttle is going to blow up," said Andrews. "Then they would be held hostage to the Russians for all access to the space station, and they don't want that. They're looking for a second domestic, meaning U.S., means of launching to the space station."

NASA obviously didn't know how soon it would need an alternative to the shuttle for supplying the ISS when, in 2000, it started the Alternate Access to Space Station program. In addition to the usual suspects (Boeing and Lockheed) NASA selected several entrepreneurial companies for funding through the Alternate Access program. One entrepreneurial winner was Andrews Space & Technology, the company Jason Andrews started just down the Seattle waterfront from his old employer. Andrews's company has designed automated supply modules that could be launched to the station atop an Atlas or Delta and returned to Earth. In positioning his company to work with NASA, Andrews clearly learned a thing or two from Citron and Kistler—his staff is packed with NASA and Big Aerospace veterans.

Kistler Aerospace was another Alternate Access awardee, and as designed the K-1 would be an ideal "end to end" solution for supplying the station—cheap, reusable and, critically for NASA, crewless. Unfortunately, with the announcement of the OSP, Alternate Access funding disappeared, leaving the strange implication that the OSP would provide the alternative access that NASA was looking for—even though it is not, as currently conceived, a cargo vehicle. As for SLI's original goal of dramatically reducing the cost of access to space—a goal singled out for praise by NASA administrator Sean O'Keefe's former employer, the Office of Management and Budget, in its analysis of NASA's 2003 budget—O'Keefe dismissed it as an unrealistic "bumper sticker" that was never achievable.

If O'Keefe was intent on setting NASA's sights lower, he succeeded with the OSP; it will do far less than the shuttle, take nearly a decade to build and cost an estimated $10 billion. The OSP also means that the shuttle team at the Marshall Space Flight Center can breathe easier about their future, since the OSP was carefully conceived to augment the shuttle, not supplant it. Throughout 2001 and early 2002, the space industry trade press was awash with news of skirmishes between SLI and shuttle factions within NASA, with reports of open bickering at conferences and trade shows. As it stands, it is obvious which camp prevailed. It is, further, no surprise that Marshall was named the lead NASA center for OSP development.

Limited performance, huge price tag and the triumph of self-interest over space-faring vision: for NASA, the OSP is familiar territory.

The Orbital Space Plane won't be operational until 2008 at the earliest (and then only for crew return, not crew or cargo delivery). Considering that the fate of *Columbia,* and that Russia's straitened financial circumstances will probably continue to limit production of its Soyuz and Progress spacecraft, finding other means of supplying and boosting the ISS looks to be an issue for NASA for some time to come. Even if the OSP solves part of NASA's ISS access problem, it is worth asking why the agency would stop at finding a single "alternate" means of accessing the ISS, rather than investing what amounts to relatively little money to pursue other options to at least the demonstration stage. It is also worth asking why NASA believes that the OSP is considered "assured access" when its cargo capacity is limited to the equivalent of hand luggage. Fortunately, such questions are being asked, and the Alternate Access program, or some variant, may yet return with NASA funding.

The failed constellation boom notwithstanding, there is still a market awaiting the K-1; NASA, military and commercial space customers—all of which wrestle with tight margins on huge investments—would no doubt be glad to reduce their launch costs, as the K-1 promises. But even if the company never gets another dime of NASA funding, or if the company folds altogether, if nothing else Kistler Aerospace has solved the brother-in-law problem, hands down—a lesson that Jason Andrews and other novice entrepreneurs have applied to their own ventures.

"It was very difficult at first for Kistler to get partners on board, and since Kistler people [at NASA] have said, 'Well, yeah, all right, I'll work with a small company,' " Andrews said. "That's been a real key element to our success. People are calling us instead of us calling them. That's new." That said, NASA's approach to the entrepreneurial sector hardly amounts to a warm embrace: "NASA gets very excited about the idea of new companies," Andrews ruefully concluded, "but when it comes down to actually building hardware, especially if it's going to the space station, there's still a real pucker factor."

As for Kistler Aerospace, if it appears that the company's future now hangs almost completely on NASA, that doesn't much bother Bob Citron. "NASA is a tough bureaucracy," he said matter-of-factly. "But you know, entrepreneurs put a lot of blame on a lot of things. They'll say, 'Bureaucracy is in my way,' but that assumes you've got to fight the system to make things happen. That's the wrong way. You've got to have NASA in your pocket."

If anyone knows a thing or two about what it takes to work with the space establishment, it's George Mueller. In 1963, Mueller was hired by James Webb—the most powerful leader in NASA's history—to get the Apollo moon-landing program back on track, for off-track it was. When Mueller replaced D. Brainerd Holmes as head of Manned Space Flight, NASA's time frame for a moon landing had slipped to 1971, an unacceptable situation since President John F. Kennedy had made a very public promise to land a man on the moon and return him safely to Earth before 1970. Webb tasked Mueller with pulling the moon-landing timeline back into the sixties. In the broader picture, Mueller had been charged with ensuring the survival of America's space program and saving face for the entire nation.

And so, at the age of forty-five, Mueller became head of the Office of Manned Space Flight and responsible not only for Apollo but also for its critical stepping-stone, the Gemini program. In the mid-1960s, the Office of Manned Space Flight controlled three-quarters of the entire NASA budget, and Mueller was not shy about using his power to get results. In his biography of James Webb, the historian W. Henry Lambright described Mueller as a "first-rate engineer and as much a workaholic as Webb. . . . [He called] the technical shots as he saw them, and seldom [worried] if his decisions or manner of making them ruffled the feathers of subordinates. He fastened on the lunar deadline and became its guardian."

Mueller proposed to remedy the Apollo schedule slip in a radical way that would become legendary within the space community, although perhaps less well known to the public at large. While working to develop the Air Force's ballistic missile program, Mueller promoted what became known as "all-up testing": the not-so-easy process of putting rocket stages together and launching them together—literally, sending them "all up" into space—at the same time, without first testing each stage separately. The idea of "staging" rockets dates back to Russia's Konstantin Tsiolkovsky, who recognized the energy conundrum of getting into orbit: it takes a lot of fuel to lift a heavy spaceship through the atmosphere free of gravity, but carrying the required fuel adds to the weight problem. The solution has been to build rockets in two or more distinct sections called stages, with each stage containing engines and/or fuel. Staged rockets typically shed all but the final stage—the one containing people and/or cargo—on the way up to orbit, thus making the rocket progressively lighter as it ascends.

Staging is tricky business, and it was never more so than in the nervous early days of the Space Age. For the Apollo-carrying Saturns, each of the mighty rocket's three stages was actually built by a different company, making it particularly logical to flight-test each stage separately before sending them up all together. Yet Mueller's all-up testing had succeeded several times prior to his NASA tenure, first in 1961 with the Minuteman missile, whose all-up launch of three virgin stages cut a full year off the program's development. He did the same later with the Titan II, a two-stage ICBM that would go on to become one of the workhorses of the commercial launch industry for Lockheed.

Despite these successes, all-up testing was anathema to NASA's way of doing business, which derived largely from Wernher von Braun's hyper-incrementalism: test, test and test again, in the lab, on the ground, piece by piece and, eventually, all together in flight. But under Mueller's watch, NASA had no such luxury. The Soviets were achieving first after first as they reached for the moon, and Kennedy's mandate loomed large over NASA.

And so, on November 9, 1967—with NASA and the nation still mourning the loss of the *Apollo 1* crew—an unmanned Apollo spacecraft was placed atop an untested three-stage Saturn rocket and launched, all-up, from a new, specially built pad at Kennedy Space Center. The entire vehicle weighed six million pounds and was six stories taller than the Statue of Liberty. Had it failed, many believe the moon program would have ended right there. But the flight went without a hitch, America was on its way to the moon and George Mueller became a spaceflight legend.

To fully appreciate how risky Mueller's decision was, one only need look to the very next Apollo-Saturn flight. About two minutes after launch, the Saturn rocket began lurching like a car with a backfiring engine, the whole massive machine vibrating and oscillating in an effect rocket engineers call pogoing because it resembles the bouncy motion of a pogo stick. NASA managed to complete the flight and retrieve the Apollo spacecraft as intended, but the mission was considered a failure, since the pogoing was severe enough to have killed a human crew. Fortunately for NASA, few outside the agency took note of the failed mission, because it flew on April 4, 1968—the day Martin Luther King, Jr., was shot to death on the balcony of the Lorraine Motel in Memphis. King's assassination sparked riots across America and signaled, in many ways, the beginning of the end for an era of technology-fueled optimism whose chief beacon was the Apollo program.

As we sat in his room at San Francisco's upscale Ritz-Carlton hotel on a misty morning in 2002, those days were far behind George Mueller. Now in his eighties, he is still sharp-witted and, apparently, still tireless: the day before, Mueller had been working hard at Kistler Aerospace's Seattle headquarters, and the next day he would be in Washington, D.C. What he was doing on such a whirlwind trip—not atypical for Mueller in what must surely be the final phase of his long career—was raising money.

Mueller's reputation preceded him; he had little trouble getting appointments. The money was not yet all there, but he was optimistic, as he surely must have been to have given up what was by all accounts a peaceful retirement to take the reins of yet another impossible project at the advent of another new era for space. It doesn't take long to recognize what drives him. Quite simply, Mueller feels he has unfinished business.

"All of our expectations [during the Apollo era] were that we would be continuing the exploration of space, that we would in fact have built by now a space station and we would by now have a colony on the moon, and we would be traveling back and forth," Mueller said in his gentle yet authoritative voice. Mueller had been a vocal advocate for establishing an extensive human presence on the moon and going on to explore Mars. To that end, he was the architect of a program called Apollo Applications, which was designed to put to use the knowledge and technology of Apollo/Saturn as a bridge to new worlds.

The rapid downsizing of NASA after the moon landing meant that

Apollo Applications didn't last long, mainly because the program depended on Saturn rockets, which were terminated in the early 1970s and replaced with the space shuttle. Apollo Applications did produce Skylab, which Mueller claims among his proudest achievements. Mueller said that Apollo Applications gave Skylab a "leg up" because it was designed for the Saturn and launched as a single entity, a model of simplicity the International Space Station could never emulate. Also, as Mueller noted with a chuckle, Skylab "didn't have an opportunity to be, ah, reengineered and redesigned as the current space station has. The constraint on the space station has been the necessity to get into orbit on the space shuttle itself."

This is a sore point for Mueller for several reasons. In NASA circles and in the Kistler Aerospace literature, Mueller is called the father of the space shuttle. He presided over the shuttle's early development and, many times during its struggle for life in the late 1960s, used his clout and political savvy to save it from the congressional ax. Now, he seems to view this as a mixed legacy. The shuttle and International Space Station, whose fates are inextricably intertwined, currently consume nearly half of NASA's annual budget. More critically for Kistler Aerospace, most ideas for new space vehicle development inevitably run the shuttle gauntlet, which involves not only NASA and Big Aerospace but also the powerful politicians whose constituencies thrive on shuttle-related contracts.

Mueller fought to keep the shuttle program alive because it was part of the scheme that he believed would eventually take humans beyond Earth once more and establish a permanent human presence in space. The shuttle was to be a ferry to orbit, a space truck serving grander ambitions. Certainly, it was never meant to be the end point of those ambitions, let alone a roadblock to the future.

"There have been various attempts to improve [the shuttle], but there isn't any way you can take that design and make it into a fully reusable vehicle," Mueller said. "So you really have to start over. To me, the amazing thing is that we've left that until now, because the original shuttle should have been retired in ten or fifteen years and a new one created at that time. It's now thirty-some-odd years old and it's truly obsolete."

As he spoke these words, Mueller was very careful. He is, after all, not only an outstanding engineer manager but also a seasoned politician with a finely tuned instinct for gauging how what he says now will affect what he needs to do down the road. And while Mueller still wields clout with his old agency—amazingly so, since he left in 1969 for a successful career in industry—he also knows that he needs NASA to help make the K-1 a reality, thus taking a step toward a future he feels has been too long delayed.

But upon being asked the question that many of the more brash space entrepreneurs put forth boldly and frequently—Do we need NASA anymore?—Mueller was unable to suppress a smile and an unexpectedly candid response: "Well now, that's a good question."

The Belly of the Beast

In September 1999, two years before he resigned as NASA administrator, Dan Goldin walked into Dodge for a showdown. "Dodge" in this case was the Sheraton Gateway Hotel near Los Angeles International Airport, where Goldin was to give the keynote address at the eighth annual conference of Rick Tumlinson's Space Frontier Foundation.

Goldin and Tumlinson had publicly butted heads before in congressional and media forums, and neither man was particularly fond of the other. Despite this, at the Space Frontier Foundation meeting Tumlinson and Goldin acted more like the captains of rival football teams than bitter enemies; there was spirited ribbing and lots of laughs, with the barbs blunted just enough to keep everyone in the room.

Before Goldin stepped onto the rostrum, Tumlinson had concluded, with typical fire-breathing evangelism, an assault on the space establishment in which he described the coming "space millennium" as being possible only with a transition from "old space"—one dominated by the government—to "new space," dominated by entrepreneurs. Goldin, too, had a reputation for zealous rhetoric, and his blunt style had earned him the nickname "Captain Crazy" during his twenty-year management of spy satellite programs at the aerospace giant TRW, from which he was selected to head NASA in 1992 and would go on to become the longest-serving administrator in NASA history—the leader of the agency for a full one-quarter of its existence. But after seven years in office Goldin had failed to deliver the "new NASA" he so often spoke of, a NASA that embraced the entrepreneurial community as partners in progress.

Goldin's style was never that of the slick functionary from whose lips even the most conciliatory words seem smarmy and hollow. I had seen him speak at least a dozen times at various points during his tenure at

NASA and I kept wondering when Goldin would begin to sound like a bureaucratic automaton, effortlessly regurgitating a company line. But it never happened. Goldin was too rough around the edges, too much a regular guy. Although he'd acquired some California cool during his years at TRW's Redondo Beach enclave, Goldin never strayed far from the "this is just you and me talking" ethos of his Bronx roots.

Thus, when Goldin promised the Space Frontier Foundation crowd that NASA's goal was to help "set the stage for a permanent human presence in space," he was believable. When Goldin insisted that "strategic public-private partnerships between groups like NASA and the Space Frontier Foundation are the only way we will make the new millennium the space millennium," he seemed absolutely sincere. Audience members looked at one another in confusion. Was this *really* Dan Goldin, on whose watch the DC-X was killed and whose relentless toadying to NASA's military-industrial contractors had allowed the International Space Station—perhaps the last hope for the Apollo generation to witness some small flowering of an exciting human future in space—to grow fat in budget and gaunt in vision?

Goldin had all the right words for all the critical issues: the absolute necessity of finding new solutions to the "access" problem, the need to bring launch costs down significantly ("below $100 per pound") and therefore open the entrepreneurial market, and NASA's commitment to pursue "significant private commercial participation" in the International Space Station ("as equity partners and paying customers"). Goldin concluded, "I believe that when NASA can creatively partner with you, all of humankind will reap the benefits of open access to space." This was a space future everyone could buy into. The room broke into gleeful applause.

Despite the brothers-in-arms rhetoric of the day, only a few months later Goldin would unveil the $4.5 billion Space Launch Initiative, which would serve chiefly as a means of funneling more money to existing NASA contractors. Less than a year after mesmerizing the Space Frontier Foundation crowd, Goldin would go on full attack to quash the efforts of Tumlinson and company to save Mir, and not long after that, he would publicly question the patriotism of Dennis Tito, the entrepreneur who proposed paying the Russians for a flight to Mir.

Tumlinson would later have a curt reply when I asked him how such enmity arose in NASA toward Tito's flight, for which the Russians took start-to-finish responsibility and which actually demanded very little of NASA: "The evil guy in the whole thing was Goldin."

There have been two types of NASA leaders since the agency's inception: those who acted from day one on the understanding that NASA was a purely political animal, and those who came to this conclusion much later and, usually, much to their surprise and dismay. James Fletcher, the shuttle champion and Industrial Space Facility slayer, is a clear example of the former type of administrator, as is Sean O'Keefe, the current officeholder, a man with no scientific or technical background (a majority of NASA administrators have been engineers or have held science degrees) and who, as a career bureaucrat, has a particular reverence for the political bottom line.

The other category of administrators includes a list of forlorn dreamers with the admirable but quaint idea that NASA's mission, above all else, should be to build on the undeniable success of the Apollo program and in doing so stretch the boundaries of the human presence in space. Invariably, such administrators have been frustrated in trying to bridge the yawning gap between vision and reality. In 1970, toward the end of his tumultuous two-year tenure, then NASA administrator Thomas Paine convened a meeting where he issued Queeg-like directives calling for "a completely uninhibited flow of new ideas" and for "swashbuckling, buccaneering courage." Unfortunately, Paine's remarks came right after Richard Nixon had told him in no uncertain terms that NASA would be lucky to simply stay on the government dole: forget going back to the moon, forget Mars. Only a few months after his call to arms, with his family concerned for his health, Paine left NASA for a quiet retirement in industry.

Other administrators tried instead to work with the reality they were given, hoping the dream might follow along behind, and these administrators were more embittered than most when they left the agency. James Beggs (circa June 1981 to December 1985) began as a space station advocate in the Reaganesque spirit of that facility's potential to spawn commercial opportunities in orbit. He watched instead as the station became little more than an excuse to keep the shuttle flying, a die that was cast by his predecessor, the company man James Fletcher.

Beggs's disillusionment at NASA's bureaucratic skulduggery, which he felt largely helpless to stop, so affected him that he later became openly hostile to the agency. "NASA should not be a transportation company. . . . NASA should not be a landlord in space," Beggs wrote in a 1997 letter supporting Spacecause, formerly the lobbying arm of the National Space Society. "And above all, NASA should not be an employment agency . . . even if that means a few bureaucrats have to find something else to do." Beggs even took the extraordinary step of lambasting Dan Goldin, then the administrator, saying that despite "repeated public pronouncements of support for commercial space from [Goldin], in recent years NASA has done *in deed* just the opposite. . . . NASA, in many ways, has obstructed commercial space activities instead of encouraging them."

Despite Beggs's accusations, Goldin was surely among the company of Paine, Beggs and other frustrated dreamers who tried to wrest a compelling, coherent agenda from the hydra-headed agency and its many, many masters. To his credit, Goldin aimed gloriously high, though he fell breathtakingly short. His dearest hope was to send humans to Mars, but by losing not one but two Mars probes in 1999 Goldin arguably set that goal back by years, if not decades. He was hampered by having to manage two white elephants—the space shuttle and the space station—that had transformed NASA from a mission-driven R&D house into a clumsy purveyor of overpriced retail services. Goldin tried to change the NASA culture and do things—to use his much-repeated phrase—"faster, better, cheaper," but as the joke went in science and industry circles, at NASA the best you could hope for was two out of three.

Goldin came in fighting for change and kept his dukes up for an admirably long time. In a February 1996 speech to the American Association for the Advancement of Science, given about halfway through his tenure, Goldin was still hammering away at the shuttle/station conglomerate with rebukes like the following:

Now I don't want to be demeaning to the people that worked on the shuttle, but the shuttle has suppressed a lot of science we could be doing. We haven't landed on a planet in twenty years because we've been so excited about the service support contract on the shuttle that we haven't been doing science. We spent ten billion on the space station and didn't produce a piece of hardware, but boy did the contractors have fun. It's shameful. It's stealing from the American public.

The loss of the Mars Climate Orbiter and Mars Polar Lander in 1999—projects that bore Goldin's "faster, better, cheaper" stamp—was a blow to Goldin's reputation and, by many accounts, to Goldin personally. Afterward, the energy Goldin had directed with such vigor toward effecting real change at NASA was converted to ensuring the survival of the agency's other big-ticket start-up projects, namely the ISS and the doomed X-33. Any interest Goldin might have had in helping the entrepreneurial community—and the level of his interest remains in question to this day, at least among the entrepreneurs—disappeared altogether.

By October 2000, just before the launch of the first crew to the International Space Station, Dan Goldin had clearly experienced his political epiphany. At a Washington luncheon that month, Goldin called the ISS an "underdog" that "many tried to kill." The audience of sharp-suited aerospace executives eating hundred-dollar-a-plate lunches hardly looked like underdogs, but heads were nodding all around at Goldin's remarks. In his final days at the agency, Goldin the visionary had disappeared altogether: "The space lovers have got to get a grip!" he told a reporter from the *Huntsville* (Alabama) *Times* in November 2001. "They've got to understand that we live in a democracy, and in this democracy, people are worried about their jobs before they're worried about inspiration."

When Goldin spoke to the Space Frontier Foundation audience in 1999, such desperate words were still in the future. Yet the writing was already on the wall. Goldin would be unable to keep his promises to them; after all, by the time of the conference, the DC-X had been terminated in favor of the bloated X-33 and the campaign to bring down Mir had already begun. Given such a track record, it is a wonder that so many in the audience were ready to take Goldin at his word. Yet their enthusiasm for Goldin's promises said less about his powers of persuasion than it did about their secret hopes for NASA.

Practically all the denizens of the alternative space community had been inspired to choose their professional calling by a NASA exploit, be it the technological wizardry of the Apollo program or the promised wonders of the space shuttle. Most of them had grown up wanting to be astronauts or NASA scientists, and even NASA's most vicious critics rarely question the talent or individual commitment of its rank-and-file personnel.

For their part, even NASA employees often find themselves uncomfortably conflicted about the agency. "People come to work for NASA because they're interested in going into space," a NASA scientist told me once, with barely concealed exasperation. "People are not at NASA because it's just a job and they happen to see an ad in the paper." After pausing for a moment, he concluded, "I like working for NASA, I think it's got great people. I wish we were doing great things."

Unfortunately, NASA the agency was born an addict to the federal trough and, thus, a slave to political whim and process. But what saddens, enrages and disappoints is the fact that NASA has become an adept and all too willing confederate in maintaining its own addiction: it not only accepts its slavery, it encourages it. The entrepreneurs believe that synergy with their ideas and enterprises is a way to loosen the shackles: to make NASA's taxpayer dollars go further and to help the agency's actions live up to its PR. But despite the many proclamations of a "new NASA" dawning (proclamations that have usually accompanied new administrators, including Goldin and his successor, Sean O'Keefe), NASA is incredibly, stubbornly impervious to change.

It may seem unusual to describe a federal bureaucracy as one might describe a person, but as an organization NASA does have a distinct personality that has remained largely unaffected by time, despite many changes in the agency's leadership. NASA's consistent qualities also help explain its consistent actions over time. If NASA were indeed a person, you might find him—and NASA is definitely a "him," though women make up about a third of the agency's workforce—to be quietly intelligent, straitlaced, easily wounded, politically conservative (albeit on the moderate side) and fairly rigid when it comes to protocol and rules. In public, NASA is by turns obsequious and shy, intensely proud and intolerant of criticism, which is endured only from its political bosses—and even then, borne only with the tightest of smiles. Above all else, NASA values loyalty and self-preservation, and it will kill (figuratively speaking) to protect itself and its friends.

While politically flexible (the agency's leadership is piously apolitical or cheerfully partisan, depending on what is required at the moment), NASA is structurally rock hard. Most of its longtime managers are civil servants whose tenure and influence have remained unchanged through successive political administrations, and who know from experience that dicta handed down today can be ignored without peril, since they're likely to be rescinded or altered once political change sweeps a new boss into office.

Those in NASA below what is derisively tagged the political class protect their own with particular viciousness. Above all else, no one must be laid off, no one must be fired, no projects—whatever their merit or folly—must be canceled. NASA's feint-and-dodge strategy has kept it alive, yet this strategy has earned the agency a contemptuous riff on its acronym: Never A Straight Answer. At a deeper level, that NASA's work exists mainly to ensure that NASA's work continues prompts an even more cynical reworking of its initials: Not About Space Anymore.

To understand how such a strange beast can function, let alone prosper, one must first understand that NASA is both more and less than the sum of its parts, of which there are many and of which few are in agreement about anything at any given time. There are never-ending turf wars for internal and external funding, political control and public affection that shape up like existential title fights: Centers vs. Headquarters, Centers vs. Centers, Civil Servants vs. Contractors, Human Spaceflight vs. Robotic Exploration, even Aeronautics vs. Space.

One way to navigate this bureaucratic labyrinth is to spend some time with a NASA phone book. NASA phone books are organized in an alphabet soup of "codes," with every letter except D, O, T and V (no one could tell me why these were missing) denoting some specific responsibility. All those at NASA make their home in a code, which defines them politically and thus hierarchically in the agency. Code M, for example, is the Office of Space Flight, which includes both the space shuttle and the space station and thus wields tremendous clout (the "M" used to stand for the now politically incorrect "Manned," as the program was originally called Manned Space Flight). Code S is the Office of Space Science, which commands respect in the scientific community for its robotic explorations like Mars Pathfinder. Code M and Code S are classic rivals, since the science community has never considered human spaceflight to be much more than high-profile stunt work.

NASA code letters don't always match the task involved. The letter F, for instance, is given to the Office of Human Resources, and Procurement is Code H. In the same runic vein, Code W is the Office of the Inspector General, known commonly by its phoneticized abbreviation, "eye-gee." The IG is an internal fiscal police force that encourages NASA employees to tattle on one another by anonymously reporting "fraud, waste, abuse, and mismanagement." Most large federal organizations in the United States have IG offices, but the very public profile of NASA's IG reporting makes the agency seem like an enthusiastically penitent Catholic, constantly looking for opportunities to confess its sins.

In an April 2001 report to the House Committee on Science, for instance, the IG reported once again how badly NASA had managed its International Space Station partnership with its prime contractor, Boeing. After adding up the various ways NASA had let Boeing overcharge and underperform—to the tune of hundreds of millions of dollars—the report dryly concluded that "ISS management practices were inconsistent with the high-risk, technologically complex environment of the ISS Program." The testimony produced several media articles noting not only the IG's ongoing self-criticism about NASA's ISS management, but also the IG's rueful between-the-lines conclusion that NASA was unlikely to follow its advice on how to fix these problems.

If this sounds more than a little schizophrenic, it gets worse. The work of NASA codes is actually spread across the ten NASA centers, all of which are fiefdoms unto themselves with varying levels of political and economic influence. Like the essentially medieval organizations they are, the centers are given to frequent skirmishing with one another as well as with NASA headquarters, to which their directors theoretically report much like warlords to a king. The Johnson Space Center and Marshall Space Flight Center have direct responsibility for the shuttle/station conglomerate (which falls mostly under Code M) and so generally wield the most power, but their duumvirate is frequently challenged by the Jet Propulsion Laboratory, which commands respect for its leadership of NASA's deep-space robotics work, which is primarily in Code S. (JPL is a singular maverick, since it isn't technically a NASA facility at all, but rather it contracts to NASA through the California Institute of Technology, better known as Caltech.)

The rest of the centers fall across the spectrum of NASA vassaldom: Kennedy Space Center, where shuttles are launched, is the public darling; Dryden is the hotshot test site attached to Edwards Air Force Base, in the Mojave; the Langley, Stennis and Goddard centers quietly go about their work in propulsion, materials and structures; Ames and Glenn/Lewis are the poor step-centers, constantly whispered about as candidates for closure.

Responsible for putting a happy face on all of this is Code P, the Office of Public Affairs, whose branches in each of the ten centers are, first and foremost, loyal to those centers. The buck finally stops at the beginning of the alphabet with the King of Code A, the NASA administrator, who has the most unenviable task of all: coaxing results out of this techno-bureaucratic spider's web, and on the ever-tightening cheap.

In summary, what the phone book reveals is what NASA insiders have known for years, which is that there is no "NASA"; or rather, there are multiple "NASA"s. Which one you deal with largely determines whether you feel "NASA" is a beleaguered repository of genius fighting to carry humanity into the final frontier, or a lumbering dinosaur that has by virtue of its size and reputation managed to outlive its evolutionary window.

One can hardly blame NASA for being something of a Frankenstein's monster, since it was intentionally kludged to life as such in 1958, only a few months after the launch of Sputnik startled the Eisenhower administration into recognizing that there was indeed a Space Race under way. America at the time had no institutional mechanism for dealing with this reality, although the Atomic Energy Commission, the National Science Foundation and all three major military branches lobbied Eisenhower for a chance to take on the Soviets in space (and get the big bucks they knew would accompany that directive).

Eisenhower instead decided to build a new civilian agency around a tiny government think tank known as NACA, the National Advisory Committee on Aeronautics. NACA was created in 1915 in response to that era's technological crisis du jour, which was that America was losing ground to Europe in the development of its own invention, the airplane. NACA evolved into a crack R&D outfit that nurtured the development of the nascent American aircraft industry with both inventions and dollars. NASA, at least initially, would have much the same mandate.

In short order, onto NACA was grafted the Army's ballistic missile team led by Wernher von Braun and his compatriots who had worked on the V-2 plus a group of missile engineers from the Naval Research Laboratory. Later, NASA added the Jet Propulsion Laboratory, whose scientists had worked with von Braun's team to develop Explorer 1, launched in January 1958 as America's first satellite. Although NACA formed the core of the new National Aeronautics and Space Administration, aeronautics would quickly take a back seat (critics still snipe about its being NASA's "little A") because everyone understood that NASA was about beating the Soviets in space, not airplane development.

NASA's administrative setup tells only half the story of the agency's evolution. The other half is perhaps best described as NASA's spiritual development, which emerged from the unlikeliest of sources. The impact on NASA of a somewhat florid nineteenth-century academic paper, "The Significance of the Frontier in American History" by the University of Wisconsin historian Frederick Jackson Turner, cannot be overstated. "You really can't understand how NASA worked in its early days, or even now, without Turner," said former NASA chief historian Roger Launius. "[His paper] was a tremendous motivator."

Published in 1893, Turner's paper was a dirge for the closing of the American frontier, which occurred in 1890 when the line of settlement moving east from the Pacific Ocean met the one moving west from the Atlantic. Rather than celebrating this event, Turner saw it as a major threat to America's future: the American character, he said, was unbreakably linked to the act of pioneering and the conquest of "free land," which was in turn essential to American inquisitiveness, inventiveness and individualism. Turner's paper popularized the idea of

"American exceptionalism": the concept that Americans, as a mongrel people, were defined culturally by the adventurous circumstances of their nation's development. American exceptionalism wasn't a new concept in 1893, but Turner seized on the closing of the West to take the idea to an urgent new level: that when the American frontier closed, with it went America's opportunity for "perpetual rebirth," and thus the cross-fertilization of ideas and cultures that made it strong and unique.

When published, Turner's paper created a shock wave among historians that still reverberates today, although not for reasons Turner would necessarily be happy about. Influential though it was at the time, "The Significance of the Frontier in American History" has been roundly criticized as unabashedly Eurocentric in its portrayal of the benefits of Western conquest, virtually ignoring the not so minor detail of genocide directed against Native Americans (by which most of Turner's "free land" became available) and the huge contributions of non-Caucasian people in seminal activities like the construction of the transcontinental railroad.

While few historians today would describe themselves as Turnerians, in the 1950s, with America feeling its oats as a postwar world power, the Turner thesis struck a chord with those preparing to challenge a new enemy on a new battlefield. Copies of "The Significance of the Frontier in American History" were distributed widely throughout NASA in its early days, and it's not hard to see why Turner's would be a motivating philosophy for space exploration: more than just cold warfare, the Space Race would see the advent of American expansionism on a cosmic scale. The Space Race upped Turner's ante; not only would space provide America with an endless expanse of "free land" for the taking (and with no bothersome natives to stand in the way), but whoever got there first would determine not just the future of America, but the future of humanity. Space became not only the new frontier; it became the *final* frontier.

The moon was the first stop on this grand journey of conquest; to date, it has been the only stop. This is less curious considering that contrary to popular belief, practically the last thing John F. Kennedy wanted to do was to race the Soviets to the moon. Kennedy's vision has long been a clarion call for space advocates, both inside NASA and else-where, who say that all it would take for the conquest of space to get back on track is another such commitment by another visionary president. Kennedy's true motivation was much more practical. With Soviet power increasing, Kennedy was looking for technology spectacular to prove that a democratic, capitalist nation could produce marvels that the Communist world couldn't match.

More to the point, for Kennedy the decision to go to the moon was a desperate political gamble made in a very specific geopolitical context: he'd just botched the Bay of Pigs invasion, and Soviet missiles were looming in Cuba. The moon program was approved only after virtually every other option at Kennedy's disposal was ruled out as unworkable. Irrigating Africa, desalting the oceans and other engineering wonders were proposed before Kennedy agreed (with Vice President Lyndon Johnson's considerable prodding) to the program that culminated in the triumphant *Apollo 11* mission.

The extent of Kennedy's reluctance was revealed in late 2001 in a newly declassified tape recording of a White House planning session held only a few months after Kennedy publicly outlined his goal of putting an American on the moon before 1970 in a speech at Rice University. Although the White House meeting included Lyndon Johnson (for whom the Johnson Space Center would later be named) and several other NASA and White House staff, the exchange is mainly between Kennedy and NASA administrator James Webb.

Webb was NASA's second administrator, and by far its most influential. In guiding NASA to success in the moon race, Webb, more than anyone in the agency's history, is responsible for shaping the peculiar blend of political savvy, technological sophistication and public adoration that has sustained the agency to date. Webb had the White House's ear like no NASA leader since; Lyndon Johnson once called Webb his best administrator.

Although often labeled as a pragmatist first and visionary second, Webb tried mightily to craft a future for NASA that would extend beyond the Apollo program in bold fashion. The way to do that, Webb reasoned, was to ensure that the Apollo program was not the be-all and end-all for the agency, but rather just a dazzling glimpse of things to

come. In this, even Webb could not succeed, as the Kennedy tapes
make all too plain. The conversation is heated, particularly on Webb's
part; one can clearly hear the anguish in his forceful southern accent as
he tries to convince Kennedy that a manned moon landing should not
be the sole reason for NASA's existence, but rather chief among many
priorities in the broader context of making the United States the leading power in space and building a long-term infrastructure to serve
future cosmic expansion. Kennedy repeated over and over that where
space was concerned, his only interest was beating the Soviets: anything else was superfluous. The hot parrying went back and forth for
nearly seven minutes, as in the following exchange:

Kennedy, quietly emphatic: "Everything we do ought to be tied
into getting onto the moon ahead of the Russians."
Webb, almost shouting: "Why can't it be tied to preeminence in
space?"
Kennedy, loudly and insistently: "By God, we've been telling
everybody for five years that we're preeminent in space and nobody
believes us!"

As the debate continued, Webb tried his best to sway Kennedy
toward a broader vision of exploration and the establishment of a long-
term space infrastructure. Kennedy eventually closed down the conversation, bemoaning that where space exploration was concerned,
"We're talking about these fantastic expenditures which wreck our
budget and all these other domestic programs, and the only justifica-
tion for it in my opinion . . . is because we hope to beat them, to
demonstrate that starting behind a couple of years, by God we passed
them!"
But it's toward the end of this exchange that Kennedy's mythic reputation as a space visionary is most severely undermined. He tells Webb
that if not for the goal of beating the Soviets to the moon, "We
shouldn't be spending this kind of money, because I'm not that interested in space."
Both Webb and the current administrator Sean O'Keefe share back-
grounds as career bureaucrats who were handpicked for NASA leader-

ship by sitting presidents: Webb by Kennedy, and O'Keefe by George W. Bush (or, more specifically, by Vice President Dick Cheney, long a patron of O'Keefe's). Given this, it is not surprising that Webb is O'Keefe's professed NASA role model: he reportedly keeps a copy of Webb's biography and Webb's 1969 book, *Space Age Management,* near his desk. If that is so, O'Keefe must find pause in Webb's own assessment of the space program he brought to prominence. Webb died of a heart attack in 1992 at eighty-five, frustrated that, as his biographer W. Henry Lambright put it, "the nation he loved and served had found a new frontier in Apollo . . . but had stepped back from the opportunity and was the poorer for it."

Webb's successors would learn the hard way that presidential mandates are a risky foundation on which to build an interplanetary future. The first lesson in that reality was the space shuttle.

In March 2002, I stood on the grassy banks of the Banana River and watched *Columbia* lift off for what would turn out to be the second to last time. This was the first shuttle launch in more than two years that was not going to the International Space Station. Such a trip wouldn't have happened on *Columbia* anyway, since, as the oldest and heaviest of the shuttle orbiters, *Columbia* wasn't able to reach the station's orbiting altitude of about 240 miles above the Earth (*Columbia* also lacked a means of docking with the station).

The seven-person *Columbia* crew was embarking on what promised to be a grueling ten-day mission to give the Hubble Space Telescope a $172 million overhaul, the third of four planned service calls to keep the instrument in business through 2010. The Hubble Telescope is NASA's great and glorious phoenix, and its near-death experience provided the shuttle's finest hour to date. Costing about $3 billion, the Hubble Telescope was launched with great fanfare in April 1990. But once in orbit the telescope was found to be terminally myopic due to a manufacturing flaw in its primary mirror, making it practically useless as a research instrument.

In one of NASA's trickiest human space missions since its Apollo days, shuttle astronauts captured Hubble and fitted it with corrective optics, thereby opening the door to some truly spectacular astronomy. Dan Goldin said repeatedly that of all the activities that occurred at NASA under his watch, he was proudest of Hubble's successful repair.

Contrary to popular perception, the space shuttle is not the planelike vehicle that goes into space vertically and comes down horizontally: that vehicle is just one shuttle component, the orbiter. The shuttle is actually the sum of the four main components that result in the orbiter's getting into space: the two solid-fuel rocket boosters, the enormous, pod-shaped external tank that holds liquid fuel and oxidizer for the orbiter's main engines, and the orbiter itself—which, believe it or not, is attached to the external tank by a single, gigantic bolt. Sitting on the launchpad, the whole shuttle is about eighteen stories tall (about half the height of a Saturn V) and weighs about 4.5 million pounds, only about 6 percent of which is the orbiter and its cargo. The rest, mainly, is fuel.

In the breezy predawn of the launch day, March 1, the sky was blue-black and speckled with stars; the distant *Columbia* at its gantry looked like a floodlit castle. At about one minute before launch, the range officer announced over the loudspeaker that the Hubble Telescope was approaching the western coast of Florida and, traveling at more than seventeen thousand miles per hour, should soon be visible; as if choreographed, at T-minus thirty seconds what looked like an unusually bright star cruised lazily over the launch site, clipping the edge of the fading moon before disappearing over the brightening horizon. Just as Hubble made its exit, the launch clock ticked down to zero.

Recorded coverage of shuttle launches tends to zero in on close-ups of flame-belching engines, but before the noise and fire come the massive clouds of steam and exhaust which need some distance to be properly appreciated. The clouds are generated not only by the solid booster exhaust, but also by the vaporization of a quarter-million gallons of water poured onto the launchpad to muffle the vibration of lift-off, which has in the past shaken loose some of the twenty-four-thousand-plus heat tiles that protect the orbiter during reentry (at a cost of about $2,000 per tile, no two of which are alike).

The boiling clouds seemed to swallow the entire horizon; long after *Columbia* achieved orbit, the vaporized water hung clumped in the sky like a low, brooding cumulonimbus. Once or twice, these clouds have headed straight for the press stands, prompting journalists to scramble for cover, since the clouds are also filled with toxic rocket exhaust. A few seconds after T-minus zero, *Columbia* ascended through the vaporous boil, its engine fire brilliant as a rising sun, and then came the thunder: a tremendous roar like two dozen 747s throttling up at once, the earthshaking sound ripping across the swampland and jackhammering through the flesh and bone of everyone in the crowd.

It was all over shockingly fast. Within four minutes, *Columbia* had already jettisoned its twin boosters (which are retrieved about 150 miles offshore by specially equipped ships) and external tank (which burns up in the atmosphere) and was already more than 60 miles above the Earth and another 400 or so miles over the Atlantic. Less than ten minutes after launch, *Columbia* was in orbit.

When *Columbia* was lost in February 2003, I dug out the heavy press packet from the launch I had seen and sifted through it. Any number of gee-whiz factoids were there for the gleaning: the duration of *Columbia*'s previous missions down to the second, the total number of miles traveled (kilometric conversions cheerfully supplied), how many times each mission looped the Earth and even the landing speed in both knots and statute miles per hour. The packet contained trivia galore, like the fact that the turbopump in the orbiter's main engine is so powerful that it could drain "an average family-sized swimming pool" in twenty-five seconds.

There were mission histories, component specifications, crew biographies and a long, detailed rundown of the various experiments that would be carried out on the mission. There were photos of the astronauts and of previous *Columbia* missions leaving and coming home, offered both as glossy color prints and on several CD-ROMs.

I sat and riffled through the packet, vaguely digesting all of it: the statistics, the factoids, the photos, the voluminous compendiums of technical arcana. Yet all the while, ringing in my head, over and over, was a question asked by the maverick Nobel Prize–winning physicist Richard Feynman—the free radical of the 1986 *Challenger* inquiry known as the Rogers Commission, and easily the most accessible part of that otherwise Byzantine process for most of the world.

Upon learning that NASA's estimates for "losing" a shuttle ranged from 1 in 100 to 1 in 100,000—and quickly calculating that the latter figure meant that someone at NASA believed it could launch a shuttle every day for three hundred years without incident—Feynman put a deceptively simple query to NASA's leadership and its contractors:

"What is the cause of management's fantastic faith in the machinery?"

Feynman died in 1988. His death was a loss for any number of reasons, not the least of which was that his independence, intelligence and cut-the-spin attitude were sorely missed on the *Columbia* inquiry board seventeen years later. Even so, Feynman's question hung like a shadow over *Columbia*'s demise, just as it has haunted every shuttle mission that safely returned to Earth by dint of faith, skill, luck or an unfathomable combination thereof.

Each shuttle launch requires roughly twenty thousand people to conduct, at an average cost of about half a billion dollars per mission (not including the cost of whatever the shuttle is hauling into orbit). Much of this cost goes toward the near-constant maintenance required by machines subject to the tremendous physical punishment of orbital spaceflight, and also because the shuttle is, to put it kindly, a multigenerational vehicle; many of its parts are so old (circa 1970s) that shuttle engineers frequently troll technology junkyards and even the Internet site eBay for spares. Virtually every shuttle refurbishment effort leaks a story or two about some unexpected find, whether it's a set of vintage transistors (perhaps still in use, perhaps not) or, in one case at least, a leftover wrench in the machinery. Often enough, shuttle renovations seem to owe as much to archaeology as to engineering.

In this weird confluence of the cutting-edge and the out-of-date, the shuttle embodies the modern NASA like no other project: everything that is wonderful, and everything that is awful. The shuttle is at once a unique marvel of engineering and a dreadful patchwork of political and technological compromise. It is the most versatile space vehicle ever put in flight; it is also so ungainly and temperamental and of such Brobdingnagian complexity that predictions for its catastrophic failure (as Feynman revealed) largely depend on whom you ask. The most commonly agreed upon estimate to emerge between the losses of *Challenger* and *Columbia* was about 1 expected shuttle loss per 500 missions. As it stands in reality, with the loss of 2 orbiters in 113 missions, the shuttle's failure rate is a more sobering 1 in 57.

Very early in the *Columbia* investigation process, NASA managers and engineers confessed somewhat nervously that they might never know with exactitude the chain of events that caused the vehicle to disintegrate. That is fair enough. Planes crash, and sometimes we never know why; yet every day, people still travel around the world. Shuttles disintegrate, and we may be left only with best guesses as to what, precisely, went wrong. Yet people will continue to build and fly vehicles aimed at the stars. The boundary between the known and the unknown has always been fluid and chaotic, and the line between routine and disastrous is never thinner than at the most precipitous edges of human technology.

However, the greater the uncertainty, the greater is the responsibility to question what types of risks are worth taking—and in NASA's case, in which vehicles they are worth being taken in. In the case of the shuttle, this question has been skirted, ignored, diluted, downplayed or simply given a happy spin for far too long. It isn't about whether or not humans should risk spaceflight; it's about whether such risks are presented honestly to all involved, and whether they are taken in vehicles that are the very best that technology, management and, yes, money, can produce.

No doubt, those intimately involved in managing and preparing *Columbia*'s last mission will agonize over what they might have done differently to save the orbiter and its crew. But *Columbia*'s loss was not the result of any individual decision or technical glitch. *Columbia* was lost because NASA continues to prop up a vehicle that should have been retired when *Challenger* exploded, if not before.

More than what went wrong on takeoff, in orbit or upon reentry, much was made, and rightly so, about what was already wrong with *Columbia* before it left the ground. After a July 1999 mission to launch the Chandra X-ray observatory, *Columbia* headed to its Mojave repair facility for what was to be a fairly straightforward overhaul scheduled to last nine months and cost about $70 million. Engineers replaced thermal tiles, upgraded *Columbia*'s cockpit from the 1970s to the 1990s (using a Boeing 777 template) and yanked out more than one thousand pounds of wiring that, amazingly, wasn't even being used.

Other problems were found; other repairs were made. All told, the overhaul ended up taking seventeen months and costing $145 million. Even with the pricey makeover, a major failure in *Columbia*'s cooling system nearly ended the March 2002 Hubble mission prematurely.

After *Columbia* was destroyed, it was revealed that even after its lengthy repairs, lingering concerns remained about *Columbia*'s ability to fly; that its thermal tile system was unusually fragile; that its left wing had previously experienced unusual (and possibly debilitating) drag in flight; and that its very frame was less able to handle any undue stress than the frame of its sister craft. Boeing, which carried out *Columbia*'s repairs, acknowledged a year after the ship's overhaul that it had found 3,500 wiring faults in the orbiter, several times anything found in *Columbia*'s sister craft.

The breadth of *Columbia*'s needed work and the persistence of its problems so surprised NASA that midway through the overhaul process, the agency's leadership seriously considered taking *Columbia* out of service altogether both for safety reasons and to leave more money for the remaining orbiters. But this was never really in the cards. For starters, *Columbia* was more than just another orbiter: it was the first orbiter to fly, the vehicle that inaugurated NASA's new era in human spaceflight—the vehicle that was to spark a new human future in space. It was more than a spacecraft; it was the grand old lady of the program, a flying legend.

Few women, or men, would appreciate being labeled old at the age of twenty-eight—the number of missions *Columbia* had flown when it was destroyed. Yet even when faced with mounting evidence that *Columbia* was, indeed, too old, NASA stuck with the one original claim for the shuttle program that hasn't yet been disproved, at least not absolutely. The shuttle will never fly fifty times per year as originally advertised. The shuttle will never crack the space launch "access" problem, provide a cheap solution for launching satellites (or anything else), spark a near-Earth spaceflight revolution, catalyze a return to adventures in deep space or pay back its cost to the taxpayers who footed its bill. This is not news. It's been common knowledge since *Columbia* inaugurated the shuttle program in 1981.

However, each of the shuttle orbiters was billed to last for one hundred missions, and this is still at least a theoretical possibility for the three remaining shuttle orbiters: *Discovery* (at thirty missions, the orbiter with the most mileage), *Atlantis* (twenty-six missions) and *Endeavour* (nineteen). That two of the shuttle orbiters fell tragically short of their intended life span—*Challenger* exploded on only its tenth mission—doesn't seem to have bothered NASA sufficiently to make it do what any truly objective observer knows it should do: quickly commit to replacing the shuttles with something better—and then ground them, once and for all.

That the shuttle still works at all is the result of an incredible amount of money and manpower put in service of a vehicle that is the spacefaring equivalent of a Ford Pinto: it keeps chugging along not because it is a fundamentally good vehicle, but because it is serviced relentlessly by a swarm of technicians and has virtually every major component replaced, repaired or refurbished after every trip—not by choice, but out of necessity. As just one small example of how delicately NASA handles shuttle missions these days, *Columbia*'s March 2002 Hubble repair mission was delayed twenty-four hours even though conditions were, technically, green for launch: the unseasonably cold temperature was just too close to the margin for NASA brass to be completely comfortable that another *Challenger* disaster wasn't in the offing.

In the years following the loss of *Challenger,* thousands of new safety checks were implemented and the infamous o-rings were fitted with heating devices to keep them warm and flexible. More than two hundred parts and systems were redesigned. But all the cash and caution did not make the shuttle fundamentally safer, or better. The mission following *Columbia*'s Hubble mission, an *Atlantis* visit to the ISS in April 2002, was postponed for several days due to a major fuel leak that was fixed just in time to avoid the mission's outright cancellation. Two months later, the entire shuttle fleet was grounded for several months when fuel-line cracks were found in all four orbiters.

When the fleet was green-lighted again in the fall of 2002, the problems continued. An *Atlantis* flight that November might have met the same fate as *Columbia,* since its left wing suffered glancing blows from loose insulation foam falling from the external tank, which was the full or partial cause of *Columbia*'s demise. In fact, several previous missions had experienced the same problem.

Nor have shuttle problems been confined to recent years. *Columbia* lost about a dozen thermal tiles on its first mission, in April 1981; it lost even more on a November 1987 mission when debris struck the orbiter upon launch in an eerie precursor to the scenario that occurred on its fatal mission. The three remaining orbiters have all experienced dozens of malfunctions and near disasters, from the beginning of their operation.

Richard Feynman, in his own appendix to the *Challenger* inquiry, noted hundreds of serious failures in parts, mechanisms, sensors and other components of the shuttle orbiter. He also noted that failure estimates for the shuttle's main engines—potentially the most hazardous element of the shuttle system—skewed from more to less optimistic the farther one moved away from the center of the shuttle's NASA/contractor circle. NASA's management estimated one failure per 100,000 flights due to main engine failure, whereas the engine manufacturer Rocketdyne estimated 1 failure per 10,000 flights and an independent NASA consulting engineer gave the main engines about a 1 in 50 chance of failing.

But the larger point of Feynman's analysis was that, where NASA's management of the shuttle was concerned, probabilities and statistics amounted to just so much smoke and mirrors. If your overriding criterion for absolute success is merely the absence of absolute failure, your numbers are bound to look far better than if accidents, near-accidents, glitches, delays, malfunctions, repairs and overhauls are the aggregate basis of judging performance, reliability and safety.

Consistently, Feynman concluded, "NASA exaggerates the reliability of its product, to the point of fantasy." Above all else, Feynman lambasted NASA for trying to pass off the shuttle for what it was billed as in the first place: a space vehicle with the operability of a commercial airliner. "In any event this has had very unfortunate consequences, the most serious of which is to encourage ordinary citizens to fly in such a dangerous machine," Feynman wrote. "[NASA] must live in reality in comparing the costs and utility of the shuttle to other methods of entering space. If in this way the government would not support them, then so be it."

Such a brave path is not in the nature of NASA these days. In the wake of *Columbia's* loss, NASA and its partners immediately began spinning the line that underfunding was at the core of anything that might have gone wrong with *Columbia*. As shuttle advocates have noted, it is true that as the schedule lagged and the costs ballooned, the International Space Station sucked funding from virtually every corner of NASA, including the shuttle program.

Dan Goldin typically got most of the blame for such juggling of funds, since most of it occurred on his watch. But Sean O'Keefe, shortly after arriving to head up NASA in late 2001, canceled three planned shuttle safety upgrades; at the same time, the Bush administration sliced about half a billion dollars from the shuttle program—money slated for more upgrades, which were viewed by the administration (and presumably by O'Keefe) as extraneous. Not surprisingly, much of that money was restored in the wake of the *Columbia* disaster. Everyone, it seems, was prepared to milk the shuttle program until *Columbia* made such till-robbing blatantly déclassé.

But the argument that more money makes better shuttles is a red herring. According to an extensive *New York Times* report, during the height of the space station's budget-guzzling bender—from about 1992 to 2000—in-flight anomalies on shuttle missions dropped steadily from eighteen per flight to fewer than five. During that period, shuttle orbiters underwent modernization—often replacing equipment that belonged in a museum, not in a cutting-edge space vehicle—at an unprecedented pace. And even when milked of funding for projects like the ISS, the shuttle program has never been cash-poor. Immediately after *Challenger,* NASA received budget boosts totaling about $2 billion for shuttle upgrades; each of the last few NASA budgets has included hundreds of millions more. And *Columbia,* of course, had the benefit of the most extensive, and expensive, upgrade in the history of the program.

Showers of cash, a year and a half of first-class technical service, a team of thousands of skilled and dedicated professionals—none of this could save *Columbia* and her crew. And the sad truth is that millions, or billions, won't prevent another such disaster. One of the more hair-raising analyses of the shuttle program's post-*Columbia* safety prospects came from AirSafe, a consulting firm specializing in the analysis of commercial air-flight risks. Estimating one shuttle loss per 145 flights—an unusually conservative NASA figure from 1996—AirSafe concluded that "if the remaining fleet of three Space Shuttles continue to fly until the end of their operational lives, there is from a 40% to 99% chance that there will be at least one more fatal event."

Such a chilling prediction recalls perhaps the most famous Feynman quote from the *Challenger* inquiry: "For a successful technology, reality must take precedence over public relations, for nature cannot be fooled."

It is doubtful that the shuttle's track record and future prospects have fooled many NASA engineers; indeed, the disconnect in catastrophic predictions finds the pessimists in the engineering camp and the optimists among the managerial ranks. Yet despite its obvious shortcomings, the shuttle program remains the darling of NASA's human spaceflight program. Amazingly, less than a month before *Columbia*'s demise, NASA's associate administrator William Readdy told the *Huntsville Times* that NASA was working on a plan to stretch the lifetime of each orbiter to an astounding four hundred missions—this, when not a single orbiter has survived to even one-tenth of that life span. Such optimism wasn't quite as obvious after *Columbia*'s loss, yet NASA's leadership stood firm on plans to keep the shuttle fleet flying for nearly another two decades.

Indeed, what is the cause of management's fantastic faith in the machinery?

As Feynman undoubtedly knew, "faith" has very little to do with it. The disheartening truth is that NASA clings to the shuttle because it is terrified to give it up. The shuttle was so hard to get in the first place that the entire agency is, in essence, tailored directly to the task of keeping it flying even at the expense of advancing NASA's Turnerian dream (now somewhat faded) of expanding humanity's space frontier.

The shuttle was announced to the world in 1972 and first flew nine years later. However, as a building block of NASA's future, the shuttle was being heavily marketed well before the *Apollo 11* moon landing. It was part of a magnificent plan to build a human infrastructure in orbit, to establish a human presence on the moon and, in short order, to send a human mission to Mars. In the glow of Apollo's triumph, such things not only seemed possible; they seemed inevitable. Yet just a few years after that magnificent achievement, the shuttle would stand alone as NASA's only human spaceflight project in the post-Apollo fiscal blood-bath that nearly tore the agency to pieces.

There are those who argue that if the shuttle was all that could save NASA from post-Apollo oblivion, perhaps NASA ought not to have been saved. Whatever its merit as a thought exercise, such bitter prag-matism is necessarily the product of hindsight: few organizations have the temerity and vision to commit institutional hara-kiri if options exist for survival. For NASA, survival became job number one very early on, since the wheels were beginning to come off the NASA loco-motive well before the Apollo program hit its zenith.

NASA's annual budget peaked in 1965 at what today would be nearly $40 billion and then went steadily downhill until 1976, when a modest reprieve brought it up to about $12 billion. Even that tiny gain would be subsequently hacked back by Congress until 1979, when the fevered peak of the shuttle's development mandated a bit more money. The budget would not post any significant gain thenceforth (the agency's 2003 budget was about $15 billion). The resulting decline in the NASA civil servant workforce was precipitous: the agency directly employed nearly thirty-six thousand people in 1966, and eight thousand fewer just six years later. The civil servant ranks would erode steadily over subsequent years; in 2002, NASA's civil servant workforce was around eighteen thousand—half the number at the agency's peak.

Undoubtedly, far fewer civil servants would work for NASA these days were it not for the shuttle. Credit for the shuttle's existence is typi-cally given to NASA's top management (particularly George Mueller and the administrators James Fletcher and Richard Truly, who oversaw the program's critical milestones) and to the Department of Defense. But behind the bureaucratic machinery stood the mighty influence of a handful of giant aerospace contractors—Boeing, Martin Marietta, Lockheed, McDonnell-Douglas and others—for whom government space work had become steady business.

The reduction of NASA's civil service ranks pales in comparison with the downsizing of its contractor ranks, even before the *Apollo 11* landing. In 1966, nearly half a million people drew sustenance from the NASA enterprise; by 1969, that number had dropped to less than 200,000. Most of the difference was in corporate contractor support, and it was in this sector that NASA found its salvation. With its economic and thus political clout, Big Aerospace made the shuttle happen. The price was that Big Aerospace would largely own human spaceflight, and NASA, from that point forward.

For some time now, it has been difficult to tell where NASA ends and its contractor workforce begins. NASA's contractor-dependent culture began to take hold in 1963 when George Mueller was called in to save the Apollo program. To get his "all-up" job done, Mueller packed his staff with personnel from the Air Force missile program that had brought him previous success. Like other American military branches, the Air Force gives relatively free rein to its industrial contractors, with budgets being a minor consideration when weighed against performance. Apollo required just such an admixture of unconstrained spending and engineering brashness; in style, its development was much more like the Manhattan Project than many would like to admit.

At first, NASA hewed closely to the NACA–von Braun ethos, which was to keep as much work in-house as possible. This worked for the Mercury and Gemini programs, but the Apollo program was of a much greater scale, and the contractors began to exert more influence over engineering and budget. Mueller's whip-cracking pace and strong administrative hand helped centralize NASA management and strengthened NASA headquarters, which had previously served mostly as a liaison office between the various centers. By 1965, just two years after Mueller came on board, contractors working on NASA projects outnumbered in-house employees by ten to one. Although there was some inevitable head-butting between civil servants and the contractor workforce, it all went smoothly enough because everyone had their eyes on the same prize: reaching the moon before the Soviets.

The rise of the contractor culture at NASA paved the way for the ascendance of bureaucrats over scientists and engineers within the agency. As NASA lost the limelight after the moon landings, it had to work harder simply to stay in business, and the agency's response was to employ more planners, more congressional liaisons, more "outreach professionals." Much of the time, the work of this growing spin-force was to get contractors lined up for one NASA project or another, since contractors employ people in congressional districts. At the same time, more people were required to support all those new bureaucrats.

The paradoxical result of NASA's growing politicization was that between 1960 and 1967, the number of professional administrators at NASA grew three times as fast as the agency as a whole. By 1972, the year the shuttle was finally given the go-ahead (and the year of the final Apollo mission), administrative personnel at NASA were twice as likely to receive a promotion as scientists and engineers.

The shuttle is emblematic of the compromise that resulted from a culture that grew in bureaucracy even as it shrank in in-house technical capability. In keeping with its growing deftness at pleasing everybody all the time, NASA pitched the shuttle as the cure for many woes and the realization of many dreams: it would be the evolutionary successor to the Saturn rockets, a return to rocket planes, the world's first reusable spaceship, a mighty military warbird and a paradigm-changing commercial vehicle. At least early on, the shuttle was also billed as the workhorse for moon bases and missions to Mars, thus continuing the march toward both von Braunian and O'Neillian futures.

NASA was not shy about putting its most powerful personnel on the campaign trail for the shuttle. Addressing the British Interplanetary Society in 1968, Mueller claimed the shuttle would work "in a mode similar to that of large commercial air transports and be compatible with the environment of major airports." Mueller also added, tantalizingly, that "the basic design for an economical space shuttle from earth to orbit could also be applied to terrestrial point-to-point transport." Wernher von Braun went a step further: "Toward the end of the seventies you will no longer have to go through grueling years of astronaut training if you want to go into orbit. A reusable space shuttle will take you up there in the comfort of an airliner."

If such wide-eyed wonderment had been converted to healthy skepticism early enough, space travel might be the better for it, since the contraption that emerged from the hype of Mueller, von Braun and others made space travel more expensive than it had been in the Apollo era and bound humans more tightly to the Earth than the moon rockets it replaced. While the shuttle orbiter looks much like a plane to the casual observer, the Federal Aviation Administration chose not to classify it as such, since the orbiter has no true capability for self-propulsion (its launch requires external rockets), nor can it power itself to landing—it glides, with difficulty. Yes, the shuttle has significant capacity for both cargo and crew, but both activities would almost certainly cost less if they were conducted separately—in expendable vehicles.

In sum, by sacrificing economy, performance and reliability (and arguably safety), NASA is able to bill the shuttle as the world's first and only reusable spacecraft—but even that claim is dubious, since virtually every major component of the shuttle is repaired, rebuilt, refitted or otherwise overhauled after almost every flight.

The shuttle's queer design—not rocket, not plane, certainly not rocket plane—results from its conceptual birth in the 1950s as something called a lifting body, a wingless, triangle-shaped craft able to generate aerodynamic lift with its fuselage alone. Lifting bodies can handle the stress of hypersonic speeds and orbital reentry better than traditional winged planes, but they have one big disadvantage: they can't actually fly. At best, they can gently drop to Earth like an oversized paper airplane. For shuttle planners, this created the unpleasant possibility that a space-capable lifting body—a package of the highest-grade American aerospace technology—might inadvertently coast down over, say, China or Russia, and be unable to maneuver its way to friendlier territory.

Thus, a set of stubby wings was added to allow the shuttle orbiter to bank so fast and at such a stark angle that, coming out of orbit at multi-Mach speeds, it could travel a thousand miles sideways in a matter of minutes. While the wings made the shuttle orbiter heavier and less efficient, they did make it maneuverable, but just barely; the orbiter quickly earned the nickname "flying brick."

The delta wings were perhaps the most obvious example of the military's role in the shuttle's development—and, for better or worse, without military support the shuttle would not exist. Well before the moon landing, Richard Nixon was calculating how he could minimize the space program's further influence on popular and political culture, since he believed—rightly, as it turned out—that the glory of the moon program would be forever associated with his predecessors, Kennedy and Johnson, whom he loathed. Since killing NASA outright was too drastic a move for popular consumption, Nixon instead rejected Thomas Paine's post-Apollo plan with its moon bases, space stations and crewed Mars missions, leaving on the table only the possibility of developing a new, reusable vehicle to move cargo and people to and from orbit. On this modest reprieve, Paine built NASA's entire future.

In early 1969, Paine and the former NASA deputy administrator Robert Seamans, who had become the Air Force secretary under Nixon, initiated a new study on requirements for a joint NASA–Department of Defense space vehicle, even though a similar study three years earlier had concluded that NASA's and DOD's needs for such a vehicle were almost mutually exclusive. With Paine and Seamans coaching from the sidelines, the new report produced a different result. Suddenly, a cost-effective, reusable space vehicle was just the ticket to meet the needs of both NASA and the military. NASA would get a placeholder in its plans for interplanetary conquest, and the Air Force would get a means of launching and retrieving big spy satellites. It was later revealed that the Air Force also wanted a vehicle that could snare Soviet satellites in orbit, although no public documents indicate that the shuttle was ever used for this purpose.

At the time of the shuttle's conceptual development, in the late 1960s, the entrepreneurial space community did not really exist: America's conquest of space was clearly in NASA's hands. However, the shuttle's approval accelerated the agency's long slide from innovative boundary pusher to staunch defender of an increasingly shortsighted status quo and, consequently, helped bring an "alternative" space movement into being. In this effort, NASA had plenty of help from the Office of Management and Budget (OMB), which under Nixon's quiet guidance repeatedly recommended cutting huge chunks of money from the putative shuttle-development budget, thus ensuring it would never fulfill its original mandate as a fully reusable space vehicle.

A fully reusable shuttle was initially priced at about $10 billion; OMB said NASA shouldn't get more than about half that amount, and subsequent contortions in funding forced NASA to repeatedly remake the shuttle to please its fiscal overlords. Only months before Nixon finally approved the program, there were nearly a dozen shuttle concepts up for consideration, the sum of which sent a desperate message to the White House and Congress: fund a shuttle, any shuttle.

Fortunately for NASA, the shuttle was desperately wanted by the aerospace industry, which put its considerable muscle behind the lobbying by NASA and DOD. Military procurements, by far the biggest slice of bread and butter for Big Aerospace, were starting to diminish: spending would drop, in constant dollars, from a high of about $44 billion in 1968 to about $7 billion in 1975. In such a dire fiscal context, the shuttle contract looked to the aerospace industry much as a steak looks to a hungry dog.

As NASA, the DOD and Big Aerospace steamrolled the shuttle toward White House approval, claims for the shuttle's performance went from outrageously optimistic to barely believable: at one point, the shuttle's industry champions said that a schedule of ninety-five flights per year was possible and would bring the cost per flight down to about $350,000, or a market-changing $7 per pound of payload. NASA tempered this claim in the pitch that ultimately won Nixon over: fifty-plus flights a year, and the ability to handle most if not all of America's military, civilian and commercial launch needs.

Until the eleventh hour, Nixon tried his best to avoid a big new project for NASA while still keeping the aerospace industry healthy. In 1971, several of Nixon's advisers launched an effort called the New Technology Opportunities Program (NTOP), which, among other goals, explored terrestrial options for utilizing aerospace industry expertise. Ideas like creating a national "two-way television" network (inadvertently foreshadowing the Internet) and retooling nuclear devices to be used in prospecting for oil and gas were intriguing, but none went past the "What if?" stage.

At the end of 1971, the shuttle was the only big new effort Nixon could find to keep the aerospace industry out of a major slump. Reflecting the disarray the shuttle concept was in by the time Nixon approved it, in the NASA press photo accompanying Nixon's announcement, NASA administrator James Fletcher is pictured showing Nixon a small model of the shuttle—the wrong model, as it turned out.

It quickly became apparent that the shuttle was neither flexible nor cheap to use, nor could it be launched nearly as often as advertised (the annual record is nine launches, in 1985). It is a particularly bitter irony that the shuttle was created to shatter the high-cost launch barrier by being launched more frequently and cheaply, and in doing so, to spark a commercial spaceflight revolution. Instead, the shuttle nearly destroyed the American commercial space-launch industry.

To help justify the billions of dollars being pumped into the shuttle program, NASA and its supporters pushed the American government to require military and commercial payloads to launch on the shuttle. The result was that construction of expendable rockets came to a virtual halt. When *Challenger* exploded in 1986, the shuttles were grounded for two years. The military quickly regained its feet by using what expendables it still had, but America was effectively shut out of the commercial launch market for a full three years.

While the U.S. launch industry was in a coma, telecommunications firms and other space-dependent companies still had business to do, and they began finding other options. Russia found marketing legs for its reliable suite of reformed ICBMs (Tsyklon, Proton, Zenit), and Europe found the field virtually wide open for its ICBM-inspired Ariane. The U.S. commercial launch drought ended on August 27, 1989, when McDonnell-Douglas's commercially configured Delta launched a British telecommunications satellite—the first private U.S. space launch. American rockets did claw their way back into the commercial launch market, but they have not dominated it since *Challenger*. In 2001, the worldwide commercial-launch industry generated about $7 billion in revenue. The U.S. share of this market was about 40 percent; it had been nearly 100 percent before the advent of the shuttle.

Over time, NASA managed the shuttle's public-image makeover from sleek *Star Trek* trailblazer to flying Rube Goldberg device by promoting its failures as virtues. Rather than decry the shuttle's failure as a vehicle of airlinerlike ease and economy, Kennedy Space Center tour guides now boast that the shuttle requires a purpose-built landing strip ("The longest concrete runway on Earth!") and must be flown by an elite cadre of crack pilots, because being the world's most ungainly glider, the shuttle gets only one chance to land.

Outside of NASA's immediate domain, however, the spin control quickly fades. At the National Air and Space Museum—hardly unfriendly territory for NASA—there is a shuttle display titled "A Remarkable Flying Machine." Despite the label, the last panel in the display concedes, "The high cost of Shuttle launches and changes in policy have led industry to turn again to expendable launch vehicles for access to space. Commercial satellites are being launched on privately owned and operated rockets." Tellingly, the display does not use the term "reusable" to describe the shuttle's operation; it opts instead for the vaguer "reflyable."

Reagan did throw a sop to NASA and its shuttle contractors by approving the construction of *Challenger*'s replacement, *Endeavour* (a fifth orbiter, which NASA lobbied hard for, was never approved). The construction of *Endeavour* solidified the shuttle's status as something of a necessary evil: even with its commercial and military manifests gone, the shuttle's domination of human spaceflight would continue unchecked. Worse, as NASA revved up the shuttle-dependent space station program, the agency's defense of the shuttle intensified.

In the larger picture, the space station doesn't need the shuttle. ISS personnel have traveled to and from the station in Soyuz spacecraft, a situation that will be all the more sensible if the station never grows beyond the ability to house three people (a Soyuz seats three). Automated Russian Progress vessels handle most station resupply jobs, and while the station's construction was tailored specifically for the shuttle, station components have been sent into orbit both by the shuttle and on Russian rockets.

Ironically, the one activity that was promoted from within and without as a potential saving grace for the shuttle—sending regular people into space—was fought by the agency with particular viciousness. In recent years, the intensity of NASA's opposition to civilian space travel was portrayed as fear over another *Challenger* disaster and, thus, another Christa McAuliffe tragedy (both the actual tragedy and the PR disaster that accompanied it). But more than that, it reflects the agency's struggle to maintain control of human space travel as its last, unchallenged domain.

NASA would lose that battle, too, although in a way that no one could have anticipated.

Advance Reservations Required

One of the first things Alan Ladwig ever said to me was clearly aimed at his former boss and mentor, Dan Goldin. We were in Ladwig's pickup truck, barreling through downtown Washington, D.C., along the Potomac River and heading toward the side-by-side gray slabs that make up NASA headquarters. We were talking casually about space, about what had and hadn't happened and, critically, what was needed to reengage the world in human spaceflight. Was it money? Was it political will? Was it, perhaps, vision?

At that last comment, Ladwig turned and gave me a long, smirking look—an unnerving proposition only because this meant he was less than fully attentive to the crisscrossing chaos of rush-hour D.C. traffic. "I don't have any tolerance for visionaries anymore," he growled, just as he swerved nonchalantly to miss a Land Rover careering our way. "It's not about vision, it's about practicality. You'll never see the true potential for space until you get it out of the hands of a single agency. Kennedy called space the new ocean, but we've treated it like NASA's lake."

These are not words one would expect from someone who was once seated at the right hand of the NASA administrator, but Ladwig is anything but predictable. Tall and suntanned and with a loose mane of brown hair, Ladwig looks a bit like a comfortably aging California surfer, but most of his career has been planted firmly in and around Washington and its corridors of power. By 2002, Ladwig's topsy-turvy space career had taken him full circle, from entrepreneur to NASA insider to entrepreneur again, albeit an entrepreneur with NASA firmly in his sights.

Our longest talk took place in the cafeteria on the ground floor of

NASA headquarters, where Ladwig was engaged in a series of meetings with NASA managers in an attempt to firm up the agency's endorsement of Team Encounter, a private effort to send into space a probe containing the poems, musings, artwork and, in some cases, DNA samples of ordinary people (all of which will be digitized and squeezed onto a few computer disks). Funding is from advertising sponsors and individuals who pay twenty-five to fifty dollars to have a small piece of themselves included on the flight, which is scheduled to launch on an Ariane rocket in 2005. So far, Team Encounter claims to have signed up more than eighty thousand participants from around the world, including such diverse popular personalities as the former astronaut Sally Ride and the rock star David Bowie.

Ladwig's many NASA connections clearly helped Team Encounter nip the brother-in-law problem in the bud: featured prominently in the NASA headquarters gift shop during several of my visits was an elaborate display encouraging participation in Team Encounter. This was small comfort to Ladwig: "It will be good to see the day when nobody asks you, if you have a space idea, 'What does NASA think?' And that day hasn't come yet. You still need that seal of approval."

Ladwig's space career began on the ground floor of the original space advocacy organization in the NASA era, the Committee for the Future, which wanted to buy from NASA the two Saturns left over from the Apollo program (among the committee's founders was the futurist Barbara Marx Hubbard, now on the board of Bob Citron and Walter Kistler's Foundation for the Future). The Saturns would have been used to conduct Project Harvest Moon, a citizen-sponsored lunar expedition that was to have paid for itself through sale of lunar materials and media rights. Ladwig and others pitched the idea around NASA centers in the early 1970s and picked up a fair number of supporters, including the influential Chris Kraft, flight operations director for Mercury and Gemini and director, at that time, of the Johnson Space Center.

In May 1971, the committee held a conference titled "Mankind and the Universe" that drew enthusiasts like Buckminster Fuller. Although later conferences involved the likes of science-fiction writer Ray Bradbury and *Star Trek* creator Gene Roddenberry, the Committee for the Future flamed out in the mid-1970s with Project Harvest Moon unrealized. "The flaw was that it was way too early," Ladwig said, somewhat sadly. "But it was an interesting notion, and a lot of fun."

After the Committee for the Future imploded, Ladwig's next space gig was with the Forum for the Advancement of Students in Technology (FAST), which tried to get college students involved in issues of technology policy, including space. One of the avenues for achieving FAST's goals was to help secure shuttle space for student experiments, which NASA also liked because they helped generate excitement about the shuttle program. FAST led to phase one of Ladwig's NASA career— a career that would, directly and otherwise, help reinvigorate the notion that regular people should have a place in space.

In the days when the shuttle still had a commercial mandate, Ladwig helped develop NASA's corporate payload-specialist program, considered by most people to be the first program to focus on commercial human spaceflight. "Our competitive edge over Ariane was, 'You fly one of your satellites with us, you can fly one of your employees as a corporate payload specialist,' " Ladwig explained. The corporate payload-specialist program sent an unpleasant ripple of pending change through NASA, especially the agency's elitist core. "The test pilots didn't like it when the science astronauts came in," Ladwig said, "and the scientists looked down their nose at mission specialists, which included for the first time women and minorities. And all three classes were against payload specialists, who were seen to be cheapening the glorious enterprise."

Ladwig and others believed that the payload-specialist program was a step in the right direction, both for NASA and for human spaceflight in general. The public loved it and, more important, the contractors loved it, since many of the corporate payload specialists were from their ranks. In the early 1980s, Ladwig was put in charge of the new Space Flight Participant Program, which was intended to expand the payload-specialist concept to true civilians and which, later, led to the Teacher in Space Program and its only participant to date, Christa McAuliffe.

The Space Flight Participant Program also served up the world's first space tourist, although he wasn't billed as such. When Senator Jake Garn flew with the crew of Challenger in February 1985, NASA tried to dress up his joyride by classifying him as a payload specialist, but even Garn admitted in an interview, "Technically, I don't have a payload." Shortly before the flight, a *Doonesbury* cartoon lampooned Garn's trip, showing a series of panels with the senator readying for space and rehearsing various statements to mark his flight. He finally finds the mots justes: "One giant leap towards approving the 1986 NASA budget."

Ladwig did not favor Garn's flight or the subsequent trip of another congressman, Bill Nelson, whose flight on *Columbia* touched down just ten days before the launch of the ill-fated *Challenger* mission. Ladwig ruefully noted that Greg Jarvis, a bona fide corporate payload specialist, was bumped onto the doomed *Challenger* mission in part due to Garn's and Nelson's flights.

NASA's reason for flying Garn and Nelson was brutally pragmatic: both held tremendous sway over NASA funding. Nelson, in fact, would end up being the prime sponsor for the bill to fund *Challenger*'s replacement, *Endeavour*. While such cozying up to Congress was no doubt helpful to NASA, Ladwig felt it was antithetical to the message that the Space Flight Participant Program was trying to send: that one did not have to be a specialist, a test pilot, a scientist or in a position of power to fly into space. "I fought all that," Ladwig said simply.

After the *Challenger* disaster brought the Space Flight Participant Program to a halt, Ladwig left NASA for a brief career as a space journalist. It was brief, in part because in 1990 Ladwig saw a TRW manager named Dan Goldin speak at a space conference about new ways to do business at NASA. "I thought, wow, this guy is talking about the way things should be done, not the same old way." By chance, Ladwig ended up sitting next to Goldin on a plane flight following the conference, and thereafter the two men maintained a connection that led to Ladwig's appointment by the Clinton administration in 1992 as head of NASA's policy and planning office, under Goldin. In 1998, Ladwig was made senior adviser to the administrator, with responsibilities that included disseminating Goldin's "new NASA" message to the public and in Congress.

Ladwig was of the few people Goldin trusted enough to speak for NASA in the most visible public forums, and he was as forthright as his boss. Interviewed by ABC News in September 1998, Ladwig said that NASA "gets a lot of publicity. We're on the covers of magazines. Yet our message of what we do and why we're important to the economy gets a little diffuse. When you start to see almost half your budget going into operations instead of research and development, you start to wonder if maybe we're not getting a little far from our mission."

Ideologically, Ladwig and Goldin parted ways over the first civilian spaceflight to follow the *Challenger* accident. John Glenn's 1998 flight on *Discovery* was billed simultaneously as an important opportunity to learn about the effects of spaceflight on senior citizens (Glenn was seventy-seven when he flew) and as a much-belated victory lap. It was well known in space circles that Glenn was infuriated at being grounded by John Kennedy following the triumphant 1962 Mercury flight that made Glenn the first American to orbit the Earth. Echoing the Soviet strategy with Yuri Gagarin, the White House felt Glenn was too essential to Space Race morale to chance losing him on a subsequent flight. (Gagarin died in a jet crash in March 1967, at the age of thirty-four, as part of an attempt to qualify for another space trip.)

Contrary to the popular perception that NASA called Glenn out of retirement for his shuttle flight, in fact the former senator lobbied vigorously for it, ultimately succeeding with the help of his powerful military and political allies. To Ladwig, this was Garn and Nelson all over again. "There became this feeling that he's entitled to it, he deserves it. I have a real problem with that," Ladwig said. "Don't tell me that some people deserve it and others don't, because I want to know what those criteria are. [Glenn] got to be the first American to orbit the Earth. He became a millionaire because of it; he became a senator. How long do we have to keep paying off this guy?"

Ladwig was particularly peeved about the Glenn flight because, in terms of "nonprofessional" astronaut candidates, there was someone ahead of him in the queue: Barbara Morgan, the teacher who was to have followed Christa McAuliffe into space. As NASA considered the fate of the Space Flight Participant Program in the wake of *Challenger,* there were "countless meetings, studies, conversations, et cetera" about what to do about Morgan. The aggregate result was a recommendation to Richard Truly, Goldin's immediate predecessor, that Morgan be given a definite slot on a future shuttle mission. Truly sat on the recommendation until his final day in office, when he suddenly announced his support for a flight by Morgan, leaving Goldin with the messy business of dealing with the details.

Ladwig and others had long been campaigning for NASA to get back in the saddle with the Space Flight Participant Program, starting with Morgan. "Nothing generated the public support for NASA in the post-Apollo era like Christa McAuliffe," Ladwig said, while acknowledging that nothing brought NASA lower in public opinion than McAuliffe's untimely death. Were it not for the *Challenger* disaster, the Space Flight Participant Program would in short order have flown journalists, artists, celebrities and eventually, Ladwig said, regular citizens selected simply to convey the wonder of space travel from the true layman's perspective.

Although he fought hard to convince Goldin to fly Morgan instead of Glenn, "Goldin couldn't bring himself to make that decision. It was too risky for a mere mortal to be able to fly." That Glenn's flight was immensely popular with the general public and widely covered in the media was cold comfort to Ladwig. "It was more going back and reliving the past," he said bitterly. "It's not looking to the future."

The end of Ladwig's NASA tenure was marked by his increasing frustration with Goldin. "He once told me, 'You know, Ladwig, your problem is that you're too up and down.' And I said to him, 'Well, I've learned from the master.' " Even with the acrimony between them, Goldin maintained a sense of humor; he dressed as Mr. Spock for Ladwig's farewell party, pointed ears and all.

Ladwig said his proudest accomplishment in his final days at NASA was helping to get Morgan back in line for an eventual shuttle mission, although Goldin could never bring himself to give her a firm flight date. Whatever Ladwig's influence, it seems that the sprightly Morgan, who is in her early fifties and has been in astronaut training for several years now, will finally get her chance to pick up where Christa McAuliffe left off. On April 12, 2002, while unveiling his near-term plan for fixing NASA, the new administrator, Sean O'Keefe, did what Goldin and Truly wouldn't do and announced publicly that he was putting Morgan back on the shuttle flight manifest: as of this writing, she will fly to the ISS in 2005 or 2006. Tellingly, NASA has remained committed to Morgan's flight despite the loss of *Columbia;* even more, the agency has reopened its Teacher in Space program for the first time in nearly two decades.

Morgan will carry into space the very public burden of her Teacher in Space predecessor, and her flight will doubtless be an emotional event not only for NASA but also for anyone who remembers watching with disbelief as *Challenger* burst apart in the sky on that unusually cold January morning. In the end, though, Morgan may owe her flight not to McAuliffe, Ladwig or anyone else associated with NASA, but instead to a man who bypassed NASA altogether and, in many ways, climbed into space all on his own.

On April 26, 2001, two days before Dennis Tito became the first person in history to pay his own way into space, the small group of us who had traveled across the world to witness his launch—family, friends, assorted others—took a forty-five-minute bus ride through the suburbs of Moscow for a visit to Star City and its main attraction, the Yuri Gagarin Cosmonaut Training Center, the prep school for Russian cosmonauts and anyone else who flies in a Russian spacecraft.

If the Johnson and Kennedy centers can be said to embody NASA's Turnerian ethos—neat, cheerful and gleaming with can-do techno-fetishism—Star City is their brooding, introspective cousin, a place invented in the 1950s solely for the Soviet manned space program. These days, Star City looks something like a slightly seedy community college, mildly stuffy and parklike in a rumpled way. On prominent display inside the entrance is a lavishly colorful billboard with a triptych of the Holy Trinity of Soviet/Russian space exploration: Tsiolkovsky, Korolev and Gagarin—the dreamer, the builder and the voyager, side by side.

There is a fourth figure whose name is not part of this tribute, yet his impact on the Russian space program is, in its way, as profound as Frederick Jackson Turner's influence on NASA. "Universal Resurrection is a total victory over space and time," wrote Nikolai Fedorovich Federov. "From the very beginning of its human evolution, mankind imagined that the immortal souls of their dead ancestors dwelt on the visible stars and planets of the firmament." Mixed with Soviet manifest destiny, Federov's promise of "universal resurrection" would, perhaps improbably, guide much of the grand planning for the Soviet conquest of space.

Born in 1828, Federov was one of several illegitimate children of Prince Pavel Ivanovich Gagarin, who some Russian historians claim was an ancestor of Yuri Gagarin. In his 1906 landmark work, *The Philosophy of the Common Task,* Federov laid out nothing less than an intergalactic manifesto for humankind, fusing Judeo-Christian ideas of resurrection with a sweeping extraterrestrial imperative. Federov concluded that humankind's ultimate, divine purpose was to unify and populate the galaxy, if not the entire universe. The reward for this task would be enormous: the attainment of Heaven or its equivalent. Federov's work had a wide impact on thinkers and artists of nineteenth-century Russia, including Dostoevsky and Tolstoy, who were his contemporaries, and his writings later made a deep impression on Boris Pasternak, the author of *Doctor Zhivago.*

But Federov's most lasting influence may have been on the sickly, reclusive genius Konstantin Tsiolkovsky, whom he befriended at the Chertkov Library when the pioneering space thinker was still new to Moscow (Federov paid his bills, in part, by working as a cataloger at the Chertkov). According to the historian Stephen Lukashevich, Federov saved Tsiolkovsky from suicide in his most despondent and impoverished days and encouraged Tsiolkovsky's native engineering talent and its application to dreams of spaceflight; not surprisingly, he also encouraged the inclusion of his own philosophy in Tsiolkovsky's writings, as evidenced by both the title and content of an early Tsiolkovsky paper, "The Reactive Vehicle Will Save Us from Calamities That Await the Earth."

"In all likelihood, the better part of humanity will never perish but will move from sun to sun as each one dies out in succession," Tsiolkovsky wrote. "Thus, there is no end to life, to reason and to the perfection of mankind. Its progress is eternal. And if that is so, one cannot doubt the attainment of immortality."

Nowhere is the Federovian ethos more evident than in the lobby of the Yuri Gagarin Cosmonaut Training Center, which is so unlike any NASA center that at first glance Western visitors might wonder whether they're in a space facility at all. As you enter the dimly lit foyer cradled in dark lacquered wood, the room looks much like a church vestibule. Stained-glass windows let in filtered light, but instead of Christian iconography, the stained glass bears majestic visions of planets, stars and distant galaxies. The room is not without its saint; just ahead of the steps leading to the business part of the center and backlit by the stained-glass windows stands an eight-foot-tall bronze statue of Yuri Gagarin, frozen forever in all his youthful handsomeness and embodying the what-might-have-been wistfulness of an entire nation. Gagarin's right arm is raised before all comers as if in benediction; he appears eerily Christlike.

On the walls nearby are photos of the flesh-and-blood Gagarin in his post-Vostok heyday, including a few with his mentor and protector, the bearish Korolev. These are rare images, since Korolev was actually prohibited from appearing in photographs and his very identity was kept secret until well after his death in 1966. (Khrushchev was worried that Korolev would be targeted by Western intelligence services for capture, or worse.) The room's hazy light gives the photos the faded appearance of daguerreotypes, making them seem more ancient than they are, imbuing them with an almost holy quality. It is perhaps the most surreal, moving place in all of space history, deeply Federovian in its mysticism and spiritual intimations.

The Cosmonaut Training Center was quiet when we visited, since many of the personnel were already in Baikonur preparing for Tito's launch. We were marched through several cavernous rooms resplendent with astounding artifacts of the Soviet/Russian space program. Filling a room and a half was a duplicate of the Mir space station, all six modules, only a few hundred million rubles short of being ready for a launch to orbit—an idea actually floated by the Russians shortly after Mir's demise, although everyone knew they didn't have the money. In another room was a training module for the Buran, Russia's space shuttle (which we'd see more of in Baikonur), and over here were the gumdrop-shaped Soyuz training capsules that every cosmonaut since Gagarin had trained in, including Dennis Tito.

Entering the warehouse-size building containing a training mock-up of two International Space Station modules required stepping around two buckets collecting water that dripped from a spiderweb of cracks in the high ceiling. As requested, we took our shoes off and took turns walking through the ISS modules—noting, among other things, how much more spacious they were than Mir. Among other improvements, the ISS actually has a bathroom with a door for privacy, nothing smaller than what one would find on the average jetliner (by contrast, Mir's toilet was about two feet from the dining table, with only a curtain for privacy). The ISS also has a clear floor-and-ceiling orientation, whereas Mir's general layout made sense only if you didn't have to worry about gravity-driven concepts like up and down—control panels were on the "floor," chairs were on the "ceiling," and so on.

After visiting the world's largest human-rated centrifuge, we moved on to lunch in the Cosmonaut Café—meaning that cosmonauts in training eat there, as opposed to a sardonic concern voiced by one of our group that cosmonauts had been reduced to waiting tables to supplement their income. Just outside the entrance, a group of women sold homemade candles and honey from a trestle table, reminding me of one visit to NASA headquarters where a group of women sold cheap jewelry in the lobby. Inside, the café and its fare were no worse than the cafeteria at NASA headquarters.

The café's exterior was another matter. Slabs of concrete facing on the building looked as if they had been blasted by artillery, exposing iron sheeting that was rusting and flaking. A concrete spiral staircase to the café's upper level was so eroded that strands of rotting rebar stuck out like bicycle spokes; the entire stairwell looked as if a single weighted step could cause it to crumble into dust.

Viewing the building in hazy sunshine, Ron Sellars, a friend and former business colleague of Tito's who came with his brother John to witness the launch, pulled out his miniature Sony digital video camera. He whispered narration to himself as he slowly panned across the café's shattered exterior, the heart of what was once the world's greatest space program: "Twenty million dollars could sure go a long way here."

That, in a nutshell, was the entire point of Tito's trip, at least from the Russian point of view. At the height of the Cold War, the Soviet

Union was pouring the equivalent of billions of dollars into its space program. But when the Soviet empire collapsed, its space program went into free fall. With rivers of money reduced to a trickle, rockets and spacecraft lay half-built on production lines, scientists and engineers (the lucky ones) fled to other countries, and launch complexes became high-tech ghost towns. By 2001, the annual budget of the Russian space program was about $268 million, which wouldn't even buy a single space shuttle launch.

Tito's trip was advertised at $20 million, but no one believes he paid that much. Twenty million was the price tag set by the Russians, but Tito's friends and family insist he bargained them down to about $12 million (Tito himself said he was contractually bound not to reveal the actual cost). Taking the Russians at face value, though, $20 million works out to about 7 percent of the entire annual budget for the Russian space program in 2001. While $20 million might be chump change to NASA, 7 percent of NASA's roughly $15 billion annual budget would be about $1 billion. One has to wonder if NASA would say no to someone offering roughly a billion dollars *in cash* for a seat on the space shuttle. According to several sources, Tito's payment not only covered the cost of his Soyuz flight but also paid for the next one.

Our final stop of the day was a detour to Cosmonaut Tito's monklike living quarters. Although he could easily have afforded more luxurious accommodations nearby, Tito elected to live in the same dormitory as his fellow cosmonauts: no small sacrifice, given that Tito's Santa Monica mansion is reputedly the largest private home in the Los Angeles area (in the film *Wag the Dog*, Tito's home stood in as the secluded palace of a foppish movie producer played by Dustin Hoffman).

Throughout Tito's room, evidence of an ascetic, almost academic lifestyle was abundant, from the boxes of Barilla pasta stacked next to a single hot plate to the cans of soda and cartons of orange juice in the small refrigerator. A Sony Vaio laptop and a Hewlett-Packard Deskjet printer sat on a small table near a window overlooking a pond. A pile of Russian coins was mounded on a small bookshelf that doubled as a china cabinet, complete with delicate, intricately decorated cups for tea or coffee. Also on the bookshelf, next to the coins, was a wristwatch bearing the likeness of Tito himself: a bright blue-eyed gaze and a boy's grin, happily wearing his spacesuit—just another astronaut. The image was so timeless that it could have been from the Apollo or the shuttle era, so little has that stereotypical portrait changed.

I asked Tom Stevens, another longtime friend of Tito's, whether or not he'd seen the watch before, and he laughed and shook his head. "Knowing Dennis, though," Stevens said, "he'll probably want to sell them."

William Stavros may be the only friend of Dennis Tito's who is smaller in stature than Tito himself. A genial, neatly trimmed man with an easy laugh, the Egyptian-born Stavros has known Tito since the 1960s. Back then they were fresh-faced, twenty-something engineers working at NASA's Jet Propulsion Laboratory plotting orbital trajectories for the Mariner probes that NASA sent to Mars. "When we were at JPL, Dennis was the orbital guy," Stavros said. "He knew more about orbital mechanics than almost anyone."

All the excitement, though, wasn't enough to keep either of them around. The drastic budget cuts that began to hit NASA in the mid-1960s led to sweeping layoffs between 1972 and 1975, which NASA called reductions in force, as if NASA were a great army being demobilized—and, in a sense, it was. Stavros and Tito were among hundreds of NASA scientists and engineers who saw the writing on the wall early on; like many of their colleagues, rather than waiting for NASA to hand them a pink slip, they jumped ship.

Both Stavros and Tito enrolled in advanced business programs at UCLA; Stavros finished his MBA and eventually joined the toy maker Mattel, where he moved through several top management positions to ultimately become the company's treasurer. Tito, who held astronautics and engineering degrees from NYU and Rensselaer Polytechnic, went even further in reinventing himself. Finishing course work at UCLA for a doctoral degree in finance (although never completing his dissertation work), he founded Wilshire Associates.

Tito had begun playing the market in his spare time while still at JPL. In conceiving Wilshire Associates with his then wife, Suzanne, Tito had the novel idea of applying computer modeling and mathematical analysis to the investment market. Wilshire Associates introduced the first asset/liability model for pension funds in the 1970s, a decade before most actuarial and accounting firms began to adopt such high-tech methodology. The company eventually became best known for its Wilshire 5000 Index, which measures the performance of all equity securities headquartered in the United States with readily available price data. The Wilshire 5000 Index, one of the only financial indexes that track the entire stock market (using more than seven thousand securities to adjust the index), was widely touted as a key sanity check during the tech-sector madness of the late 1990s and early 2000s. Wilshire Associates grew into one of the largest fund-management companies on the planet, directly responsible for more than $10 billion in assets and with advisory responsibility over assets in excess of $2.5 trillion.

For Tito, the transition from a career plotting spacecraft trajectories to life as an investment icon was more organic than might be readily apparent. Tito's company biography describes the core concept of Wilshire Associates as "applying science to the art of money management." Tito himself went a step further in a 1999 interview with the *Los Angeles Times,* directly linking his JPL experience with his ability to fundamentally change the way investments were tracked and managed: "You don't get to Mars by luck. It's like threading a needle [from] 10 miles away. In dealing with the stock market, you could just throw darts and end up with a portfolio that might luck out and beat the market, but you don't get to Mars that way."

It further stands to reason that Tito would be a shrewd negotiator in haggling over the details of his trip into space. Although it may not seem so at first blush, to fly into space (let alone stay there for a while) for $20 million is a pretty good deal; doing it for $12 million is practically highway robbery. Let's do some simple math and consider what flying into space on the shuttle costs per astronaut, using NASA's own cost and launch statistics for 1999. That year, the space shuttle program cost $2.47 billion, a figure covering shuttle maintenance work plus launch, landing, mission and crew operations and personnel. Thirty-two astronauts were launched on five shuttle missions in 1999: $2.47 billion divided by thirty-two works out to about $77 million per astronaut. In summary, Tito was getting a bargain.

Tito could certainly afford a high-priced vacation; Wilshire Associates made him very rich, to the tune of more than $200 million in personal assets. Tito's rise to fame and fortune was not without its bumps. After producing three children, Tito and his wife, Suzanne, divorced (Suzanne, their two sons, and Tito's then girlfriend attended his launch, while his daughter missed out on the occasion due to a last-minute scheduling problem). Through it all, Tito remained a man who by all accounts was straightforward, down-to-earth and generous with his fortune; among other things, he has long been one of Los Angeles's leading cultural patrons. "You know, we were just working-class kids from Queens," his sister, Joan Henrich, told me while en route to Baikonur. "Dennis deserves it all. He's worked for it. He never forgot who he was or what was important to him. He just was always really smart and driven, and he made that work for him."

No one among Tito's friends or family was surprised to learn that he wanted to travel into space. In the early 1990s, Tito made some tentative inquiries at NASA about visiting the ISS on the shuttle; they went nowhere. Tito occasionally attended space conferences and even lectured at a few about the importance of applying business sense to space dreams. "Most of us who left NASA, we really left," William Stavros told me as we talked just prior to Tito's launch. "I mean, we didn't suddenly become not interested in science, but we were making a clean break and focusing our attentions and energies someplace else. We were leaving that world behind. Dennis, he never stopped dreaming about space, he never stopped wanting to go."

The man who convinced Tito that he could travel into space was Rick Tumlinson. In December 1999, only a month before MirCorp went public, Tumlinson spoke in Pasadena to a gathering of the Space Tourism Society about MirCorp and its plans to revive the Mir space station as a commercial enterprise. Unbeknownst to Tumlinson, Tito was in the audience. Through a mutual connection, Tumlinson later learned that Tito was interested in MirCorp—and interested, possibly, in flying to Mir himself. Tumlinson arranged a meeting with him at Tito's office in Santa Monica.

"He knew he wanted to go, but he was very uncertain," Tumlinson recounted to me in early 2002. "He's pretty traditional, pretty straitlaced. He's not impulsive. He kept asking about the viability of Mir-Corp, how real was it, and I told him everything I could. We walked down the street and had sushi, talked, and came back a little while later. He had this real intense look and said something like, 'So this is real.' I looked him in the eye and I said, 'Yeah, we'll do this, this is real.' And he said, 'Well, I want to fly.' " Tumlinson then contacted Jeff Manber, who officially signed Tito to the spaceflight deal with MirCorp.

Tumlinson and Manber have bitter feelings about the final days of their professional association with Tito. As of September 2000, the Soyuz spacecraft being built for Tito's planned April 2001 flight to Mir was behind schedule, and another manned flight was needed in January 2001 to boost the station and conduct further repairs in advance of Tito's trip. There was no money to speed up production on Tito's Soyuz, or, more important, to pay for the January reboost and refurbishment mission. If the reboost didn't happen, neither could Tito's trip.

There was little help to be had from Energia, which was fully extended with ISS commitments. Thwarted by NASA and its partners, MirCorp's attempts to attract major new investors were failing, with one significant exception: Manber had held several discussions with a potential space tourist who seemed every bit as possible as Dennis Tito. Like Tito, this customer (whom Manber and Tumlinson declined to name) had plenty of disposable income and a burning, lifelong passion for space travel. Unlike Tito, he wanted to stay in space longer (three weeks as opposed to eight days for Tito) and would pay more for it.

There was to be another stipulation: the new customer would do a deal only if he could fly first. Manber claims that Tito's contract guaranteed only that MirCorp would fly him to Mir no later than the end of July 2001. Tumlinson said while that was technically true, the implied agreement was that Tito would be the first private citizen to pay his way into orbit. "When I shook Tito's hand, we knew what the deal was," Tumlinson told me. Tumlinson also knew that Tito's ego was firmly engaged from the very beginning of their discussions. "He wanted to be first. I knew that, we all knew that."

After much heated internal discussion, MirCorp's Walter Anderson backed Tumlinson's position. But as far as Tito was concerned, it was too late. The deal with the other would-be space tourist never materialized, but word of the deal had leaked to Tito, and he was livid. Tito's ire didn't make things any easier when Manber called later and asked him to free up some of his escrow funds in exchange for a guarantee of flying first. This funding would, Manber and others believed, have given MirCorp enough liquidity to keep Mir on orbit long enough to fly Tito in April—thus freeing up the rest of his money, and buying the company (and station) more time. Tito declined, perhaps driving the final nail in Mir's coffin. As late as November 2000, Tito was technically still scheduled to fly with MirCorp, but he knew Mir's fate was sealed. Already, he was quietly working directly with Energia and Rosaviacosmos (as the Russian Space Agency is most commonly referred to) on a deal that would redirect him to the ISS.

NASA wasn't necessarily supportive of Tito's Mir vacation plans, but it didn't target him for intense criticism, at least not at first. But on November 16, 2000—the same day Vladimir Putin signed the formal de-orbit order for Mir—Tito leapt firmly onto NASA's radar screen by announcing that it was "highly likely" he would be traveling into space more or less on schedule and, as a bonus, staying at the posh ISS. At first, Rosaviacosmos (which had jurisdiction over who flew on Russian rockets to the ISS) would not confirm Tito's claim, which made international headlines. However, Rosaviacosmos chief Yuri Koptev later admitted that his agency, in conjunction with Energia, intended to fly Tito to the ISS on an April 2001 taxi flight (this is ISS-speak for a routine resupply or crew-change mission) aboard a Soyuz spacecraft, which would, conveniently, have its third seat open.

It was a risky move. In approving Tito's flight, Koptev ignored the advice of legal advisers who said the consequences of problems during Tito's flight, particularly a launch failure or on-orbit catastrophe of any kind, could cripple the Russian space program. According to the other ISS partners, Koptev's decision to fly Tito circumvented ISS partner protocols, since Rosaviacosmos was the partner of record, not Energia— even though the Tito deal rested firmly with Energia, not Rosaviacosmos. On this last point, Koptev claimed that Russia had the right to fly whomever it chose on its spacecraft to its part of the station, and for whatever reason it chose; NASA and the other ISS partners claimed that ISS protocols required partner consensus on who visited the ISS. Koptev argued that this was a suggestion and not a mandate.

Publicly, NASA had no immediate response to the news of Tito's ISS flight plans. As it turned out, the first public noises of opposition came from the European Space Agency, which began grousing to European media about Tito's looming visit. In a February 2001 press conference, Jorg Feustel-Buechl, ESA's director of manned spaceflight, said Russia had no right to send "amateurs" to the ISS, particularly before its construction was completed. But the real ugliness began that March, when Tito and his Russian crewmates, Talgat Musabayev and Yuri Baturin (both of whom had been training with Tito for nearly a year), showed up at the Johnson Space Center for additional training on the American sections of the ISS.

At the training center, Tito and his crewmates were met by Robert Cabana—the NASA manager who had previously expressed such keenness to see Mir de-orbited. Cabana said he was ready and willing to train Musabayev and Baturin—but not Tito. Musabayev, the mission's commander, protested that Tito's seven-hundred-plus hours of cosmonaut training in Star City—essentially, the same training that NASA astronauts received there—more than qualified him to fly to the station, particularly considering that he wouldn't have to do anything except stay out of the way. The argument became increasingly heated, then Musabayev dropped a political bombshell: the entire crew would train, he announced, or none of them would. Cabana didn't budge: "In that case, we will not be able to begin training, because we are not willing to train with Dennis Tito." With that, Musabayev and his team went back to their hotel.

Cabana's words were saved for posterity because, unbeknownst to him, a *Newsweek* reporter was traveling with Tito and his crewmates at the time. The story that appeared the following week in the magazine was the first of many that would leave NASA looking less like a government agency trying to exercise appropriate caution and more like an intractable bully. It wasn't long until a full-scale political and media war erupted, with the nasty rhetoric culminating in Goldin's public implication that Tito was anti-American for choosing to fulfill his dreams against NASA's wishes and—the real capper—with U.S. senator Barbara Mikulski, a longtime Russia basher, branding the Russian space agency as "pimps" for selling the Soyuz seat to Tito.

Tito mostly steered clear of the war of words, instead focusing on his singular interest: achieving his lifelong goal of going into space. In media interviews, he was polite and boyishly enthusiastic, stressing again and again that he was pursuing a purely personal interest through appropriate channels with his own time and resources. "I had this dream over 40 years ago and at that time, I didn't have a penny to my name," Tito told the *New York Times*. "We have one life to live on this Earth. . . . I am optimistic that I do have the right stuff to make this flight." Tito's can-do comments echoed NASA's early entrepreneurial spirit and endeared him to the American public. The net result was a portrait of a likable, slightly eccentric dreamer—much richer than you or me, certainly, but still very much a Regular Guy.

As the date of Tito's flight grew closer, the tone of NASA's rhetoric became increasingly shrill and uncomfortably personal. Speaking to CNN after the training flap, the NASA astronaut and manager William Readdy went so far as to imply that Tito was endangering the very future of U.S.-Russian détente: "We are not going to let him be the wedge to drive us apart," Readdy proclaimed. Goldin howled that Tito's trip was putting "incredible stress" not only on the ISS crew but on all of NASA, a position somewhat undercut by concurrent statements from Goldin's own deputies and ISS crew members. Speaking to CBS Radio a month before Tito's flight, ISS crew member Susan Helms said, "We are absolutely going to welcome anybody who is on the other side of a hatch that we open from any visiting vehicle and we're not going to worry about the things that happen on the ground and the discussions that have taken place." Astronaut Ken Bowersox, backup to Bill Shepherd as commander of the first ISS mission, went a step further and lauded Tito's trip as "supporting the [Russian space] program and that helps us." Dan Tam, handpicked as Goldin's special assistant for ISS commercialization, told *Aerospace America* in its April 2001 edition that "as long as Tito doesn't mess up what the ISS crew members need to do, it's really between Tito and the Russians." Tam even went a step further and actually praised Tito's initiative: "I think he has the right idea." (Tam was quietly moved out of the commercialization role shortly afterward.)

Meanwhile, Tito praised Russia, the ISS crew and even NASA, whose ruffled response to his flight Tito said was "understandable."

Tito's hottest public comment was in response to Goldin's charge only a
few weeks before his flight that Tito was not an American patriot—
unlike the *Titanic* director James Cameron, who had also asked NASA
to fly him to the station but agreed to wait until the agency was ready.
(Cameron, actually, is Canadian.) Even Tito's response to Goldin's insult, which would boil the blood of most Americans of Tito's generation and conservative bent, was mild and measured. Tito told a Russian
journalist, "With all due respect to Mr. Goldin, I don't think that he
is in the position to determine who is an American patriot and who
is not."

Privately, friends and family said, Tito was livid over NASA's public
assault on his motivations and, particularly, his character. "Dennis
never wanted to go to war with NASA and he was upset that they
decided to go to war with him," said Tom Stevens. "But he wouldn't
take the bait. He wanted to go into space, and that was what dictated
everything Dennis said and did. He wasn't going to get into a pissing
match with NASA because that wasn't going to help him get there." Ed
Marzec, a Wilshire Associates attorney who also joined the launch trip,
agreed: "NASA tried to make Dennis look like some selfish asshole who
was stomping all over the American space program," Marzec said to
me. "They failed not only in America; they failed everywhere. I mean,
in Russia, they think he's the new Armand Hammer."

The combination of Tito's cheerfulness and NASA's loutishness
worked wonders for Tito, transforming him in the public eye from a
bored playboy millionaire to a scrappy poor-kid-made-good, struggling against a bullying government agency to realize a lifelong dream.
As Tito's flight drew nearer, the press was awash with praise for Tito's
entrepreneurial pluckiness and criticism for NASA's boorishness. Writing in the *National Review,* Andrew Stuttaford sniped that NASA was
"little more than the postal service in a space suit" and said that the
"surly and self-important way in which the agency has handled Dennis
Tito" was yet further evidence that "the old, marvelous improvisa-
tional NASA . . . was dead." Even James Oberg, a veteran American
space analyst who is often critical of Russia's space program, noted of
Tito's upcoming flight that "over the past century or two, millionaires
have opened the public's access to dozens of activities which now
entertain and thrill millions." In an Associated Press article published

on the eve of Tito's launch, Oberg went on to blast NASA for its ostensible concerns for safety during Tito's trip, which he viewed as a thin disguise for sour grapes: "NASA kept repeating that the Russians were teaching us all they knew about space station safety, and now suddenly NASA is proclaiming it knows more than Russia."

Unable to triumph with the kind of backdoor strong-arming that downed Mir (and facing a Russian space program determined not to give in again to NASA following the loss of that space station), NASA had no choice but to salvage what it could from the situation. On April 24, four days before Tito's scheduled launch, NASA and the other ISS partners granted "an exemption" for Tito's flight to the ISS that amounted to little more than a face-saving formality for NASA—a last attempt to seem in control. In its news release announcing the exemption, NASA offered no rationale for the decision, as if assuming that it would be common knowledge that the agency had been backed into a corner and forced to concede its just and true position. Even through the stilted language of the NASA release, one can practically feel the agency's helpless, angry resignation.

But by that point, the position for NASA was no doubt crystal-clear. Short of having the resident ISS crew lock the station's entry hatches (particularly unlikely, since the ISS was, at the time, commanded by the veteran cosmonaut Yuri Usachev), there was nothing NASA or its partners could do to stop Tito from becoming the station's first private lodger.

The Baikonur Cosmodrome is a three-and-a-half-hour plane flight southeast from Moscow. The lush, fairy-tale forests of Mother Russia gradually give way to a harshly beautiful vista reminiscent, in its arid splendor, of Utah. As you fly farther south, small lakes dotting the increasingly parched landscape below appear to be rimmed with snow; these are the result of alkaline buildup along the water's edge, a legacy of failed efforts to siphon the lakes and use their water for cotton farming. Still farther south, the landscape becomes even more dessicated, slowly becoming a crackled scrub desert virtually unmarked by human presence.

As we flew over this scenery in our chartered jet, a Russian journalist in our group looked out the window, assessed the landscape and announced that we must be over Kazakhstan. For the next couple of hours, the scenery grew harsher and more desolate. Even at thirty thousand feet above the ground, you could practically feel the life leaching out of the land, and with every mile we seemed to be traveling farther from the known world.

Despite the bleak topography, in the context of space exploration the parched steppes of Kazakhstan are the most significant place on Earth, the terrestrial starting point of more space "firsts" than anywhere else. The history of the Baikonur Cosmodrome (typically called simply Baikonur) has its roots deep in the loamy geopolitical soil of the Cold War; its construction was approved in 1955, simultaneously with the approval to develop the SS-6 intercontinental ballistic missile, or ICBM, which was aimed squarely at the fears of Western political leaders.

The reasons for locating what would become one of the Soviet Union's primary ICBM launch sites—and its only manned space-launch site—within Kazakhstan were, at the time, fairly straightforward. More or less conquered by Russia in the nineteenth century, Kazakhstan was located at the far south and west of the Soviet empire, meaning that missiles test-launched from there could complete flight trajectories of thousands of miles entirely within the USSR, and thus be more easily hidden from prying Western observers. Kazakhstan was also a lightly populated land, and the Kazakhs were a nomadic people; even today, Kazakhstan's largest city, Alma-Ata, has the provisional feel of an outsized caravan. A lack of large, permanent settlements in Kazakhstan meant that the Soviets could, by domestic decree and/or physical force, shoo the nomadic Kazakhs across the vast territory of Kazakhstan (nine times the size of France) whenever it suited them, a practice not dissimilar to the American policy of "relocating" nomadic Native American tribes in the late nineteenth and early twentieth centuries whenever their lands were found to hold oil, gold or other valuable resources.

The combination of Soviet mandate and Kazakh portability made Kazakhstan the perfect choice for a nation interested in building a large rocket facility, and particularly valuable for a nation without easy access to a hospitable oceanfront site (like the swampy Florida coast) to provide a margin of error for the fiery malfunctions that plagued both the Soviet and American space programs in their early days. Also key to establishing the Baikonur Cosmodrome where it is today was the ability to stealthily transport materials and people to build and support launch facilities. The Baikonur site had a network of roads and rail lines ideal for this purpose.

The launch complex was built over several years, under the tightest cloak of security. Even the name of the cosmodrome was an intentional misnomer. There exists today a dusty, ancient town called Baikonur far to the northeast of the cosmodrome, which was actually built near the small community of Tyuratam, a name by which the cosmodrome is still occasionally referred to. This name game was a deliberate attempt to fool the Western intelligence community, which the Soviets reasoned would eventually get word of the new launch complex and try to sniff out its location. The ruse never quite worked: U.S. tracking stations in Turkey were sophisticated enough to be able to detect even the earliest ICBM launches from Baikonur, giving the West an accurate bearing on the cosmodrome's location early in its development.

Throughout the Cold War, the Baikonur Cosmodrome was the target of more U.S. reconnaissance than was anywhere else in the world except perhaps Cuba. In one of the Cold War's tensest incidents, the CIA pilot Francis Gary Powers was on a mission to photograph Baikonur on May 1, 1960, when a Soviet antiaircraft missile knocked his high-flying U-2 spy plane out of the sky, forcing him to eject into the waiting arms of the Red Army. Powers's public trial in the Soviet Union and his eventual return to America—in exchange for an incarcerated Soviet spy—capped one of the most humiliating incidents of the era for the United States. Strangely enough, one of the cover stories used by Powers and the other elite pilots of the ultrasecret U-2 corps was that they were conducting meteorological research for NASA.

Baikonur and its missiles were mandated in no small part because of American aviation advances like the U-2. When Powers was shot down, the Soviets had known about the U-2 flights for nearly four years but had been unable to stop them. Powers's plane was hit only because engine trouble forced him to lower his cruising altitude, giving the Soviets their chance. The Soviets had no aircraft nearly as sophisticated as the U-2; however, as evinced by the missile that shot down Powers, their rocket technology was just as sophisticated as the Americans' if not more so. Thus, Khrushchev built Baikonur, hoping to narrow the military gap by developing a fleet of continent-spanning nuclear missiles.

The first successful ICBM flight from Baikonur was in August 1957, the hulking missile flying straight and true to a target in the Kamchatka peninsula far to the northeast. In October of the same year, as the leaves began to turn in Moscow, Washington and London, Soviet engineers loaded a basketball-sized metal ball onto a modified SS-6 ICBM under Sergei Korolev's watchful eye. The ball weighed 183 pounds and was chockablock with relatively simple electronics, including a small radio so that its makers could keep track of its location. (Western fears that this rudimentary telemetry device was designed as an early, if crude, means of targeting ICBMs were later proved to be unfounded.) The SS-6 fired into the bright morning sky, speeding straight up toward the vacuum of space, desperately burning fuel to reach higher than its design limits, finally exhausting itself and then gently lobbing its cargo into the microgravity of low-Earth orbit, exactly as the chief designer had planned.

Hours later, millions of people around the planet watched, stunned, as the metal ball known as Sputnik (Russian for "traveling companion") zipped overhead, winking like a star in the sun's reflected light as it cut an arc through the night sky and traveled an elliptical orbit that took it around the Earth every ninety-six minutes. For its radio transmission, the ball trailed four spring-loaded whip antennae that gave it the look of a supercharged spermatozoan. The analogy was profound in its contrary implications: mankind had blasted a protean simulacrum of itself not into a fertile new place, but into the one place where only a simulacrum could survive unaided. Here was notice that technology had, quite literally, gone beyond us.

More Baikonur firsts would follow, most notably the launch in 1961 of the first human into space, the eternally photogenic Yuri Gagarin. Sputnik and Gagarin were perhaps the greatest triumphs of the Cold War for the Soviets, and they made Baikonur the empire's darling. In short order, the spaceport became nothing less than a city unto itself, sprawling over four thousand square miles of the desolate Kazakh steppes and peaking in the 1980s at nearly one hundred thousand inhabitants.

The collapse of the Soviet Union was not kind to Baikonur. As the Space Race slowed to a leisurely walk, the spaceport's population dropped to half of its peak as the power of a vast empire drained away. By the early 1990s the spaceport's facilities were literally in shambles: rusting pipes, pockmarked launchpads, entire sections of the cosmodrome deserted. In 1994, there were reports of riots among military personnel assigned as caretakers to the crumbling cosmodrome. Buildings and even spacecraft were stripped of metal that was sold for scrap. More and more space activities were being pulled out of Baikonur, most particularly ICBMs, which were being consolidated within Russian territory at military-only launch sites like Plesetsk and Kapustin.

This left Baikonur with a mostly "scientific" space manifest, which was extremely cold comfort. Since the late 1980s, financing for Soviet scientific missions in space had virtually dried up, and Soviet scientists in general were seen less and less at international gatherings of their peers—one of the earliest signs to the rest of the world that the Soviet Union was on the brink of collapse. Since all Soviet space activities were at that time under the auspices of the military (unlike the United States, which created NASA as a civilian agency), civilian space interests took a back seat to the Russian military's broader goal of simply avoiding disintegration. The cradle of the Space Age, Baikonur was dying a slow, humiliating death.

Baikonur today is anything but dead. On the contrary, it is arguably more vital to the future of Russia than ever before, a fact that has not been lost on the Republic of Kazakhstan. In 1991, Kazakhstan declared its independence, and Russia's space leadership suddenly found that its only launch site for human missions to space was in another country. After declaring its independence, Kazakhstan initiated negotiations with Russia over the future of Baikonur which were tense from the very start and which dragged on until 1994, when an agreement was signed with terms almost disproportionately favorable to Kazakhstan. For starters, Russia would lease the Baikonur Cosmodrome for the equivalent of a steep $115 million a year. Later, the Kazakh government began charging a per-launch fee atop the annual rent and required the Russians to get Kazakh approval prior to any Baikonur launch.

This lease arrangement has not been without its moments of high drama—case in point, the July 1999 shutdown of Baikonur following the crash of a Proton rocket carrying a Russian military satellite, which showered the landscape with tons of highly toxic fuel and dropped a four-hundred-pound chunk of steaming rocket onto the lawn of a local home. The Kazakh government immediately banned all launches until Russia paid the equivalent of $270,000 in damages, which the Russians grudgingly did.

In addition to being politically and economically expedient, such blood-from-a-stone negotiating on the part of the Kazakhs is also no doubt a bit of sweet revenge. During the height of the Soviet era, Baikonur rockets exploded frequently with deadly consequences, and even today sundry rocket parts—charred hull fragments, crumpled fairings, spent motors—are strewn like flame-roasted sculptures across the Kazakh landscape. During the Space Race, launch catastrophes would rarely slow the pace of the Soviet space juggernaut, let alone cause a complete site shutdown. In the post–Cold War political land-scape, Russia's need for Baikonur means supplication to the Kazakhs as never before (a point proved in early 2002 when the Russians, express-ing clear optimism about the future of their space program, extended their Baikonur lease through 2044). These days, Kazakhstan sometimes shuts down Baikonur just because, at long last, it can.

About half an hour before we were cleared to land on Baikonur's sin-gle runway (we needed both Russian and Kazakh military clearance), the vast spread of the cosmodrome came into view. "You know, ten years ago they would have been shooting at us by now," quipped Tito's friend Ed Marzec. Everyone laughed, but we were all transfixed as the spaceport materialized from the fierce landscape below. From our van-tage point, Baikonur looked like a vast mining town in the American high desert, in Nevada perhaps; dozens of squat, dumpy buildings the same sand-gray color as the desert, connected by pockmarked roads. The few structures indicating any relationship to space exploration were visible only on the periphery of our limited view out the plane's windows: launch gantries, at least half a dozen of them, scattered across the blowy plain, looking like fragile toys hardly capable of launching huge rockets high above the Earth.

After clearing Kazakh customs, our group piled into two vans for the forty-five-minute drive from the airstrip to our hotel, passing bunches of anonymous blocky buildings and the occasional wild camel grazing contentedly by the side of the roadway. The hotel itself was astonishingly upscale; a beautiful lemon-yellow lodge called the Sputnik Hotel, complete with an indoor pool, spa, an on-call massage therapist and even a disco. The hotel was built in 2000 by Starsem, a multinational aerospace company that manages commercial launches of Russian Soyuz rockets and their satellite payloads ("Soyuz" is both the name of a rocket and a spacecraft that sometimes sits atop that rocket). The hotel was part of a reported $35 million that Starsem pumped into Baikonur in late 1999; other Starsem-financed improvements included launchpad upgrades and new facilities for satellite preparation. Starsem's cash is only part of the infusion of Western capital into Baikonur that has made the spaceport once again among the busiest in the world. Another multinational satellite company, International Launch Services (ILS), which is led by American investors, has built a satellite processing facility at Baikonur.

The newer, swankier Baikonur has occasionally produced some bizarre side effects for its Western patrons. Officials of the woebegone constellation company Globalstar, which used Baikonur to launch its satellites on Russian rockets, discovered that Baikonur held such a hypnotic allure for its crews that the company began mandatory rotation every three weeks. "We actually had to force people to leave," a Globalstar representative, Dennis Barr, told *Space News*. "People who stayed there too long ended up with personality disorders. Part of the personality disorder was that they didn't want to leave."

If you spend any time in Baikonur, the meaning of Barr's comment takes hold fairly quickly. On the ground, it feels less like the end of the world than a world unto itself; a place of sand-whipped, Zen-like peacefulness, a quality not uncommon in remote desert communities but even more surreal here, considering Baikonur's Sturm und Drang history. After checking into our rooms, we were given a tour of the cosmodrome's small residential and commercial district, which looks much like Moscow without the urban coil (no smog and crowded streets, but also nowhere to get an espresso). The wind pounded in booming gusts, and low scudding clouds raced overhead; it felt more like fall than spring, with dust and scrap paper swirling in miniature tornadoes and tumbleweed skittering across the pavement. Our van parked at one end of a broad square next to a massive stone statue of Lenin around which a group of children laughed and chased one another on Rollerblades.

Children are not an unusual sight at Baikonur, which is complete with Soviet-style apartments, a limited but serviceable shopping mall and a school. Our cosmodrome guide, Yelena (whose family of five had lived at Baikonur for several years; her husband worked in satellite processing for Starsem), told us that the long expanse of pavement stretching from the statue to the beginning of Baikonur's main promenade was once the scene of elaborate military celebrations and ceremonies in the heyday of the Soviet space program. The promenade itself was quiet when we visited; Yelena told us that most of those who might be strolling on a typical early evening such as this were preparing for tomorrow's launch.

At that, someone joked that with the cold evening approaching, everyone was probably at home having a few drinks, but Yelena said seriously, "Not much drinking here. You drink, you can't work. You don't work, well, you must leave. This is good work, so people don't want to leave, so people don't drink too much." Not everyone was enchanted by Baikonur's grubby transcendental spareness. Strolling down the promenade, with a few streetlamps beginning to hum to life, one of our group remarked quietly, "Wouldn't you just kill yourself if you had to live here? This must be the asshole of the world."

On April 28, the day of Tito's launch, we rose early for a short tour of the business part of Baikonur, starting with an enormous building of gleaming white paint and fresh-from-the-trade-show high-tech equipment where Soyuz rockets are assembled and prepared for launch. Another assembly building held sundry space vehicle parts and machining equipment, along with a stunning surprise: a fully assembled Energiya booster, atop which sat the one and only flight-tested Buran shuttle. The Energiya was Russia's successful answer to the Sat-

urn moon rockets, but it came too late. It was used to launch the one
and only flight of Russia's space shuttle, the Buran ("snowstorm"),
which made its lone, completely automated and perfectly executed
orbital flight in 1988.

Although the Buran and NASA's shuttle look identical to the unin-
formed observer, the Buran fully embraced its status as a glorified
glider, and thus its orbiter has no integrated engines; that is why its
launch required the powerful Energiya. No engines meant greater payload capacity, and many have argued that NASA's shuttle would have
benefited from this configuration. But the Buran was never a real competitor to the shuttle, since after its first flight the program ran out of
money and the flight-tested Buran and its siblings (which were at various stages of completion) were mothballed. One Russian space engineer
half-joked that Russia's space program was broken by its perceived
need to match America in building a shuttle, since the Russians knew
full well that NASA's shuttle had turned out to be a lemon even as they
pressed ahead with the Buran. He followed that with another half-joke:
"International Space Station project," he chortled in gnarled English,
"is Russian revenge."

Both the Energiya and the Buran sitting in the Baikonur hangar
looked virtually flight-ready. When I put this question to our facility
tour coordinator, a taciturn engineer named Sergey, he replied cryptically: "The Energiya and Buran programs are no longer open. But these
vehicles are not models." I later asked Jeff Manber about the status of
the Buran and particularly the Energiya, which if functional would be
the only operational Saturn-class launch vehicle left in the world.
Manber laughed, and said, "Well, truthfully, they're probably a few
hundred million away [from launch readiness]. But that's the thing
about the Russians. They don't throw anything away. They're keeping
the Energiya as preserved as possible until the day when money
returns."

That day couldn't come soon enough for the Russians, since about a
year after our visit, the roof of the Buran building collapsed during a
repair job. While the Buran and Energiya suffered only minor damage,
eight workers were killed.

As if to underscore the Russians' thriftiness, after leaving the
Energiya-Buran building on our way to Tito's final preflight check, we

passed what appeared to be an equipment scrap yard, filled mostly with rusting and unidentifiable parts and machinery. It had the odd beauty of a modern-art exhibit, all protruding edges and lumpy totemic mounds. Amid the tangle of twisted iron sat two gigantic half cylinders that looked like do-it-yourself Quonset huts, reportedly used for some kind of storage. They were neatly placed side by side and perfectly symmetrical, as if cut from a single huge tube of metal. The perception turned out to be correct—this was the bisected upper stage of the last N-1 rocket, the rocket that was to have lofted the first men to the moon but never got off the launchpad in one piece—the Götterdämmerung of the Space Race for the Soviet Union, the failed nemesis of the Saturn V. It's arguable which rocket had the more ignominious fate—the Saturns, which ended up as museum pieces; or the N-1, which became a storage shed.

At the final media briefing in advance of Tito's flight, those in our group, along with a few specially invited businesspersons (Jeff Manber and Walter Anderson were somewhere in the crowd) were the only Americans in attendance. American media were banned from Baikonur on the occasion of Tito's launch, a bitter payback to NASA for its pugnacious opposition to his flight. America would learn about the flight of the world's first self-financed space explorer—ironically, one of the few certifiable American "firsts" for human spaceflight—from foreign media.

Tito and his crewmates sat behind a Plexiglas wall facing a room crammed with cameras, microphones and bodies; this was clearly a rock star moment and the Russians had staged it as such. Largely ceremonial questions were asked in Russian by a panel of stony-faced Rosaviacosmos and Energia personnel, and while all three cosmonauts participated, it was clearly Tito who drew the most attention. Although he had learned get-around Russian in his year-plus adventure preparing for the flight, a female translator whispered the questions in Tito's ear, in English. Slowly and deliberately, Tito replied to each question in Russian, sometimes pausing to ask his translator for the mot juste. The process took about twenty minutes. Then, at last, Tito and the crew were certified to fly.

After yet another ceremony outside the press briefing building, attended by what looked to be about five hundred people (a large crowd for a Baikonur launch), there was a mad dash for vehicles. No one wanted to miss what came next—one of the great rituals of human spaceflight, started by none other than the great Gagarin himself. Two RVs carried Tito, his crewmates and other essential launch personnel toward the waiting Soyuz; we were driving immediately behind the vans until very close to the launch site, when the RVs took the low road while we kept going. Then, in the distance, Tito's van stopped and we stopped as well, getting out and bracing our cameras for the Kodak moment to come.

A few hundred yards away on the road below us, the cosmonauts got out of the van and clustered around the right rear tire of their vehicle for a ceremonial pee, just as Cosmonaut Gagarin did forty years earlier out of urgent necessity, defying even the mighty Korolev to answer nature's call only moments before climbing into his Vostok capsule for his historic flight. Impromptu potty breaks were also a hallmark of the early American space program. Delayed for hours inside his cramped Mercury capsule on May 5, 1961, Alan Shepard famously disobeyed NASA orders for him to "hold it" and, while the Redstone rocket was still on the launchpad, turned his spacesuit into the world's most expensive diaper. While Shepard undoubtedly had a less comfortable trip than a dry suit might have afforded, there was no other ill effect; later spacesuits were fitted with high-tech diapers.

The business of launching humans into space is different in Russia (or Kazakhstan) from in the West, often in unintentionally humorous ways. As we drove up to the Soyuz pad, Karl Hanuska, a Reuters reporter fluent in Russian, began to giggle uncontrollably. "Oh, that's great," he cackled, his eyes nearly tearing up with laughter. "That's really great. Kaboom!" He pointed out the window to an acronym stenciled rather formally on several buildings around the launchpad and, sure enough, there it was in big blue letters: KBOM, which we discovered was actually the acronym for the Rosaviacosmos division responsible for general construction at Baikonur. We all had a good laugh until the white fléchette of Tito's Soyuz came into view, cradled like a delicate piece of ivory in its spindly launch gantries. A few minutes later, we found ourselves standing squarely on Pad One, walking the cracked and weathered concrete on which Gagarin had walked four decades earlier as he prepared to become a human warhead atop an undependable ballistic missile.

In interviews afterward, Gagarin admitted that he wondered whether or not he'd actually get into space, what it would be like and, most particularly, whether or not he'd come back alive (he nearly didn't when a chunk of equipment on his Vostok capsule refused to separate during reentry). The Soyuz certainly looked much more like a missile than something Buck Rogers would climb into: a gleaming, unnervingly slender tube only a few stories high, adorned near its base with four strap-on engines which would drop away after the initial boost phase and which gave the rocket the look of a big white lawn dart. Even though the Soyuz holds the record free and clear as the most frequently flown human-rated vehicle in history (and arguably the most reliable), it seemed unaccountably small to loft three men and several tons of equipment up through miles of crushing atmosphere and into the vacuum of space.

The repercussions if something disastrous happened to Tito were well known not only to Rosaviacosmos and Energia, but also at the highest levels of Russian leadership. On the bus ride to our hotel the night before, we sat rapt as Alexei Mitrofanov, a prominent ultra-nationalist member of the Russian parliament who had spearheaded the political effort to keep Mir aloft, told us that Tito's flight was symbolic of Russia's efforts to reassert itself as a major player among space-faring nations. But it was not a step without risk and the potential for disaster. Recognizing this, Mitrofanov, who possessed the moist bonhomie of a young Zero Mostel, told us soberly, "If things do not go well, there will not be another person to go to space like this for many years. For us, I think, it is now or never. So I hope it is now."

Despite such words of portent, on the ground less than two hours before launch the Russians seemed very laissez-faire about the whole operation. Although we were merely observers, our group loitered so close to the Soyuz that any one of us could have picked up a rock and banged it off the rocket's metal hull with little effort—a far cry from the keep-your-distance agenda of shuttle launches. Liquid oxygen fuel hissed from the fuselage of the Soyuz, creating for a few minutes the surreal effect of snow as the liquid boiled to gas, froze in a thin coating of frost on the cool hull, then flaked off and fluttered over the launch site in the morning desert breeze.

We never actually saw Tito climb aboard the Soyuz, since at T-minus one hour we were shuttled to an open-air viewing area in order to give us the best chance of jostling for a good position for the launch, much as you would arrive early to get a good seat at a parade (like most parades, the viewing area at Baikonur is standing-room only unless you bring your own lawn chair, which several people did). The Soyuz, about half a mile away from our new position, looked like a cigarette held at arm's length, somehow appearing larger from this distance than from up close.

The crowd gathered and thickened, short Russian phrases were barked over a tinny loudspeaker and then, without any noticeable countdown, a cloud of brown dust boiled up from around the base of the Soyuz as it slowly rose off the ground, a bright yellow-orange candle flame at its tail. The dust quickly settled as the rocket rose; without the mountain of smoke and steam that accompanies a shuttle launch, all we saw was an upside-down torch improbably rising over the flat landscape and gaining speed by the second.

There was at first very little noise, only a dull scratchy roar out over the steppes that was easily overwhelmed by the cheering crowd. But as the Soyuz accelerated, the sound of hundreds of thousands of pounds of thrust grew in volume, just as gut-shaking as a shuttle launch, drowning out all human noise altogether. Then, within seconds it seemed, the Soyuz was nothing more than a glowing ember high in the clear morning sky. Through binoculars, I just barely caught the faint starburst signifying that the Soyuz's four strap-on boosters had separated from the main fuselage, which now was merely the faintest speck to all but the most powerful lenses in the crowd. Seconds later, even that was gone.

In the context of NASA's worst fears, Tito's time on the International Space Station was uneventful: he stayed out of the way and helped where he could, but mostly he looked out the window and listened to his opera CDs. While NASA couldn't stop Tito's launch, it did manage to inflict some pain on the Russians in negotiating his exemption, including getting Rosaviacosmos to sign off on a you-break-it, you-buy-it agreement stating that NASA would be reimbursed for any damage Tito might cause during his visit and also releasing NASA from any liability should anything go wrong for Tito on the station (Tito immediately agreed to both proposals). Goldin also vowed publicly to send Rosaviacosmos a bill for lost productivity on the station due to Tito's visit. No such bill was ever sent.

In November 2001, NASA, the Russians and the other ISS partners agreed upon guidelines for future flights of "spaceflight participants" (resurrecting NASA's old terminology) to the ISS. In addition to the predictable requirements that spaceflight participants not have criminal records or serious substance addictions (past or present), the guidelines also included prudish language prohibiting participants who had "dishonest, infamous, or notoriously disgraceful conduct" in their personal or professional history. Ironically, such language would probably have disqualified most of the original Mercury and Gemini astronauts, not to mention a few from Apollo; the test-pilot ethos practically mandated the ability to party hard, break some rules and still be able to operate supersonic vehicles.

But the rules were more or less a consolation prize for NASA and ESA, since Tito met all the requirements before the requirements were even set. What's more, the high-profile nature of Tito's flight, intensified by NASA's public harrumphing about it, would ensure that the evolution of "space tourism" would from that point forward happen squarely in the public eye. Only days after Tito was launched, the Russians said they had plenty of takers for future trips; to prove their point, less than a month after Tito returned to terra firma, Rosaviacosmos confirmed that a Russian rock star and a South African businessman were in the early stages of training at Star City. The rock-star deal would fizzle but the businessman, Mark Shuttleworth, would launch to little fanfare in April 2002 to become the first space traveler from Africa.

In May 2001, MSNBC published the results of an online readership survey asking the question, "What do you think about putting paying passengers in space?" Of the more than 66,000 people responding, 81 percent said either that they wanted to go themselves or saw nothing wrong with others going. How was it that NASA so misjudged the public response to Tito's flight? "I think it's real straightforward," the former NASA manager Lori Garver said. "Early on, Mr. Goldin said [Tito's flight] is not something that's going to happen, and he's not a person who likes to go back on his word. Unfortunately it wasn't his call. That's almost harder to swallow."

Even harder for Goldin was the hero's welcome Tito received upon returning to terra Americana. Audiences were not hard to come by— newspapers, magazines and television shows lined up to interview Tito, and virtually every space organization on the planet (NASA being a conspicuous exception) feted him as quickly as it could get time on his busier-than-ever calendar. Overnight, Tito became the poster boy for a new era in human space travel, the man who had knocked down the gates for the rest of us who might like to someday visit the final frontier. As for Goldin's insinuation that Tito was un-American for going around NASA's back and flying with the Russians, that most American of organizations—the Boy Scouts—put this accusation to rest once and for all by giving Tito its top honor: the Americanism Award.

But the truest testament to the impact of Tito's flight may have been found in a small moment that, for most, probably went unnoticed. Shortly after his return, Tito appeared at a congressional session on the subject of space tourism. From the crowd, the cameras, Tito and his entourage (I counted one PR manager, two assistants and a new leggy girlfriend) and the clearly starstruck attitude of the congresspersons present, you would have thought a Hollywood celebrity was in the room.

As the hearing was getting organized, a group of teenage boys from a local science club gathered at the table where Tito was to sit and offer his thoughts on space tourism, along with Rick Tumlinson and Mike Hawes, then deputy manager of the International Space Station. Rumor had it that Hawes had been sent to represent NASA at this hearing precisely because he and a handful of others had tried to convince Goldin that even if NASA didn't want to encourage Tito's flight, it should at least not openly oppose it. For being so contrary (made worse because he'd been right), Hawes had been sentenced to represent NASA in the House hearing. He shuffled into the room looking understandably hangdog as he prepared to bear the brunt of what everyone anticipated, correctly, would be another scathing diatribe by members of Congress on the subject of NASA's incompetence.

The hearing's fourth participant was Dr. Edwin Aldrin, better known to the world as "Buzz" and one of the most famous space travelers in history. Aldrin has been a hero to fans of human spaceflight ever since he and Neil Armstrong decided, with the computer on their *Apollo 11* landing craft having failed and their spacecraft plummeting toward the lunar surface, to try a manual touchdown, deciding then that it was glory or death. As the stark lunar surface loomed ever larger, Aldrin fired off streams of numbers to Armstrong, who then made split-second adjustments to various thrusters and landed, having overshot the target calculated by the computer that failed them with only seconds of fuel margin to spare. After Armstrong's famous small step, Aldrin then became the second human to set foot on another world.

Now seventy-something, still handsome and fit, Aldrin has been a tireless advocate of opening space to the masses by writing books, giving lectures and sponsoring space development projects through his Starcraft Enterprises and ShareSpace Foundation. Most of those attending the congressional hearing were of a generation that remembered Aldrin's historic journey; despite a well-publicized bout with alcoholism, to most in the room he had never ceased to be anything less than a hero.

The science-club boys were obviously of a post-Apollo demographic, but they seemed no less starstruck as they nervously moved as a pack toward the table where Aldrin, Tito, Tumlinson and Hawes were sitting. With pens and paper at the ready, it was clear they hoped to get an autograph. The boys then walked right past Aldrin and up to Tito, who looked genuinely surprised.

Aldrin watched the scene briefly, smiled and then returned to his notes. Tito recovered himself, and then signed with a flourish like the hero he had unexpectedly become.

Business in a Vacuum

The most curious aspect of Dennis Tito's space odyssey was not his struggle to get into space or even his time on the ISS, but rather what he did, and didn't do, afterward. Most of what Tito said in the wake of his adventure was predictable, i.e., it was a fabulous experience and everyone should have a chance to do it. As for NASA, Tito had no hard feelings: "I have nothing but great pride in NASA," he gushed at one fete, while adding that he didn't believe efforts like the Space Launch Initiative were going to do much to reduce launch costs or promote space commerce.

Tito also warned that the Russians were beating the pants off the United States in the "new Space Race," the one for commercial dollars and, particularly, commercial passengers. Russia hosted the world's first paying space tourists well before Tito and, in doing so, notched up more space "firsts." In 1990, the Tokyo Broadcasting System paid the Russians a reported $12 million to send the journalist Toyohiro Akiyama to Mir for nine days, making him Japan's first space traveler. (Akiyama's trip was reportedly an unpleasant one; he was space-sick much of the time.) Britain's first astronaut, the civilian scientist Helen Sharman, was selected in an industry-sponsored competition to spend time on Mir in 1991, which she did with aplomb, complaining only about the wonky showers.

Given that track record and his own experience, Tito not surprisingly believed that the Russians would continue to run away with the space tourism market for the foreseeable future. "I don't see anything happening on the American side," he told a crowd at a sold-out lecture at the National Air and Space Museum in early 2002, a comment he echoed on other occasions. What Tito meant was that he didn't see NASA booking paying passengers on the shuttle anytime soon, which

seems right enough. A year after Tito's flight, NASA's Robert Cabana—
the man who barred the way to Tito's training at the Johnson Space
Center—told Reuters that the agency had "no plans to fly nonprofes-
sionals on the ISS" (paradoxically, despite the fact that his JSC showdown caused a ripple of ill will between NASA and the Russian space
program, Cabana was subsequently assigned to represent NASA in
Russia).

Such developments were not unexpected, yet they are critical to
understanding what changed and what didn't in the post-Tito universe. Tito's flight did not send a flood of millionaires rushing to fill
seats on Soyuz. To date, despite dozens of rumored customers, only
Mark Shuttleworth, the South African, has enjoyed the same experience as Tito. There have been a flurry of "reality" show concepts aimed
at sending winners to the ISS, but none of these have yet taken hold
firmly; even in the entertainment business, $20 million (or even
$12 million) is a lot of cash. The loss of *Columbia* put another roadblock
in the path of those who would like to send private citizens to the ISS,
although few believe this roadblock is permanent.

The most obvious change wrought by Tito's flight was its proof that
NASA was not invulnerable where its last undisputed monopoly was
concerned. But Tito's flight did not send the monopoly crashing to the
ground; at most, it merely highlighted its problems more clearly. Albeit
indirectly, Tito's various postflight comments put the onus firmly on
NASA to tear down barriers to commercial space activities like tourism.
Yet in doing so he reinforced the very concept that his pioneering flight
had, in effect, shattered—that the only voice that really mattered when
it came to human spaceflight was NASA's. In this light, Tito's unkindest
cut of all might have been his answer when asked if he would recommend any commercial human-spaceflight activity to his Wilshire Associates clients as an investment opportunity. "Absolutely not," was
Tito's terse response.

With painful clarity, Tito's words highlight the conundrum that has
long faced the entrepreneurial space movement: the people with the
money and the people committed to building a vibrant entrepreneurial
space industry usually aren't the same people. There are the Walter
Andersons of the world who put their dollars where their cosmic inter-

ests are, but as the effort to commercialize Mir made clear, their isolated millions cannot compete with the annual multibillion-dollar taxpayer stipends given to government space agencies and Big Aerospace
monopolies.

In this sense, Dennis Tito is right; the ability to turn human space
activity into something more than a government-directed enterprise
rests not with private industry, not with maverick millionaires, and not
even with the alternative space community. The agent of change must
be NASA.

The first step toward loosening NASA's grip on human spaceflight is
introducing competitive enterprise into the NASA culture—long the
subject of many "paper NASA" efforts, but solidly boxed out of the
"real NASA" for predictable reasons. For the most part, the world's
largest space agency has attempted to address the call of the free market
through what it describes as "commercialization" and "privatization,"
although one is hard-pressed to find a consistent definition of what
either term means where NASA is concerned.

Over time, responsibility for commercial activities has been handed
around NASA like an unwanted pet. Offices and departments bearing
the mantles of commercialization, commercial products, space manufacturing and space pharmaceuticals (although not yet space tourism)
have been opened, closed, reborn and dismantled dozens of times,
depending on how loudly the congressional call is sounding for such
activities. The result has been predictably scattershot, but there has
been a consistent theme: what action NASA does take in terms of space
commerce is focused on controlling such activities. Too often, NASA's
idea of enabling space commerce is NASA, Inc.

Despite systematically quashing proposed entrepreneurial competitors (like MirCorp), killing potentially beneficial commercial co-
ventures (like the Industrial Space Facility), and clumsily marketing
vehicles singularly unsuitable for commercial use (the shuttle), NASA
has long labored to demonstrate that it is a player in the commercial
space game. However, when it comes to actually engaging in commercial enterprises, NASA has failed resoundingly. Potential and even
once-committed partners (like Ortho and 3M during the ISF era) have
been driven away by NASA's legendary paperwork requirements, inex-

plicable cost structures and ever-shifting launch schedules. Proposals for outright commercialization of existing NASA assets—such as the several serious propositions to buy, lease or replicate shuttles that came along in the 1980s and 1990s—have typically been dismissed out of hand as simply heretical.

One might rightly ask whether a government research agency should be forced to engage quite so directly in developing the commercial sector. Yet NASA's original 1958 charter specifically directs the agency to "seek and encourage, to the maximum extent possible, the fullest commercial use of space." Those who drafted this declaration knew that the organization they were creating would, if successful, dominate space for some period of time—just as it was clear that the Manhattan Project, and later the Atomic Energy Commission, would dominate the use of nuclear power during and immediately following World War II. This was acceptable, since both spaceflight and atomic energy were initially tied directly to national security interests. But from the start it was understood that there would come a day when commercial applications would succeed military applications, or at least be able to complement them.

It is incumbent upon NASA—and upon those who have influence on its work—to create a space market, not control it. This has happened to some degree in the realm of unmanned spaceflight, even if it took place largely by legislative force after *Challenger*'s loss, and despite the fact that this market remains dominated by a handful of heavyweights who work hand in glove with NASA to block competition. Yet NASA has not been forced to commercialize, even in part, human spaceflight: the stakes are perceived to be too high, and NASA has convinced lawmakers and most of the private sector that it alone has the ability to safely send humans into space.

That the world buys this argument, despite exploding shuttles and lost astronauts, is largely the result of circular reasoning on a grand scale: since no one in the United States other than NASA has ever been allowed to send humans into space, there is no reason to believe anyone else can do it. This applies indirectly to the Russians as well, who continue to withstand tremendous political pressure from NASA and its allies not to launch paying passengers on their spacecraft (as for independently developed commercial human spaceflight in Russia, the investment capital simply isn't there).

It is essential to note that NASA's original charter does not exempt human spaceflight from the agency's mandate to encourage the commercial use of space. Thus, even if NASA intends to remain a leading player in human spaceflight, it is obligated to change its operational ethos from subsidized monopolist to subsidized customer. To put it mildly, this is a daunting philosophical and operational leap. At the macro level, NASA has no intrinsic interest in being a free-market player, in part because spending significant organizational energy on such vulgar concerns as competitiveness and return-on-investment sullies the splendid von Braunian enterprise. Also, because successful commercial space activities would inevitably compete with NASA and its contractors, such activities are first and foremost perceived as threats.

"When I was in policy and plans, a lot of entrepreneurs would end up in my office because nobody else would talk to them and they knew I was in favor of commercial opportunities," Alan Ladwig told me when we talked in March 2002. "It was like, 'Well, is there some policy way you can help us out?' There were plenty of things you could do from a commercial point of view, but they might end up taking a NASA person's job. That was the perception."

The most cogent rationale for NASA's singular inability to conduct itself in the free market was offered by the former NASA deputy administrator Hans Mark as part of an amazing 1985 speech given only months after he left the agency to head the Texas State University system. Mark was a well-known space station advocate who had been knee-deep in NASA's attempts to make commercial hay of both the station and the shuttle. His remarks, presented at a private event not far from NASA headquarters in Washington, D.C., speak volumes about why NASA as a commercial proprietor is destined to be a fruitless proposition:

Now that I'm out of government, I can say that I found nothing funnier than bureaucrats wringing their hands in town around here about how to commercialize something where most of them, myself included, don't have the slightest idea on how to make investments, how to judge markets, how to do all those things that are important for commercialization. It is ludicrous. It is crazy.

Crazy or not, the cry of "commercialization" is sounding louder than ever these days as NASA is shaken by one embarrassment after another. Most such mortification has come as a result of how NASA has managed (or in many cases, not managed) its relationship with its two biggest contractors, Boeing and Lockheed, in their lockdown of the ISS and shuttle contracts. In such dealings, the basics of sound business practices have been glaringly absent, extending from a lack of rudimentary management ability to inconsistent and inept accounting procedures and purchasing philosophies. Above all, lacking is a sense that as a publicly funded agency NASA feels a duty to find the best value for its taxpayer dollars and to provide stimulation to a market that is, after all, supposed to be free and not closed.

A random sampling of recent fiscal mishaps large and small is sufficient to understand why the Office of Management and Budget listed NASA, among twenty-four federal agencies audited for adequate business practices in 2001, as one of the two organizations that not only flunked but actually performed worse than in the previous year (the Federal Emergency Management Agency was the other). The problems are so staggering and so numerous that many critics have concluded that NASA's Big Aerospace partners have been taking the agency and its management naïfs for a long, expensive ride. "If you look at Boeing, the space station is a cash cow," said Keith Cowing, a former space station scientist who edits the web sites SpaceRef and NASA Watch. "If they fuck up, they get more money. If they fuck up in Seattle [Boeing's headquarters for aircraft construction], they don't sell planes."

This blunt assessment is all too clearly borne out by fact. In November 2001, a NASA Inspector General audit report concluded that Johnson Space Center "did not justify the restructuring of the ISS contract" that shoveled an additional $404 million to Boeing while getting "little assurance that Boeing did not overstate the value" of the claimed cost increase that resulted in the extra payment. As if that weren't generous enough, the IG reported that Johnson also willfully eliminated its option to recoup any fees paid to Boeing, even if Boeing's "technical and cost performance is ultimately unsatisfactory."

In a withering report issued in June 2001 by the Senate's Committee on Governmental Affairs, the list of NASA's business failings included "mismanagement of information technology and security, poor management of contracts and major projects, and poor financial management." That report gave examples of fiscal folly in copious and excruciating detail, including the fact that the Marshall Space Flight Center spent $10.79 on the same eight-pack of AA batteries that cost the bargain shoppers at Langley only $1.47. On a somewhat larger scale, the report noted that "despite continued cost overruns and schedule slippages" Boeing was awarded $16 million by NASA in "unearned incentive fees." (NASA did, eventually, get that money back.) The Senate report's coup de grâce was to note that NASA made "a $590 million error" on its financial statements that went undetected "even by the auditors whom NASA had paid to check the agency's books." Similar issues would haunt the auditors in question—Arthur Andersen, now famous for their sleight of hand with Enron's books—for years to come.

The NASA Inspector General struck again in August 2001, this time at Lockheed over the validity of a 1998 agreement to consolidate a number of NASA-wide "space operations" under Lockheed and thus supposedly save NASA a whopping $1.4 billion over the course of a decade. However, the IG stated that NASA could not "substantiate" $62 million in alleged cost savings over the first two years of the agreement—savings the agency had confidently reported to Congress. The IG concluded that because the initial reported cost savings couldn't be justified, any overall savings promised by the contract must necessarily be in question. The Consolidated Space Operations contract is worth a healthy $3.6 billion to Lockheed and various subcontractors.

As if all this wasn't bad enough, yet another report slammed NASA in the one area in which it might reasonably be assumed to return a strong bang for the buck: informing the public about its work. NASA spends millions each year on awards, events, public forums, web sites, sponsorships, school field trips, press materials and educational activities intended to keep some flame of inspiration alive in the general populace. NASA's proposed 2004 budget for such activities was a very healthy $170 million.

But NASA scored dead last among twenty-three federal agencies in George Mason University's May 2001 report titled "Which Agencies Inform the Public?" Among other depressing thoughts, this means that the organization which inspired humanity by sending the first humans to another world was less connected to the public than the Nuclear Regulatory Commission or the Small Business Administration.

Although the above examples were extracted from the time of the Goldin administration, such embarrassments did not cease under Sean O'Keefe, although O'Keefe's hatchet-man mandate made it easier for him to bring the ax down quickly where needed, at least initially. In September 2002, O'Keefe canceled a new shuttle launch system that was more than five years behind schedule and, if completed, would have cost more than double its original price estimate. Approval of the so-called Computerized Launch and Control System was largely based on promises that it would save 50 percent over the cost of the quarter-century-old system it was to have replaced; instead, an internal assessment concluded that using the new system would cost $15 million per year *more* than the old system. The cancellation undoubtedly stung at the Kennedy Space Center, where more than one hundred civil servants worked on the project, and even more so at United Space Alliance, which had nearly five hundred contractors on the job. In shutting down the project, O'Keefe also wrote off the $273 million it had already consumed.

Still, even O'Keefe has not been immune to NASA's powerful political gravity. In the early months of his tenure, which began in December 2001, O'Keefe talked tough about possibly closing NASA centers as a means of trimming NASA's sails. Within NASA, center closures are spoken of only in whispers, much as one would speak of a dread disease, yet they have long been encouraged by fiscal purists who view NASA's ten-center system as a redundant and outmoded relic of the agency's bigger-budget, bigger-mission days. Critics have argued that NASA should not be exempt from such closures any more than the military, which was forced—albeit by an independent panel—to close dozens of bases in the 1980s and 1990s, a process that in turn refocused military priorities and spending.

The influential *Space News* offered a helpful push just before O'Keefe took office, writing in an editorial that both Glenn/Lewis and Ames "could be closed without a significant impact on NASA's major mission." The editorial repeated other familiar recommendations: privatizing the Wallops Flight Facility in Virginia, and eliminating substantial program overlaps between Marshall, Stennis, Goddard and JPL. Such moves could save hundreds of millions of dollars each year.

After a spring 2002 tour of each center, however, O'Keefe toned down the closure rhetoric and thereafter talked only quietly about the possibility of consolidating certain work at certain centers, saying on several occasions that each NASA center was "vital" to the agency's future. On his tour, O'Keefe undoubtedly received a lesson or two in the economic and political clout of the center structure; for example, the Marshall Space Flight Center (which Cowing described to me in early 2002 as "NASA's spleen," calling much of its work unnecessary and redundant) pumps an estimated $700 million into the Alabama economy each year and employs, one way or another, more than seventeen thousand people.

One of the key people assigned to help O'Keefe during his early months on the job was Courtney Stadd, who initially came to NASA in December 2000 as head of the newly victorious Bush administration's "transition team"—which meant, loosely, that he was charged with preparing the agency for the expected resignation of Dan Goldin. The transition-team job was supposed to last six weeks; by the time Goldin finally left nearly a year later, Stadd found himself as O'Keefe's chief of staff and White House liaison.

Among the military brass and number-crunchers filling the ranks of O'Keefe's NASA, Stadd was something of a curiosity: a dyed-in-the-wool, let's-colonize-the-galaxy space advocate from way back in the 1970s when he was an early member of Gerard O'Neill's L5 Society and helped O'Neill market *The High Frontier*. In the spring of 2002, sitting in Stadd's spacious suite next to O'Keefe's office at NASA headquarters, I asked him what O'Neill would think of the current state of affairs in human space travel.

"Oh, he'd be enormously frustrated," Stadd said with a faint smile. "He was frustrated even back then by the lack of willingness of the agency to break out of its straitjacket. He saw the limitations of the

space shuttle. He hoped by now we'd at least have initial habitats floating out there in space and using the surface of the moon for further
development."

"But," Stadd added quickly, "he was a physicist and not an econo-
mist or a sociologist, and I think those other elements shape the future
as much as anything. He was also a rationalist, and after I settled him
down again, he'd understand why we are where we are, and why we
hadn't gotten as far as we might have."

In conversation, Stadd is by turns smooth (on politically sensitive
topics) and giddy (on virtually anything having to do with spaceflight).
Yet he is unable to hide the fact that he seems conflicted about the stifled state of human space development, and he clearly recognizes that
NASA is at the heart of the problem. "For better or worse, and from an
entrepreneurial standpoint it's often for worse, NASA is the big enchilada. That ain't good," he admitted. "It's not good because it means the
government has an inordinate say over the business climate."

Among other jobs when he joined NASA, Stadd was tasked with
cleaning up the agency's commercial strategy in the waning months of
Goldin's tenure. He continued in that role under O'Keefe. Stadd was not
short on practice, having been given essentially the same job in the
mid-1980s with the Department of Transportation (DOT), which was
told by the Reagan administration to straighten out its regulations to
make it easier for commercial space companies to set up business (this
was in the wake of the *Challenger* disaster). The DOT today is the point
of regulatory clearance for commercial launches in America, having
centralized a process that previously required separate go-aheads from
nearly a dozen government agencies. This is due, in part, to the work
of the Federal Aviation Administration's Office of Commercial Space
Transportation, a DOT office that Stadd helped create.

With a clear expression of world-weariness, Stadd described the
seemingly eternal miscommunication between NASA and the entrepreneurial sector in terms of "angular truth"—meaning that "it's all based
on your frame of reference. I show you a picture of a launch, and that's
a truthful moment. But if you're standing over here, it's slightly different."

Stadd speaks from painful experience. In the 1980s, he was part of an entrepreneurial launch company known first as Arc Technologies and later as Starstruck. The company was an early example of a space enterprise funded by the newly minted "siliconaires" of the high-tech revolution, a trend that continues to this day: Jeff Bezos, founder of Amazon.com, and John Carmack, cocreator of the apocalyptic video games Doom and Quake, are among the newer siliconaires turned space entrepreneurs.

Starstruck's big idea was to power small rockets with "hybrid" fuel in order to send small payloads into suborbital space, cheaply. In the parlance of space propulsion, "hybrid" is a generic term for any number of liquid-solid propellant combinations whose general appeal is their combination of solid fuel cheapness and liquid fuel safety. Starstruck planned to apply hybrid technology to the sounding rocket business—not an unreasonable entrepreneurial proposition, since sounding rockets (used frequently for atmospheric research) are less complicated and less costly than orbital rockets. Once the sounding rocket business had taken hold, Starstruck planned to expand its hybrid business to larger vehicles that would, eventually, compete with Big Aerospace for commercial payloads.

For the most part, the business of sounding rockets flew, so to speak, beneath NASA's political radar screen—or so Stadd and others at Starstruck thought at the time. "NASA was not exactly friendly to it," Stadd said of the agency's response to Starstruck's attempts to secure NASA's endorsement. "They viewed us as a bunch of upstarts, what did we know? We wanted to break into a business which was quasi-monopolized by the Bristol Corporation of Canada."

Therein lay the barrier. In addition to being a quasi-monopoly, the Bristol Corporation was also a longtime NASA contractor. Even though NASA itself had no major stake in either hybrids or sounding rockets, as Stadd and countless others have learned, NASA protects its tried-and-true contractors like it protects itself. The result for Starstruck was that NASA protected the Bristol Corporation by refusing to offer its endorsement of Starstruck, which left the company vulnerable to the brother-in-law problem. Unable to attract enough investment or even NASA's tacit approval, Starstruck eventually folded.

Stadd is the first to admit that most space entrepreneurs face similar scenarios today. He also acknowledges that there is no easy fix for changing a culture that was set up from the beginning to be insular and highly defensive of its interests—not to mention being wholly unequipped to foster commercial activities. "Simply establishing an office called commercial space doesn't help," Stadd said. "If you're an entrepreneur, you can check a box, 'Yes, I visited with NASA's commercial people,' but from the standpoint of advancing the industry, we have to infiltrate the culture. The easy thing to do is an office, a box, a title. The hard thing is, how do you get people in line jobs around NASA to have a presumption in favor of the commercial sector first? That's going to take a while."

Stadd may be understating the case considerably, since judging by its public spin, the public might easily conclude that NASA is conducting as much free-market commerce as it can handle. As a commercial "brand," NASA wields tremendous if sometimes inadvertent power. Any random scan through airline catalogs, magazine advertisements or corporate brochures reveals an inordinate number of products claiming to have been either developed or endorsed by NASA (that many such claims are false is arguably even greater testimony to NASA's brand power). Remember that NASA is, after all, a government bureaucracy and then try to make a comparable list of products sporting endorsements from, say, the highly competent National Oceanic and Atmospheric Administration or even the National Science Foundation. Despite exploding shuttles and multibillion-dollar financial fumbling, the popular consciousness has retained an image of NASA as a competent, cutting-edge organization. If the synthetic stuff in your sleeping bag is good enough for NASA astronauts, isn't it good enough for you?

In terms of influencing the development of commercial products, NASA can point proudly to some successes, having either pioneered or funded the development of hundreds of processes and products that have found their way into medicine, manufacturing and other industries (contrary to popular myth, however, NASA did not invent Tang, Teflon or Velcro; all of these items existed before they became associated with NASA). But what NASA touts as actual commercialization or privatization is typically neither, as evidenced by NASA's hand in creating the United Space Alliance (USA).

After USA was created in 1998, NASA pointed to it as an example of cost savings provided by successful privatization and competition. While the process that produced USA was competitive in the broadest sense of the word—proposals were solicited, received and reportedly reviewed—few doubted that Boeing and Lockheed would come out on top, since they'd had most aspects of the shuttle's operation locked up for some time. As for any actual competition sparked by the creation of USA—a major motivation for the creation of a private shuttle management entity—the RAND Corporation described USA's de facto monopoly on shuttle contracts as "profoundly noncompetitive"; though all of the major shuttle contracts are managed under USA's guidance, they all belong to either Boeing or Lockheed. In other words, one way or another Boeing and Lockheed simply manage Boeing and Lockheed, albeit in a dazzling array of configurations.

Even granted that a dollop of competition was involved in the United Space Alliance's creation, the common definition of privatization is taking a publicly developed asset (like the shuttle) and giving, selling or leasing it to private entities to do with as they see fit. However, USA only operates the shuttle; NASA still owns it and controls its use. This comes as no surprise; as the congressional space analyst Marcia Smith said to me when I spoke to her in March 2002, "If you were to truly privatize something like the shuttle, the private sector can decide to walk away from it." Therein lies the rub. "NASA is dependent on the shuttle, since they need to have something to go to the space station," Smith said. "In the sense that the private sector owns it, they could charge whatever they wanted and the government has to buy it."

Even more to the point, if the shuttle were actually privatized in the truest sense of the term, someone other than NASA could walk in with the requisite half-billion dollars and "buy" a shuttle mission from USA. In the worst-case scenario for NASA, a truly private owner might listen to critics and ground the shuttle vehicles altogether, perhaps cannibalizing them for technology to create newer, better spacecraft.

Strangely enough, when NASA truly hands technologies and operations to the private sector, as opposed to creating a quasi-private caretaker like the United Space Alliance, the results tend to be positive. In 1962, the year John Glenn became the first American to orbit the Earth, NASA spun off control of commercial communications satellites to the then new Comsat Corporation, an act that sparked the global communications revolution and created an industry generating more than $100 billion in annual business worldwide. Likewise, global positioning—in which signals are bounced between satellites and ground receivers to provide precise terrestrial locations for objects and/or people—began as an exclusively NASA-military proposition and was offered to the private sector beginning in the 1980s, albeit in a version with somewhat less accuracy (the better to prevent folks deemed unfriendly from having cheap access to military-grade technology). Since then, GPS has become integral to automotive companies, the long-haul transportation industry and sporting goods manufacturers. In the future, it may be common for cell phones to have GPS capacity.

Satellite imaging is another technology that NASA and the military successfully handed over, at least in part, to private industry. The private sector did so well with imaging that the government returned as a customer. Shortly after the September 11, 2001, terrorist attacks in America, the U.S. government paid Space Imaging, which operates the for-profit Ikonos imaging satellite, nearly $2 million on a monthly basis for exclusive rights to all Ikonos images being taken of Afghanistan and surrounding areas. The purchase, meant to supplement the government's imaging and keep Ikonos pictures out of unfriendly hands, was considered to be a landmark event for the imaging industry. In another era, the government might have exercised what is commonly known as "shutter control" on Ikonos, simply terminating the company's ability to operate for as long as it saw fit. Instead, the government chose to use Ikonos just as the company's private customers did.

While NASA voluntarily spun off communications satellites, GPS technology and orbital imaging, commercial space launches had to be wrested away from the agency, so central were they to NASA's very identity. Even so, the result was similar: the American commercial launch industry survived *Challenger* only because the Reagan administration's Commercial Space Launch Act, enacted in 1984 and amended several times since, laid the groundwork for prizing commercial and military payloads from NASA's viselike grip, with the agency finally forced to let go after *Challenger*. Letting go of human spaceflight, however, has been a step too far for NASA; any commercial benefit that might result from such a move remains, as such, only speculative.

At the same time, there has been no shortage of analyses, assessments, testimonies and reports about how the future of human spaceflight might shape up under a truly commerce-driven agenda. The popular human spaceflight market seems particularly ripe for development. A 1998 study done by NASA and the Space Transportation Association concluded that the "space tourism" market by around 2020 could generate up to $20 billion annually. This conclusion was based not only on interest in America, but on the conclusions of extensive surveys done in Japan, the United Kingdom and Germany.

In 2001, Andrews Space & Technology released its "Future Space Transportation Study," a NASA-funded report to assess the market for "S-commerce" as an impetus for the development of new reusable launch vehicles. The Andrews report drew many of the same conclusions as previous studies: that the potential for space-based private enterprise is huge, that players are already assembled in the entrepreneurial sector to exploit its potential and that prospective customers for a panoply of space-based enterprises are practically waiting for such enterprises to be offered to them.

In the end, however, the Andrews report reiterated the chicken-and-egg problem of space enterprise, whatever its form—that most space-business concepts won't fly, literally or figuratively, if the cost of sending stuff into space is much above $1,000 per pound. With launch costs below that threshold, however, possibilities for space commerce abound—a conclusion not only of the Andrews report, but also of the landmark study of which the Andrews report was a direct descendant.

In 1994, the aerospace titans Boeing, General Dynamics, Martin Marietta, McDonnell-Douglas, Lockheed and Rockwell released the results of an exhaustive study of current and potential space markets in a three-inch-thick tome called the "Commercial Space Transportation Study." The companies involved in the study dubbed themselves the Alliance and briefed Dan Goldin that April; the report is said to have played a significant role in subsequent commercial thinking about the International Space Station, and is also credited with helping to spark the Space Launch Initiative.

Like the Andrews study, the Alliance report sought to use a tantalizing end—big business in space to produce big bucks on Earth—to justify an expensive means: the development of a new space transportation system to succeed the shuttle, in which Alliance companies would undoubtedly have a major stake. Echoing the early drumbeating that led to the shuttle's approval, the Alliance report said that a commercially focused, next-generation space transportation system would bring launch costs down to levels that would actualize the multibillion-dollar commercial potential that had long been predicted for space.

The Alliance report was far-reaching and uninhibited, albeit often in scary ways. The report suggested that the moon would make an excellent nuclear waste dump and called lunar nuclear waste disposal "a huge market attainable . . . using near term technologies" if launch costs could be brought down to around $500 to $600 dollars per pound. The report did acknowledge that the explosion of a rocket containing tons of radioactive waste could have a few unpleasant consequences here on Earth, and gingerly noted that there might be "political issues" to be overcome: would you want a rocket-powered nuclear garbage scow launched from your neighborhood? (The report failed to mention that the moon-as-nuclear-dump scenario was the foundation of *Space: 1999,* a cult favorite 1970s television series in which an explosion of all that nuclear waste blasts the moon and those occupying its base out of Earth orbit and onto a multi-episode tour of the galaxy.)

On page after page, the Alliance report ticked off one fascinating space commerce possibility after another. With launch costs down to around $400 per pound, an orbiting television and film studio would be a worthwhile business proposition (this is one of the ideas MirCorp was promoting for Mir). If launch costs came down to about $100 per pound—close to the ultimate cost target of Kistler's K-1—many things became economically viable: various forms of manufacturing and product testing, facilities for "high-density and high-intensity agricultural production," and even orbiting health and medical facilities, which would be particularly useful for heart disease and burn victims whose recovery would, according to doctors consulted for the Alliance report, benefit greatly from a microgravity environment.

Space theme parks—that is, a theme park actually in space—were also a $100-per-pound proposition as was faster-than-overnight package delivery, with the report citing a thumbs-up to this idea from the FedEx and Emory executives interviewed for the study. The Alliance report even gave a nod to O'Neillian space settlements at this level, although despite lower launch costs such facilities would require "other very large in-space business activities" to support their creation.

What is perhaps most unexpected about the Alliance report was its outright dismissal of NASA as a major player in the future of most of these activities, saying that the agency's modus operandi is "totally incompatible with commercial ways of doing business." The Alliance members were, after all, NASA's biggest commercial contractors; in fiscal year 2000 alone, they collectively consumed close to $5 billion in NASA funding. But the Alliance report concluded that, at best, NASA should act as an administrative regulator or, particularly in the case of science payloads, as a customer, not as an owner or operator (and since the Department of Transportation already regulates nonhuman commercial launches, even NASA's potential regulatory role would be significantly diminished). As for the shuttle, it didn't figure at all in the Alliance's thrilling space future.

But NASA doesn't really need the shuttle to stay firmly in the commercial space game. In many respects, the agency wields more power than ever in the commercial space arena, despite having no real commercial mandate and, as of this writing, not even a dedicated department or office of commercialization at NASA headquarters. What it has instead is arguably more important, because for all intents and purposes, NASA owns the International Space Station—which, since the demise of Mir, is the only place in orbit left to do business in at all. Whether true commercialization, rather than a NASA-dictated version of such, can claim a foothold there could determine the course of space commerce—and by extension, human presence in space—for decades to come.

In early 2001, NASA attempted in the quietest possible way to slip a budgetary time bomb under the door of the Bush administration, which had barely begun to unpack its boxes after coming to power.

The International Space Station, only partially completed, was going to need an extra $4.8 billion over its cost cap just to reach minimal operability. Unfortunately, this was only the latest jolt of ISS sticker shock, since the cost of finishing the station had already jumped by $4 billion in the prior three years.

The sheer magnitude of the station program makes the shuttle look like a mom-and-pop operation. In support of no more than three crew members who occupy a space about the size of a three-bedroom house (although if finished, the ISS will have three times as much room), NASA and its partners employ an estimated one hundred thousand people in sixteen countries, including the employees of about five hundred contractors. An average ISS crew, which stays at the station for several months, consumes about six thousand pounds of supplies. Assembling, supplying and crewing the station have, as of the end of 2003, required sixteen shuttle flights and about twenty flights by various Russian spacecraft. It is by any standard an incredible undertaking.

It is also, by any standard, incredibly expensive; the station has blown every budget attached to it, as far back as its days under Reagan. Most estimates for the ultimate cost of the station, including future operation, are comfortably over the $100 billion mark. Thus, far from being received quietly, the 2001 news of a new cost overrun of such magnitude exploded all over the media and set off any number of congressional hearings, several of which involved Sean O'Keefe while he was still deputy director of the Office of Management and Budget. In that capacity, O'Keefe was called before a congressional committee in May 2001 to offer recommendations on how to deal with the station's ballooning budget.

The question O'Keefe was being asked to address during the May 2001 ISS hearing was, in essence, what could be done to halt the station's seemingly chronic fiscal hemorrhaging? O'Keefe began by praising the station as a marvelous technological accomplishment, then he laid into NASA for the Keystone Kops management practices that allowed the station's budget to spin out of control. He said that while funds had been added repeatedly to the ISS budget, that budget had continued to skyrocket even as development lagged. Pointedly, O'Keefe noted that "adding funds may have made it too easy to avoid consideration of lower cost alternatives and make tough decisions to manage within the budget."

The result, O'Keefe summarized dismally, was that NASA estimated it was six years and $14 billion from completing the station—the same estimate the agency had made in 1996, five years earlier. O'Keefe's recommendations on how to bring the station and its costs under control included a number of sensible strategies to improve budgeting and oversight procedures, most of which are taught in the average Management 101 course. But O'Keefe's broader conclusion about NASA's management practices was more direct: "Business as usual must come to an end."

Before the year was out, O'Keefe the NASA administrator found himself mired in a squishy political bog created by trying to live up to that credo. The Bush administration refused to increase the station's budget cap, as O'Keefe knew would be the case, since O'Keefe himself had encouraged such a recommendation (which was furthered by the budgetary reshuffling that took place after the September 11 terrorist attacks). Thus NASA was forced to find other ways to keep the station moving forward while living within its means.

Under O'Keefe's management, the ISS went on a crash diet. Shedding hundreds of millions of dollars in a matter of months, the ISS quickly began to bear an uncomfortable operational resemblance to Mir, if not Skylab. Gone was a dedicated habitation module, seen as a crucial component to make the station truly livable (right now, the ISS allows for only three residents, each of whose private space is roughly equivalent to a single-person tent). Also gone was a commitment to complete the station at all; a three-person capacity, dubbed "core complete," might be as much as NASA and its political masters would spring for.

The result was a mob of angry constituents from every corner of NASA's world: international partners livid that they were being asked for more money to keep the ISS afloat while seeing crew time for their astronauts slashed, scientists concerned that less station capability would mean less station science, politicians who were sure that they'd paid for more than what they were getting and a press corps that lapped it all up and spat it back to an increasingly incredulous public.

The station's future looked even bleaker after the loss of *Columbia,* since some of the most critical remaining modules—including Japanese, Russian and European research facilities—need the shuttle to get into orbit (although in theory, they could be retooled for other launch vehicles like Russia's Proton, although at such expense that they might be better off starting from scratch). Thus, without the shuttle, completing the ISS might not be possible, even if NASA did commit to it.

Then again, a smaller station might be all the ISS partners can handle for the near future. In early 2002, NASA halted development of a crew return vehicle prototype called the X-38, replacing it later that year with the Orbital Space Plane concept. However, the OSP won't come into use until 2008 at the earliest, and since the Soyuz capsules currently used as escape pods seat only three, sole reliance on them as a means of emergency crew return necessarily limits the permanent station crew to that number. Also gone until at least 2008 is any reasonable hope of conducting substantial scientific research on the ISS, since it takes about 2.5 crew members just to keep the station running properly on a daily basis.

Scientific research, of course, is the most loudly trumpeted reason for the ISS's existence. Yet the value of the station for science has been the point of major debate since the days of the Industrial Space Facility. In his book *Voodoo Science: The Road from Foolishness to Fraud,* the physicist Robert Park (who also writes an excoriating, and usually funny, e-mail newsletter on science, politics and the media called What's New) says bluntly that the space station "cannot be justified on scientific grounds" and lumps it in with cold fusion and perpetual-motion machines as an example of high-tech hocus-pocus trying to pass as scientifically valid enterprise. If you want to conduct microgravity experiments, Park and others have long argued, you're better off doing so in an unmanned environment (like the Industrial Space Facility) that is free of the inevitable jostling by humans padding around in a space station.

In 1991, the influential American Physical Society (APS), which represents more than forty thousand physicists, issued a policy statement which is still in effect and which states that "the potential contributions of a manned space station to the physical sciences have been greatly overstated." Along with the American Physical Society, the American Society for Cell Biology and the American Crystallographic Association—whose members presumably would provide many of the biological and chemical experiments for the ISS—are among the scientific groups on record as being skeptical that much valuable science will happen aboard the ISS. As a *Florida Today* article put it, "It is still hard to find a scientist outside of NASA who expects much progress from station research."

But perhaps the most disappointing aspect of NASA's response to its ISS budget woes was its obstinate refusal to consider commercial solutions as a real alternative to making the station more viable. There was an effort along these lines in late 2001, when for the first time in its history, NASA actually put on paper a comprehensive, agency-wide commercial development plan. NASA's "Enhanced Strategy for the Development of Space Commerce" proposed a variety of sensible options for commercial development of the ISS, including turning over management of some activities and operations of U.S.-owned sections to organizations not a part of the government.

The document said many of the things entrepreneurs had been longing to hear and included vows to "simplify management processes for commercial activities" and to stamp out "inappropriate competition between for-profit commercial companies and non-profit or governmentally subsidized entities." The report even acknowledged human space travel as a "new area of commerce," although it stopped well short of recommending that NASA begin selling seats on shuttle missions or even encouraging more Titos and Shuttleworths as paying ISS guests.

This particular version of NASA's commercialization manifesto didn't make it past the draft stage, mainly because it was almost immediately ripped to shreds by the very sector it was intended to encourage. "What they're talking about is NASA actually engaging in commercial business," Pat Dasch, then executive director of the National Space Society, told CNN in response to the report. "But we don't think NASA engaging in commercial deals is what this is supposed to be about. NASA is not supposed to be bailing itself out."

Rather than finding ways to turn over key activities to the private sector or use NASA resources to facilitate entrepreneurial opportunities, the plan's language seemed crafted mainly to give NASA as much control over current and future space commerce activities as possible. "NASA will dedicate Agency resources in partnership with private sector ventures when these activities advance substantial NASA mission objectives," read part of the plan's vision statement. The plan also sought to ensure "that the conduct of commercial space activity is consistent with the safe and successful accomplishment of the NASA mission." Such statements in and of themselves aren't necessarily unconstructive, but Dasch (who joined the board of the Space Frontier Foundation after stepping down from her NSS post) saw them as portents of how NASA was planning to handle the commercial development of the International Space Station. Once again, it was to be NASA, Inc.

Even Courtney Stadd, who led the development of the maligned ISS commercialization policy draft, said NASA was not necessarily the best organization to direct business in a vacuum. "We're unreliable," Stadd admitted. "Ask me how much I'm going to charge you to use equipment on the space station. I don't know. Challenge me: ask me, within certain limits of reliability, if you have a particular experiment today what are your chances of getting up to the space station in a certain period of time? Can't do it. So the fundamentals that are necessary for stability and predictability in the business environment are not there."

NASA has made an effort to lay down cost limits for setting up shop on the ISS, yet even so the answer to the question "How much does it cost to do business on the International Space Station?" is somewhat nebulous—"a lot" is probably the best answer. In 1998, NASA released its Commercial Development Plan for the ISS and subsequently set up a web site to encourage entrepreneurs to jump aboard. "The strategy," according to the web site's introduction, "is for NASA in partnership with the private sector to initiate a set of pathfinder business ventures that can achieve profitable operations over the long run without public subsidies. These business pathfinders will break down public sector and market barriers in the near term and lead to economic expansion over the long run."

Taken in the most generous spirit, this is an admirable sentiment. Yet the execution of this concept has been anything but smooth. First there is the issue of the ISS's rather prohibitive pricing for commercial activities. To secure a standard research slot on the ISS for a year costs a cool $20.8 million. For this price, a paying customer gets a dedicated niche on the station, 2,900 kilowatts of power, 86 hours of astronaut time, and 2 terabits of data download. That tidy sum does not include possible charges for additional crew time (at $15,000 per hour), additional power (at $2,000 per kilowatt-hour) or any number of other "premium" services that could drive the cost considerably higher. Most critically, the price does not include transportation to and from the station via the shuttle, for which you can expect to pay the customary $10,000 per pound—in each direction. (On the subject of transportation, NASA graciously allows that shuttle transport is priced separately to allow for "privately offered space transportation services in the future.")

If you, as an entrepreneur, have somehow managed to convince your investors to swallow such costs (plus the cost of developing whatever you're planning to send to the ISS in the first place), there is still the small issue of when your undoubtedly time-sensitive research will actually be sent aloft. Alas, in the Terms and Conditions of Use, you will find the following troublesome disclaimer: "Timing of use is negotiable, subject to the inventory of resources and accommodations available at each phase in the ISS program." *Columbia,* of course, made this caveat even more unnerving.

The response to the nominal opening of the ISS for business was not exactly overwhelming. As of 2003, the ISS commercialization web site listed only three signed agreements with industrial partners (remember, the ISS commercialization policy was released in 1998). Of these three partnerships, however, one is actually a ground-based test of NASA hardware designed to mimic a weightless research environment, and another fizzled out due to lack of funding before anything became spaceborne.

To date, only one agreement on the ISS commercialization site has actually flown: an experiment created by the Baltimore-based biotech company StelSys to learn how human liver cells process medicine in space, which flew for a short ISS stint in late 2002. Notably, the StelSys experiment was virtually indistinguishable from the two dozen or so other small biology-themed experiments it accompanied at the time, all of which were conducted by government-funded entities (universities, research labs, NASA centers) and which were of the general scope and complexity of experiments flown on the shuttle for years.

If nothing else, however, the StelSys flight proves that NASA will actually fly and operate commercial experiments on the ISS, albeit for a very hefty price (the total cost of the StelSys package was not made available). What NASA will not do is to give a leg up to promising commercial enterprises that do not have the ready cash to use the station. NASA says as much when addressing the issue of "unsolicited proposals" for ISS commercial projects that are speculative and would require NASA support at some level. "NASA has no currently appropriated funds, nor does NASA anticipate future appropriations, for funding commercial ventures," the ISS commercialization site states.

Into this category fall the vast majority of endeavors—launch vehicles, entertainment and media ventures, popular human spaceflight—that stand the best chance of opening the near-Earth space marketplace. The lesson of ISS commercialization efforts to date is that NASA is playing a familiar theme by appearing to do what is required by the letter of the law (the law in this case being the 1998 Commercial Space Act, about which more will be said later), but in action protecting NASA interests as much as possible to the exclusion of anything resembling a threat to the status quo. Courtney Stadd confessed that NASA is not "in the business to subsidize or otherwise ensure the success of the entrepreneurial folks."

For all their feisty independence, it would be untrue to say that most space entrepreneurs wouldn't love to get substantial NASA subsidies to kick-start their work. Said Mitchell Burnside Clapp, "If NASA simply said, 'Look, we're going to set aside a billion dollars a year to launch sand into orbit'—or mass of any description—and announced they'd pay, say, five hundred dollars a pound for that, you could begin to get real investment. That's a market, even if it's jury-rigged. The point is that once the capabilities exist, a real market would have a chance."

Many entrepreneurs point to the Kelly Air Mail Act of 1925 as an analogous precedent for what NASA might do to open up commercial opportunities in space. Under the Kelly Act, the U.S. government gave critical subsidies and incentives to the fledgling commercial airline industry in the form of contracts for delivering mail, even though air-mail at the time was virtually untried and definitely unprofitable. Helping the entrepreneurs along was a highly competent federal R&D office that provided critical technology assistance to nurture commercial airline entrepreneurs into a position to be competitive for Kelly Act money. It is worth repeating that this R&D office was the National Advisory Committee on Aeronautics, or NACA—the organization from which NASA was created.

In the years immediately following the Kelly Act, most mail was still ground delivered, but the promise of future business and infusions of government cash was sufficient for entrepreneurial companies to attract enough investment to get their ideas off the ground. An updated version of the Kelly Act in 1934 broke up burgeoning aircraft industry monopolies that had already begun to consolidate the air transportation and service business under as few roofs as possible. United Aircraft and Transport, for instance, was broken into three parts: Boeing, United Airlines and United Aircraft, with the latter eventually consumed by its core business, the engine developer Pratt & Whitney.

Unfortunately for space entrepreneurs, the space industry today is far more evolved than the air-travel industry was in 1925, and space business monopolies have been solidified by government actions rather than broken up by them. In the meantime, NASA has demonstrated a talent not only for strangling nascent competition in the cradle, but also for using subsidies to bring more mature entrepreneurial ventures close enough to kill them as subtly as possible.

Where ISS commercialization is concerned, the thorn in NASA's side is Russia, whose space leaders have demonstrated a remarkable willingness to put ISS-oriented commercial schemes into action. The Soyuz that delivered Dennis Tito to the ISS also made what was billed as the world's first fast-food delivery in space: a six-inch pizza with salami and tomato sauce, the product of a million-dollar promotional campaign courtesy of Pizza Hut (in 1999, Pizza Hut slapped a forty-foot-long banner on the Russian Proton rocket that carried the ISS Zvezda module into orbit, reportedly paying another million dollars for the privilege). In fact, Tito's Soyuz was a veritable delivery van of promotional products: in addition to the pizza, the ISS gained copies of *Popular Mechanics* (courtesy of the magazine), talking picture frames from RadioShack and two hundred LEGO toys that were later given to kids around the world as part of an advertising campaign (the LEGO deal was organized by Space Media, Inc., a subsidiary of Spacehab, the company founded by Bob Citron).

It may come to pass that Russia's commercial zeal will help bail out the ISS, and NASA. There has been serious talk about a commercial "module," to be built by Spacehab and Russia's Energia, serving as temporary living quarters should NASA and its international partners not find the money to build the station's living space beyond its three-person capacity. Once the ISS is completed, each of the partners will be able to do with their portion of it what they choose, be it state-sponsored or privately funded. Given their aggressive promotion of the tourism market, it comes as no surprise that the Russians are planning to give over a significant portion of their ISS real estate to commercial activities; the Spacehab/Energia module, appropriately dubbed Enterprise, is intended to be the flagship of those efforts. Once in place, Enterprise would be open for microgravity research, multimedia activities (filming products in space, shooting weightless sequences for movies, etc.) and, yes, tourism.

One might reasonably conclude that NASA would jump at the chance to have somebody else pay for the development of a habitation module for the ISS. Instead, the agency's response to the crew-capacity crisis was to explore government options for making this happen, asking Italy and France, among others, if they might see their way to coughing up more public funds (the answer thus far has been no). NASA's relative lack of interest in a commercial habitation module has meant rough going for Spacehab and Energia in their attempts to secure the money needed to get their module built and launched. Once again, it seems, NASA would prefer to be owner, not customer.

In mid-2001, MirCorp announced that it had struck a deal with Energia—seemingly ever ready to consider entrepreneurial space schemes—to design and build a small private space station that would open for business much as a commercialized Mir would have done. Originally, the station was to be called Mir 2, but when this name proved a bit too tender for the Russians, MirCorp gave it a more prosaic moniker, Mini Station 1. The station would be developed to accommodate up to three humans for stays of twenty days, with an operational lifetime of fifteen years. Its development cost is estimated at about $100 million.

The Mini Station 1 was announced with great fanfare, lots of media coverage and not a negative peep from NASA. I asked Jeff Manber how this had come to pass, and he had the most unexpected response: "This could not have happened without NASA's consent." Manber said he and other MirCorp leaders calmly met with NASA's leadership and essentially got their assurance that the agency wouldn't try to shoot down the private station concept as they had torpedoed the privatized Mir. Another private space station effort, led by the Las Vegas hotel mogul Robert Bigelow, actually received some technical advice from engineers at Johnson Space Center in the full light of day (Bigelow, a colorful figure who also very publicly funds UFO research, is hard to miss).

Where the convergence of NASA and entrepreneurial interests is concerned, there are other glimmers of hope. In 2002, Alan Ladwig moved from Team Encounter to a venture that is closer to his heart in terms of getting regular people into space. Zero Gravity Corporation, of which Ladwig is now chief operating officer, is looking to take a bit of space tourism business back from the Russians by becoming the first company to offer private parabolic flights over American soil.

Parabolic flights fly in roller-coaster arcs, producing moments of gravitational equilibrium that, for a few seconds at a time, simulate microgravity weightlessness. Astronauts and cosmonauts have trained for space on parabolic flights for decades. Several companies have contracted with the Russians to sell these thrill rides, but NASA has long controlled parabolic flights in America, making them available only to NASA astronauts (or film studios that strike deals with the agency).

Founded by the X Prize impresario Peter Diamandis and the former shuttle astronaut Byron Lichtenberg (who cofounded the X Prize with Diamandis), Zero Gravity is following Bob Citron's staffing model and packing its ranks with top-level ex-NASA talent, the better to thwart the brother-in-law problem. In Zero Gravity's case, that means not only Lichtenberg and Ladwig, but also Robert Williams, who managed NASA's KC-135 zero-gravity training program at Johnson Space Center for two decades and is probably the single most experienced person in the parabolic-flight business outside of Russia. By early 2003, Zero Gravity had already received FAA approval of its specially refurbished 727 for parabolic flying, inked its first major contract (with a Hollywood studio) and begun signing up individual customers.

More intriguingly, Ladwig has talked with NASA about outsourcing some of the agency's parabolic training to Zero Gravity—which, if it happens, would be an entrepreneurial first where human spaceflight is concerned. Because of Sean O'Keefe's cost-trimming drive, Ladwig said managers at NASA headquarters have been warm to the idea of Zero Gravity's taking on NASA parabolic training as a baby step in the novel idea of competitive outsourcing for human spaceflight activities. But there is resistance. "The managers of the KC-135 program at [Johnson Space Center] are—surprise, surprise—not overly enthusiastic on the notion of giving up the [parabolic training] function to the private sector," Ladwig wrote to me. "It's all about control."

In June 2003, NASA announced that Courtney Stadd would be leaving his NASA post to move back into the private sector. During his final days at NASA headquarters, Stadd sent me several long and enthusiastic e-mails about an endeavor on which he'd been working "overtime" yet which he admitted "sounds boring and irrelevant to entrepreneurship and exploring the high frontier."

Indeed, on its surface, the details of NASA's gradual transition to a fiscal practice known as full-cost accounting are fairly yawn-inducing unless you're a professional number juggler. Traditionally, the cost of most NASA activities—be they shuttle launches or ISS module development—reflects only the contractor cost, not the integrated cost of the NASA civil servants and various administrative and facility costs that provide the framework in which NASA-contracted projects actually get done. Such costs are dispersed across a variety of internal budgets and occasionally even dumped into the budgets of unrelated projects, thereby making these costs virtually invisible, sometimes even to NASA's accounting staff. The advantage of such an accounting system is clear: in the eyes of auditors and Congress, it makes any given project seem more affordable than is actually the case.

Full-cost accounting would pull the veil off such fiscal hocus-pocus by including the cost of every item, service and facility required to bring a given project to fruition, regardless of where such costs lie. Needless to say, price tags for most NASA projects and programs would appear even fatter than they do now.

"At the very least, [full-cost accounting] should flush out what the real costs are and thereby promote a more honest debate re trade-offs the government is willing to make in terms of going with a given commercial offer vs. standard government approaches," Stadd wrote to me in June 2003, although he is under no illusion that this will necessarily bring NASA to heel where its general aversion to competitiveness is concerned. "Will the rice bowls in government seek other ways to play a shell game? Of course," Stadd said. "Whether any of this will make any difference, only time will tell."

Questions without clear answers abound where NASA and its relationship with the private sector are concerned. But regardless of whether or not full-cost accounting ushers in a new era of fiscal credibility for NASA or Zero Gravity Corporation gets a NASA contract or MirCorp's Mini Station 1 ever gets built, the fact that such diverse activities have at least tacit NASA approval signals one of two things: either the "rice bowls" within NASA don't find them threatening (for now), or the agency's leadership is beginning to recognize that when it comes to the future of the space frontier, if it's not going to lead, it had at least better get out of the way.

Welcome to the Revolution

Whatever their form, revolutions begin less often with a gunshot than with a conversation—with an idea expressed in a way that is so compelling as to seem inevitable. Every spring on Capitol Hill—that crazy mixture of idealism, Machiavellian wheeling and dealing, and big, big money—the alternative space movement gets a chance to further its own revolution, its opportunity to plant a seed of doubt about the way things are done now and to encourage even the faintest interest in a new idea that might, someday, change the whole future of humans in space.

This is MarchStorm, the annual raison d'être of a nonprofit lobbying organization called ProSpace, which since 1995 has worked to pry open the final frontier with the slippery lever of political influence. The alternative space movement has taken its case to Capitol Hill from its earliest days, through private meetings with sympathetic space fans in Congress and testimony before key committees and occasionally through old-fashioned, ad hoc ranting and raving. But as the movement has matured, so has its political approach. ProSpace and MarchStorm are perhaps its most evolved incarnations.

To date, the biggest ProSpace coup has been the Commercial Space Act, which ProSpace is given broad credit for having pushed through Congress toward its signing into law by Bill Clinton in 1998. Among other mandates, the Commercial Space Act required NASA to demonstrate progress in fostering commercial opportunities for the International Space Station and to commit to studying options for privatizing the space shuttle. It also required the U.S. government to buy available space science data from the private sector, thereby putting some official weight behind entrepreneurial efforts in a variety of areas, from launch vehicles to satellite imaging.

The Commercial Space Act is not bulletproof; for example, the NASA administrator can, at his or her discretion, simply decide that a NASA payload must go up on the shuttle. One provision of the Commercial Space Act is comical in that it was required at all. The act made it legal for commercial space vehicles to reenter Earth's atmosphere and return space payloads to American territory, obviously a key prerequisite for reusable launch vehicles. Prior to this provision, any commercial reusable launch vehicle in America could go up, but it couldn't legally come down.

In theory, the Commercial Space Act can be interpreted to mean that even human spaceflight must be provided by the private sector, subject to the appropriate clearances and regulations. The act's language makes no distinction between inanimate and live cargo, stating simply that the term "payload" means "anything that a person undertakes to transport to, from, or within outer space, or in suborbital trajectory." In reality, though, where human spaceflight is concerned the Commercial Space Act merely sets some rules for the future, since the shuttle remains the only vehicle on offer.

Whatever its limitations, the Commercial Space Act did make the NASA/Big Aerospace leviathan move. There is an ISS commercialization effort under way (at least on paper) and NASA has made semi-serious efforts to consider true shuttle privatization, although the agency has recently backed away from that term and now talks instead about "competitive sourcing" (and even that phrase was used with less frequency after *Columbia*). Through its MarchStorm agenda and its frequent one-on-one and roundtable interactions with Washington lawmakers, ProSpace views its role as keeping up the pressure on NASA in these and other endeavors. In the larger picture, it seeks to nudge Washington further forward in loosening NASA's stranglehold on the space marketplace.

The person who connected me with MarchStorm and ProSpace was Eric Dahlstrom, whose experience down the rabbit hole of the space establishment is emblematic of why MarchStorm draws so many former NASA employees, contractors and consultants. Dahlstrom, in fact, has been each of these at some point in his space career. Now in his mid-forties, Dahlstrom—somewhat shy yet possessed of a barbed sense of humor—runs a Washington-based space consultancy; his wife,

Emeline, is a marketing manager at Space Adventures, the astro-tourism company that helped to organize Dennis Tito's time in Russia. The Dahlstroms were among those who watched Tito's pioneering flight in Kazakhstan. Thrilled at seeing so much Russian space hardware up close, Dahlstrom wore a wall-to-wall grin during most of the trip.

Dahlstrom's space advocacy work began while he was still a university student studying astronomy. Fresh from reading *The High Frontier,* Dahlstrom founded the Maryland Alliance for Space Colonization and soon met the future space advocates Peter Diamandis and Todd Hawley in a caravan of young O'Neillians who journeyed to Florida to see *Columbia*'s maiden voyage in 1981. Diamandis would go on to found the X Prize and Zero Gravity Corporation, and he and Hawley would help found the well-regarded International Space University with major funding from Walter Anderson and promotional support from Rick Tumlinson.

For his part, after *Columbia*'s first flight and all it promised, Dahlstrom found himself far less interested in being an astronomer and fairly desperate to be part of the next big NASA project on the horizon: a giant space station that was being billed as the jumping-off point for journeys back to the moon and missions to Mars. In 1984, the year Reagan announced Space Station Freedom, Dahlstrom was still a graduate student at the University of Maryland when he found an unlikely entry point to this grand venture, a job at NASA's Goddard Space Flight Center developing the specifications for a database to help manage the station's engineering design process.

Dahlstrom laughs about this now, as well he might, considering that he had practically no experience in database design when he got the job—he and a friend applied with a "Why not?" philosophy, never really expecting to be successful. In retrospect, however, Dahlstrom recognizes this event as his first instruction in the cynical, counter-intuitive nature of NASA politics.

"They set up the whole project to fail, consciously," Dahlstrom said to me when we talked in July 2001. He was obviously still bewildered by the episode. "They were told to try out some low-cost options for the project, like grad students, but the NASA managers were just waiting for it to fail so they could go back and say, 'OK, we need to include our higher-priced engineers.' "

Thrilled to be working on the space station whatever the reason, Dahlstrom set out to make good on what he had been hired to do. He boned up on every aspect of the station's early evolution and spent late nights cramming on database development as if preparing for a battery of university exams. In the end, Dahlstrom said proudly, "We kept the NASA guy happy," to such an extent that the database requirements Dahlstrom's team put together made Goddard's proposal the prototype for the system that would go out to industry for a quarter-billion-dollar contract, won by Boeing—which then decided it didn't need Dahlstrom. "They said, 'You're not in information systems,' " Dahlstrom said with a wry smile. "I couldn't argue with that technically, but by that point I had written or at least helped write the requirements that made Goddard the model on which Boeing's contract was based. It was a little discouraging."

From there, Dahlstrom's space career would increasingly resemble a Carrollian journey through the projects and problems that made NASA what it is today. While not wanted by Boeing, Dahlstrom's hard-won space station expertise was nonetheless a valuable commodity that he ultimately put to use for Lockheed as part of its space station work at NASA's Langley Research Center. At Langley, Dahlstrom was involved in five redesigns of Space Station Freedom before the program was officially canceled in 1993 and reborn that same year as the International Space Station.

Even before that transition, Dahlstrom saw that political requirements were literally shaping the station's future. "In one of the redesigns we asked, 'What is the space station designed to do?' and they gave us a list of sample science missions and it was sorted by congressional district. This was the headquarters view of what science was supposed to be done."

It may have been easier for Dahlstrom to weather the political storms he dealt with at Langley, since an earlier experience had already broken his faith in NASA. Just prior to joining the Lockheed space station team, Dahlstrom worked with the Planning Research Corporation on a contract with NASA headquarters to assess the likelihood of another *Challenger* disaster and to develop a series of "lessons learned" from the incident that could be applied to future large-scale projects.

By the time this report, titled simply "Lessons Learned from *Chal-*

lenger," was submitted in March 1988, any number of analyses had already been conducted on the disaster, including the famous Rogers Commission report that contained Richard Feynman's searing comments. But while the Rogers Commission had a presidential mandate— meaning that its findings could not be easily ignored—it was an external assessment of the *Challenger* disaster, not an internal analysis. The study Dahlstrom contributed to was to be the agency's official in-house soul-searching, and therefore was to serve as the real foundation from how NASA would move forward not only with the shuttle program, but also with the space station. The report was initiated in the first place specifically to apply *Challenger*'s lessons to the station's development.

The sixty-six-page report that Dahlstrom and his colleagues submitted to NASA headquarters (through the agency's safety division) did not include any new investigations, but instead synthesized the work of the Rogers Commission, several congressional inquiries and even other internal NASA studies. For the most part, the report mirrored the recommendations of the Rogers Commission: tighter communication, management, testing and reporting standards were necessary to avoid such a disaster in the future. But recommendations were less what consigned "Lessons Learned from *Challenger*" to a dark corner of a deep NASA vault than its statement of underlying issues, many of which reached far into the NASA psyche.

Motivation and skills were said to be lacking in large and critical segments of both the NASA and contractor workforce, to such a degree that the report questioned whether or not NASA was even capable of running anything as complex as the shuttle program (let alone whether or not it could handle a space station). Importantly, the report also found that NASA's "forgiveness policy" for employees was so harsh as to be virtually nonexistent, and therefore was a major contributing factor in explaining why so many people knew so much about what might go wrong with *Challenger*'s launch on that particular day and yet failed to bring their concerns to light.

The term "forgiveness policy" generally describes the degree to which an organization's management creates an environment in which employees feel free to be critical of the organization and its work. On that point, NASA has a long-standing reputation for intolerance. While discussion is encouraged, open dissent is simply not condoned; more to the point, it is often punished.

In May 2001, the *Cleveland Plain Dealer* reported that a NASA Glenn engineer had been fired for publicly criticizing cutbacks to an innovative energy-storage device intended to replace the pricey, inefficient batteries on the space station. A year earlier, the *Huntsville* (Alabama) *Times* had reported a similar situation in which Marshall's radiation safety officer was fired because he "blew the whistle on lax practices that could have led to worker injury or death." In both cases, the whistleblowers were contractors and therefore easily muscled out. NASA civil servants, who short of committing a felony are employed more or less for life, simply get shifted to less desirable jobs if their names are attached to any public disapproval of the agency.

While Dan Goldin's NASA was particularly notorious for its censure of whistle-blowers, fear of reprisal for criticism has been an agency hallmark for years and thus is deeply ingrained in the NASA culture. The issue surfaced again in the wake of the *Columbia* disaster. Only weeks after the event, it was revealed that several shuttle engineers had exchanged a flurry of e-mail messages fretting over whether or not damage sustained to *Columbia* on launch—which resulted when a large chunk of insulating foam broke off the external tank and hit the orbiter's left wing—would cause catastrophe upon reentry. Such discussion is common among shuttle engineers during missions, since virtually every mission has its share of glitches and faults. But what made the e-mail exchanges notorious is that they were withheld from NASA's top managers, including Sean O'Keefe, even though they expressed obvious concern about the possible damage *Columbia* had sustained— and even though they laid out scenarios for catastrophe that, in the end, turned out to be remarkably close to what postcrash investigators believe may have happened.

By most accounts, O'Keefe is not a manager inclined to shoot the messenger. Nonetheless, at an early congressional hearing, the usually composed O'Keefe was uncharacteristically defensive about the e-mail exchange, retorting that he was "not privy to every single discussion that goes on within the agency." That response, however, did not explain why the e-mails were also kept from William Readdy, the associate administrator for spaceflight and the person who, logically, might have been expected to have all important scenarios and analyses at hand.

What's more, O'Keefe's repeated statement that experts had assessed the potential for the "foam" problem to seriously damage *Columbia* and concluded that there was no risk in landing was not borne out by the e-mail exchange he wasn't privy to. Indeed, the most telling remark in the *Columbia* e-mail exchanges came from a United Space Alliance engineer: "Why are we talking about this the day before landing," the engineer wrote, "and not the day after launch?"

The most pessimistic response to these and other criticisms is that *Columbia* couldn't have been saved even if NASA had concurred, from its top brass to its rank-and-file engineers, that the spacecraft faced a fatal breakup. *Columbia*'s crew was unequipped to conduct even the most basic survey of any potential damage to the vehicle, let alone repair it. There was no onboard means of fixing any external damage to the orbiter, and *Columbia* was both unable to reach the International Space Station and incapable of docking with it in any event. The riskiest option was rushing another shuttle, probably *Atlantis,* into orbit; while the *Columbia* inquiry board admitted that such an option was viable, it noted that the preparation for such a rescue attempt would have needed to be letter-perfect to have any reasonable chance of success.

But if nothing else, NASA is an organization with a stellar track record of dealing with emergencies. The most shining example of NASA's seat-of-the-pants resourcefulness occurred in 1970, when the *Apollo 13* mission was rocked by the explosion of an oxygen tank on its way to the moon. Heeding the now famous rallying cry "Failure is not an option," NASA engineers created an ingenious strategy to get the three-man crew safely back to Earth by making novel use of materials already at hand on the Apollo spacecraft (duct tape, socks, plastic bags) and by using the spacecraft in ways it was never intended to be used. It was an extraordinary demonstration of ingenuity and persistence on the part of all involved, and may have been NASA's finest hour as a unified, focused organization.

To be fair, it was obvious to all involved that *Apollo 13* had experienced a potentially fatal accident, whereas it was unclear whether or not *Columbia* had suffered significant damage upon launch or whether any such damage was sufficient to cause a catastrophe. And to ask whether or not *Columbia* could have been saved is not to question whether or not, with full and unambiguous knowledge of a potentially fatal problem, NASA would have tried to save her—in such a context, NASA most certainly would have put in an effort worthy of the agency's best days.

Rather, the issue stemming from the e-mail exchange is why, when so many people throughout the NASA system apparently knew that there was a significant possibility for disaster, no one made a stronger, more concerted effort to unequivocally confirm whether or not this was the case. On the heels of revelations about the e-mail exchange, it was reported that a senior NASA engineer told NASA management on January 21—only five days after *Columbia*'s launch—that NASA should "beg" other agencies (mainly the American military) to use satellites to image *Columbia*'s left wing, a request that belies O'Keefe's statement that no one believed the orbiter to be in serious trouble. That request wasn't the first. At least one, and possibly two groups of NASA engineers made similar requests, all of which were declined.

For all of O'Keefe's pronouncements that NASA's in-flight analysis of *Columbia*'s health was thorough and complete, one of the first reports presented to the *Columbia* inquiry board pointed out that NASA had no way to track, catalog and analyze a diverse array of safety concerns from within its vast and scattered empire, or to ensure that such concerns landed on the right desks—this, despite the fact that an independent assessment team had recommended three years earlier that the agency put just such a system in place. The recommended system wasn't delayed; it was dismissed as unnecessary.

Piece by piece in the *Columbia* investigation, a picture emerged of an organization in which isolated pockets of analysis and concern were only occasionally threaded together into anything coherent, and where none of the findings produced the kind of all-out, top-level, just-to-be-sure analysis of the spacecraft that might have produced undeniable evidence that an *Apollo 13*–like situation was, or was not, at hand. But the NASA of *Columbia* was not the NASA of *Apollo 13;* with *Columbia,* the fragmented and fearful nature of the organization prevented a genuinely open, truly unified approach to the problem—in turn deferring early, aggressive measures to determine quickly and concretely how bad things might be and what might be done to prevent disaster. Options for saving *Columbia* may have been limited, but whether or not they were nonexistent is a "What if?" exercise the agency, and the public, will probably be engaged in for years to come.

Sadly, "Lessons Learned from *Challenger*" foreshadowed all of this. The finding that perhaps struck deepest into the fractured heart of NASA was the report's conclusion that the agency suffered from a "lack of a program team spirit," since personnel "primarily identify with their own organization, element, project, or function rather than with the program as a whole." In other words, while the world watched in horror and sadness as an agency called NASA lost *Challenger,* the report concluded that if such a unified organization had existed in the first place, such a loss might never have happened. More disheartening is that NASA's internal culture seems to have changed very little in the seventeen years between *Challenger* and *Columbia,* with the result being another disaster, and another inquiry.

If you chance across a copy of "Lessons Learned from *Challenger*," consider yourself lucky: I spent two weeks scouring the NASA archives in Washington for an "official" copy of this document with the assistance of several very helpful and knowledgeable NASA staff, to no avail. Even if you can find the report, it will probably not be the one Dahlstrom and his colleagues first submitted, which Dahlstrom describes as Version 0 (one of the last remaining original copies of which, Dahlstrom believes, is in his possession). Dahlstrom said the report was heavily edited for content before its eventual submission to NASA headquarters, and even then, Dahlstrom claims, only about twenty-five copies were made, sans the usually requisite tracking numbers (the easier for the report to simply disappear). Not only was the final report cleansed of most of its accusations about the problems behind the *Challenger* disaster, Dahlstrom said, it pointedly refused to connect them in any way, even constructively, to the future development of the space station.

"The safety guys were supposed to make sure these things didn't happen in the future, but they didn't want to hear anything had ever been wrong," Dahlstrom recalled. "At the meeting where we presented the original version of the report, the safety guy in charge threw it in the waste basket. He said he was sick of hearing bad things about NASA." After keeping his copy of the original "Lessons Learned" document to himself over the years, Dahlstrom decided to publish it on his own web site on January 28, 2001—fifteen years to the day since *Challenger* blew apart and with its lessons largely unheeded. "I just thought it was time that people had a chance to see it," he said, still clearly upset by the experience.

The primary lesson Dahlstrom himself took away from his time inside the belly of the NASA beast was a familiar one: that money and politics—not scientific discovery or spacefaring vision—were what made the agency's world turn. In 1993, once Space Station Freedom became the ISS, the Langley space station office was disbanded, putting Dahlstrom out of a job. Dahlstrom continued his employment with Lockheed doing design work on small satellites and then worked for the space industry analyst Futron and later a London-based space "edutainment" company before launching his own consultancy in the mid-1990s.

When he connected with ProSpace in 1997, Dahlstrom found an organization that spoke his language. Space business-as-usual had to end, and the only way for that to happen was through organized resistance that struck at the heart of the space establishment.

When I last caught up with Eric Dahlstrom, it was the day before MarchStorm 2003 and the space community was still raw with *Columbia's* loss. In the wake of *Columbia,* NASA was scrambling for technical answers while simultaneously running for political cover. While NASA bent over backward to appear forthcoming in the investigation, behind the scenes the agency's leadership was working the halls of Congress to ensure that come what may, the shuttle would get more money. For its part, ProSpace was developing its annual MarchStorm agenda so that while NASA might run, it couldn't hide.

MarchStorm 2003 was a decidedly different affair from my last experience with the group, in 2001, when I had donned my best business
suit, pinned a ProSpace name tag to my lapel and joined the March-
Storm legions to see what grassroots space lobbying was all about.
Then, there had been a marked exuberance in the assemblage of men,
women, students, retirees (whose occupations had included teaching,
accountancy, medicine and, predictably, engineering). A deeply ingrained passion for space exploration had driven them to use precious
vacation time and pay their own way to Washington to advance the
cause of what they believe to be humanity's destiny as a space-faring
civilization. Having arrived at a year loaded with such predictive
import for space, the feeling seemed to be that if a *2001* future hadn't
actually come to pass in 2001, then at least that year could mark a turn
in such a direction.

The MarchStorm 2001 lobbyists fairly swaggered onto Capitol Hill
with a sort of triumphant moral authority. This sense of possibility was
even reflected in the MarchStorm training session, which by tradition
took place the day before the first round of lobbying. With time being
a precious commodity for most participants, MarchStorm training is
necessarily a quick-and-dirty activity, and it varies from year to year
based on the results of each effort and feedback from each MarchStorm
group. Training in 2001 was a quirky combination of Sturm und Drang
motivational speeches, introductory political lobbying and introductory community theater, in which small squadrons of recruits were put
in front of the group and prodded one by one to the room's equivalent
of center stage, there to deliver a brief who-are-you-and-why-are-you-
here spiel. All the while, a MarchStorm veteran paced the middle of the
room, serving as each speaker's personal Henry Higgins as he critiqued
the delivery of recruits and advised them on the finer points of body
language, diction and choice of words.

Some of the lessons clearly took hold. Over the course of three days,
sixty MarchStorm lobbyists delivered the ProSpace message to nearly
three hundred legislative offices. Some of these meetings produced
more meetings, in which bills were sponsored, hearings were scheduled and bipartisan discussions were organized on how best to inte-
grate private-sector interests into the public-sector space program.

Training for MarchStorm 2003 was a more serious, more focused affair than that of 2001. In fact, "training" was virtually nonexistent: everyone, including rookies (most of whom had never before set foot on Capitol Hill), spent most of the day on the critical issues of the Pro-Space agenda and the broader political context into which they would be expected to insert this agenda during their lobbying. It was a smaller turnout than usual, forty-four volunteer lobbyists in all—the ranks being thinner, most believed, due to the downturn in the economy and the unmistakable sense that with war with Iraq looming, many felt that being at home might be safer than being on Capitol Hill.

There were a couple of short presentations, but no big speeches; there was barely a lunch break. There was, in sum, an almost grim sense of mission. Above all, though, there was a clear understanding among veterans and rookies alike that *Columbia* had created a window of possibility for change that would not stay open for very long. There was also an understanding, although never stated as such, that NASA would use *Columbia* to tighten its shuttle/station interests. Pressing for change would become progressively more difficult with time, not easier.

The one person I expected to see at MarchStorm 2003 but didn't was Charles Miller, the man who had conceived MarchStorm in the first place. As it turned out, Miller decided to stay away from MarchStorm 2003 out of concern that his involvement would constitute a conflict of interest. At the top of the MarchStorm 2003 agenda was a three-step plan to restore the Alternate Access to Space Station program, the success of which could have positive ramifications for Miller's company, Constellation Services International, the other entrepreneurial semifinalist (along with Andrews Space & Technology) in the Alternate Access program before it was subsumed by the Orbital Space Plane.

Miller is a boyish forty-something and a denizen of the alternative space community. Like Dahlstrom and so many others, Miller grew up on heavy doses of Asimov, Clarke, Heinlein and, later, Gerard O'Neill. Their influence on him is clear. "A thousand years from now, we won't remember who was president or who gave the speech that motivated the first man to walk on the moon," Miller once told me. "We won't remember any of the wars. But we will remember this as the time when people first ventured off this planet. It's like when the first fish came out of the ocean; this is the same scope, the time when human life ventured out into the cosmos. It's the same level of significance."

While still an undergraduate, Miller showed an early talent for space activism, founding chapters of O'Neill's L5 Society at his alma maters Chico State University and Caltech, where he recalls being chastised for poaching on the territory of the Carl Sagan–founded Planetary Society, which considers Caltech its spiritual home. On the basis of such activism, in 1988 Miller was tapped to be second in command of the National Space Society (NSS), formed a year earlier through a merger of Gerard O'Neill's L5 Society and the organization NASA created to counter it, the National Space Institute (NSI).

At the time of the merger, both organizations had become moribund, since neither the von Braunian nor the O'Neillian agenda seemed to be going anywhere. If NSI was a bit too stodgy for the O'Neillians, L5 was a bit too wacky for the space establishment. "The L5 Society had a lot of enthusiasm and energy, but they had kind of a flaky image, like a bunch of hippies," recalled the decidedly clean-cut Miller. "That image was good to get people excited, but it wasn't very effective in actually making things happen when it comes to political action and legislation. They were trying to shed that skin."

Predictably, the L5 and NSI contingents butted heads over numerous issues, including the passage in 1990 of the Launch Services Purchase Act. Conceived and lobbied for primarily by hard-core L5 chapters in Tucson and San Diego, the Launch Services Purchase Act required NASA to buy launch services from commercial companies, thereby helping to kick-start the post-*Challenger* commercial-launch market in the United States. Much like the later Commercial Space Act, the Launch Services Purchase Act did not exempt NASA from using the shuttle for any payloads the agency deemed to require the shuttle's "unique capabilities," nor did it prevent the agency from being noncompetitive in its purchase of launch services.

Within the National Space Society, the NSI faction sided with NASA in its opposition to the Launch Services Purchase Act, while the L5 crowd obviously supported the legislation. As it turned out, the L5ers were surfing on a wave of change that was practically inevitable after

Challenger; they couldn't have backed such legislation at a more oppor-
tune moment. Even so, the Launch Services Purchase Act is widely
hailed as the first piece of legislation enacted as a result of grassroots
space advocacy.

Although admitting that it was "not my fight," Miller said NASA's
vicious opposition to the Launch Services Purchase Act left him doubt-
ing just how effective NASA could be in implementing a more exciting
agenda for space. His faith was further shaken when NASA bungled an
opportunity given to it by President George H. W. Bush to lay out plans
for a return to the moon by humans, with an eye on going to Mars.
Miller left his NSS administrative post in 1991, and although he
remained on the NSS board of directors for some time afterward he was
beginning to believe that changing the space status quo would require
more direct, and more radical, methods.

Miller's first plunge into space agitation was wonderfully controversial. Flying in the face of the 1967 Outer Space Treaty, which bars
national claims on extraterrestrial bodies, in early 1992 Miller publicly
proposed the establishment of an American lunar "colony" of at least
fifty people by the year 2010. "At the moment you have fifty people on
the moon, the United States leaves the Outer Space Treaty, or at least
that part of it prohibiting national sovereignty over extraterrestrial
bodies, and we claim the northern hemisphere of the moon," Miller
explained, still clearly excited about the idea. "That would get the
public's attention."

What Miller called his Luna 2010 proposal reached a popular audience in a *Wall Street Journal* article provocatively titled "Will American Flag of the Future Display 50 Stars and a Half Moon?" The article
included Miller's twist on the famous Reconstruction offer of forty
acres and a mule, which would grant one hundred lunar square miles
to "any citizen who can survive [on the moon] for six months." Miller
was more than ready to be one of those who took their chances.
"There'll be people that die," he told the *Journal*. "That's one of the
risks. You're gonna be on the frontier." Ultimately, Miller proposed, the
United States would indeed annex this lunar real estate as a fifty-first
state.

That Miller got play at all in the influential *Journal* was due in part to the company he was keeping at the time: none other than the libertarian icon and nemesis of the Clinton administration, Newt Gingrich. When Miller was first shopping his lunar homesteading idea around the halls of Congress, Gingrich was a rising Republican Party star with a reputation as a keen student of American history and possessed of a well-known fascination with the Turnerian potential of space to extend America's manifest destiny. Gingrich had also been taken with Gerard O'Neill's *The High Frontier,* to the extent that he once drafted legislative requirements for making an O'Neill-style colony a bona fide American state (he worked on that document with the futurists Alvin and Heidi Toffler, well-known O'Neillian confederates).

Miller had several meetings with Gingrich and two commercial space allies, Congressmen Dana Rohrabacher and Robert Walker, and at one point this group planned to float Miller's lunar settlement idea through Congress as a trial balloon to shake up NASA's status quo approach to space and, possibly, shake some more money out of the White House for more ambitious space activities. As Miller described it, "They felt NASA's approach to commercialism smacked of socialism."

The Clinton wave that swept George Bush the Elder from the White House also drew Gingrich's focus from plans for lunar conquest, and the meetings about the moon ended. "That was really my last stab at the old paradigm, because this was going to be a government-led initiative through and through," Miller said. "I still had the blinders on. I was still looking for the president to make the big speech."

Disillusioned and worn out, in 1992 Miller left Washington and went home to California, where he crafted vague plans to join the tech boom in Silicon Valley and put space on the back burner. He didn't stay out of politics, or space, for long. At the urging of his adoptive mother, who is part Native American, Miller began to help several local Native American tribes with their lobbying efforts in Washington. The tribes had hired a professional lobbyist but were coming home empty-handed after making what were, on the scale of typical Washington funding proposals, very modest requests. "All they really wanted was a little money to set up offices and hire someone to help coordinate things like common tribal services," Miller said. "They weren't getting anywhere, and I said to them, 'You have to take your case straight to the legislators, in person, face-to-face.' We did that and got seventeen million a year."

Flush with success, Miller realized he'd also found a method that might work for the kind of space reforms he'd long dreamed of. Thus was born, in 1995, the first MarchStorm. Calling a few friends and handing them a one-page agenda, Miller and the seven other original MarchStorm foot soldiers made some appointments on Capitol Hill and came away with what eventually became a $25 million line item in NASA's budget for reusable launch-vehicle studies (for better or worse, MarchStormers also played a role in pushing for the X-33 program). Miller founded ProSpace in 1996 largely to provide infrastructural support for MarchStorm, whose agenda each year has become increasingly ambitious and increasingly effective, now counting some of Congress's most influential members as allies.

As its name suggests, MarchStorm is planned much like a military campaign, from the precision with which congressional briefings are secured (and more important, held) to the strategic objectives developed by the ProSpace leadership that each MarchStorm volunteer gets drummed into his or her head. There is even a mandatory uniform: suits and ties for men, dresses or pantsuits for women.

While MarchStorm volunteers are trained to deliver a clear message involving four or five primary legislative objectives, they are not trained to anticipate much resistance on the part of the people they are briefing. ProSpace president Marc Schlather says this has happened occasionally, but for the most part congressional staffers are at worst disinterested or niggardly with their time, and at best knowledgeable and enthusiastic. And 99 percent of the time, MarchStormers brief staffers, not actual legislators. Schlather said this was not necessarily a bad thing. "Space just isn't much of a priority on the Hill," he observed. "That means when it's time for a vote or decision on something space-related, most members [of Congress] turn to their staffers and ask how they should vote. The staffers often wield tremendous power. They're sometimes the most important people in the process."

Schlather and others are very clear as to what they are up against in engaging the space establishment on its home turf. Schlather, who also chairs the Senate's executive roundtable on space, provides much of the continuity between the annual MarchStorm events, using the intelligence gathered during each MarchStorm (volunteers must file briefing reports that analyze potential interest in the ProSpace agenda) to press those members of Congress who seem most inclined to push for change, whether for ideological or political reasons. Frequently, Schlather will remind MarchStorm lobbyists that passion is not enough. He tells them that even the message itself is not enough. "You need to talk about what's important to them, not what's important to you," he says.

In that light, what Schlather and others do not convey to the March-Storm participants is that for every half-trained eager-beaver volunteer ProSpace can send onto the Hill, Big Aerospace has a veritable platoon of well-paid, expertly trained lobbyists who lavish Congress with money and influence. Boeing and Lockheed alone spent more than $19 million on lobbying in 2000, which paid for dozens of professional lobbyists who work the Hill daily. In addition to such indirect wheel-greasing, since 1994 aerospace companies have given more than $6 million directly to members of Congress, including about $2 million that went straight to lawmakers with significant responsibilities for NASA programs and spending.

Even those members of Congress who don't drink directly from the Big Aerospace fountain are likely to find themselves eating at NASA's table, for NASA was organized from day one so that all fifty states would have NASA contracts (predictably, the heavy-hitting states—Texas, California and Florida—get the most money). The money involved is often substantial. In fiscal 2000, NASA spent $12.5 billion—a full 83 percent of its total budget—on outsourcing to agencies, companies, nonprofit organizations, universities, schools and other organizations. Of that $12.5 billion, 74 percent went to industry. And of that amount, about 53 percent—or nearly $5 billion—went to Boeing and Lockheed through one corporate permutation or another (including the United Space Alliance).

NASA funding is also used as a political tool at the highest levels, to wit, the agency's decision in early 2002 to consolidate its shuttle repair work at Florida's Kennedy Space Center, in the process decimating the longtime shuttle repair operation in California. Many saw this as another gift to the president's brother, Florida governor Jeb Bush, and more payback for the way California had voted in the 2000 presidential election, in which it picked Democrat Al Gore by a large majority. NASA touted the move as potentially saving up to $30 million per shuttle repair job, but that was mere speculation. At the time of the consolidation, the Florida facility had only done one shuttle overhaul. The rest had been done in California.

NASA money often flows to congressional districts in more roundabout ways, quite often through a budgeting mechanism known as earmarking. On the surface, earmarking seems practically illegal, or at least patently unethical: lawmakers blatantly attach funding requests for pet projects to large federal budgets like NASA's, and such projects are often the deciding factor in whether or not the lawmaker in question approves the agency's budget. Money is usually added to a given agency budget for earmarks, but it is almost never sufficient to defray their impact on the budget; thus, core agency programs always suffer.

Few federal agencies are so enthusiastically targeted for earmarks as NASA, whose budget over time has become a veritable smorgasbord of pork. Since 1998, Congress has drained nearly $2 billion from NASA budgets through earmarking, which the *Orlando Sentinel* bitterly noted was enough to "fix all deferred maintenance on the shuttle's ground facilities ($800 million), fund a costly science mission to Pluto ($500 million) and still have almost enough left over to pay for a shuttle flight ($500 million)."

In approving the FY 2002 NASA budget, the Bush administration approved a record 136 earmarks totaling more than half a billion dollars. In many cases, earmarked money didn't go to programs that were exactly crucial to NASA's mission, such as the $30 million to renovate classrooms at a South Carolina university (in the district of a key senator). Earmarks add to the difficulty of ProSpace's job, since they virtually ensure that every legislator has some sort of vested interest in NASA.

The ProSpace agenda has been remarkably consistent over time. Its driving theme is that the private sector should be given the opportunity to create a market in orbit consistent with the way markets are created on Earth, free from government-sponsored ownership. Lest this seem too blatantly libertarian for many tastes, ProSpace does see an important role for the government as regulator, customer, motivator and incentive provider.

The legislative tools ProSpace proposes for economic change in

space are similar to those employed on the ground, and in this regard
ProSpace and MarchStorm have pushed three interrelated efforts for
several years running. The first is a federal tax-incentive plan to stimulate investment in space transportation companies, which would operate under the aegis of the Department of Transportation's Office of
Commercial Space Transportation, not NASA. Second is a bill to create
a tax-free "enterprise zone" in orbit to give companies more help in
finding their feet.

The third ProSpace push is for a bill called the Spaceport Equality
Act, which, despite its curious title, would simply recognize spaceports
as transportation hubs and give them the same right as airports to issue
tax-exempt construction bonds. Far from being hubs of sci-fi exotica
akin to Mos Eisley in *Star Wars,* actual spaceports are simply places
where rockets are launched: Vandenberg Air Force Base, for instance, is
a spaceport. There are active spaceports in Alaska, California, Florida
and Virginia; ten other states have spaceport plans in the works.

All three pieces of legislation have congressional sponsorship (although only the Spaceport Equality Act has both Senate and House
sponsorship), and all three have been introduced into Congress, where
they have thus far made only limited progress. This is not unusual,
since legislation can take many years to make its way from proposal to
law. As did the 2001 and 2002 agendas, MarchStorm 2003 sought to
reacquaint some congressional offices with the legislation while introducing it to others for the first time. In either case, the goal was to sign
on additional sponsors, in the time-tested knowledge that more names
meant a stronger chance of eventual approval.

Along with the incentive plans, the MarchStorm 2003 agenda also
sought support for study and demonstration funding for two other
long-standing ProSpace interests: identifying near-Earth objects (asteroids and comets, mainly) whose paths might cross the Earth's at the
wrong time, and for solar power generating from orbit. In both efforts,
ProSpace has been successful in drawing support from NASA and the
National Science Foundation. In the case of what is typically called
space solar power, even the influential Electric Power Research Institute—the research arm of America's electric-power industry—has
come on board in recent years as an advocate.

But the big MarchStorm push for 2003 was for restoring the Alternate Access to Space Station program. ProSpace had advocated the idea of alternative means of reaching the ISS well before NASA created a program to address this issue, and it was as shocked as the rest of the space community that the Alternate Access program was summarily cut at just the moment when, logically, NASA should have been accelerating it. As originally conceived, NASA was to have purchased services from an Alternate Access winner in 2006. Had such a timeline been upheld, NASA would have had a cargo option for the ISS independent of Russia's Progress spacecraft and the European Space Agency's (ESA) proposed automated transfer vehicle, scheduled for its first flight in 2004. As it stands, if Russia finds itself unable to build more Progress ships, or if ESA's Ariane V explodes again (the rocket has a short but troubled history), another shuttle grounding would make even a fleet of Orbital Space Planes useless, since they could not supply air, water and other ISS crew essentials.

The fact that ProSpace supports Alternate Access at all is unusual, considering that the program was part of the Space Launch Initiative, an effort ProSpace vigorously opposed. ProSpace believed that SLI would produce an outcome much like the doomed X-33—lots of nifty technology that would never fly—and felt the money would be better spent on pure R&D and on funding a variety of existing private-sector launch vehicle concepts to the flight demonstration stage. ProSpace's greater fear was that SLI would end up as a shuttle replacement program rather than a template for both government and private-sector spacecraft, a charge NASA hotly denied while also saying that the program would almost certainly produce the shuttle's successor; positions that, if not contradictory, were certainly hazy in their distinction.

By early 2002, ProSpace's relentless attack on SLI had gained attention, if not necessarily concrete results. Just days in advance of March-Storm 2002, in which killing SLI was still very much on the agenda, *Space News* reported that SLI contractors "descended en masse" to brief lawmakers on the merits of the program. Only a few weeks after MarchStorm 2002, the House Science Subcommittee convened a hearing on SLI and grilled several NASA representatives about the program's ulterior motives.

In late 2002, ProSpace was unexpectedly granted its wish to see SLI changed, although not in the way the organization had hoped for. Out was a push for a better spacecraft, in was the Orbital Space Plane, and gone altogether—and without the slightest bit of fanfare—was Alternate Access: the program's budget was zeroed out after fiscal 2003, and the money already allocated for 2003 was diverted to the OSP (perversely, this happened just before the *Columbia* disaster).

Unlike the broader SLI program, the Alternate Access program was forced upon the agency in 2000 by the Office of Management and Budget (ironically, this approval occurred when Sean O'Keefe was still OMB's deputy director). OMB viewed Alternate Access as a low-cost exercise in creating options to keep the government's multibillion-dollar space station investment operational in the event that current resupply options went away, either permanently or temporarily.

The fact that Alternate Access was imposed upon NASA, rather than invented by it, is just one reason that NASA had little trouble giving it up, although the program did not go down without a fight. "I can't put a pretty face on this," said a NASA insider, speaking anonymously. "It was a fight between people for control." The people in question, said this source, were managers and engineers in the Office of Space Flight—which owns the shuttle program—and NASA's Code R, the technology development wing that was overseeing SLI.

These same sides had wrestled over the very definition of SLI, and the battle for Alternate Access had much the same theme, namely, whether or not any successful efforts might jeopardize the future of the shuttle. The answer, unequivocally, was yes. The Orbital Space Plane (OSP) alone was fine, since it would still require the shuttle for cargo operations. The OSP plus a separate cargo vehicle would, in effect, make the shuttle expensively redundant. Once again, the shuttle cabal—stacked with vested interests and corporate interests—won the day. As long as SLI's ambitions for new vehicles were limited to the OSP, the shuttle would be guaranteed business for many years to come. Alternate Access, therefore, had to go.

ProSpace saw Alternate Access as both a sensible insurance policy for the ISS and an opportunity for the private sector to grab a piece of the government's human-spaceflight market, which is why the organization supported it while trying to kill efforts for a next-generation, government-owned space vehicle. The summary elimination of Alternate Access drew an unusually sharp response from ProSpace's leadership. Schlather even broke what he describes as his "cardinal rule" when dealing with NASA on Capitol Hill. "You never want to bash NASA, because you often never know, even if there isn't a vested political interest, what kind of emotional interest someone might have," Schlather said. "So our position has always been that we don't hate NASA, we love NASA. We just believe that NASA could be a much more effective agency."

However, for ProSpace, the combination of the *Columbia* disaster, the advent of the OSP and the killing of Alternate Access resulted in the gloves coming off. In briefing materials for MarchStorm 2003—materials delivered to congressional offices—ProSpace described the removal of Alternate Access as "reckless and negligent" and called upon Congress to direct NASA to restore the money diverted to the OSP and reinstate the program through its demonstration phase.

Much was made of the dire possibility of a foreign-only option where a lack of Alternate Access was concerned. Whatever the individual political leanings of ProSpace leaders and MarchStorm lobbyists, MarchStorm agendas are tailored to appeal directly to the dominant political zeitgeist on the Hill each year. Thus, it was noted in the briefing packet that without Alternate Access, NASA might well be forced to rely not only on Russia, but also on Germany and France (which, respectively, make the ESA transfer vehicle and the Ariane rocket to launch it)—three nations that in early 2003 were well on their way to being added to the Bush administration's "axis of evil" for their staunch opposition to the American-led war on Iraq.

In 2003 as before, MarchStorm volunteers headed onto the Hill in groups of two or three, about twenty teams in all, and were told that they would typically have fifteen to thirty minutes to make their case, depending on the interest level and schedule of the congressional staffers they met with. Shortly before MarchStorm 2003 began, I remembered Charles Miller's description during MarchStorm 2001 of the average congressional staffer: "They're typically twenty-two- to twenty-five-year-old liberal arts majors, bright-eyed and bushy-tailed, idealistic and terribly paid," he said. "They want to make a difference. Most of them won't stay in Washington, in the political life, and there's a 20 to 30 percent turnover each year. If they've been in their job longer than about a year, their idealism has probably been blunted a bit."

MarchStorm 2003 invaded the halls of the House and Senate just a week before the American military stormed into Iraq, and the pending war clearly dominated most people's thoughts. The machinery of government hummed along apace, although the sense of business-as-usual felt more than a little bit forced with thoughts of war so prevalent. In this jumpy context, it is therefore telling not only that MarchStorm did not lose any appointments, but also that legislative staffers seemed clearly engaged in the MarchStorm message when their minds might understandably be on other things.

During MarchStorm 2001, I had learned firsthand that a staffer's attention was not guaranteed even when the nation was on a peacetime footing. That year I had been teamed with Howard Beim, a talkative middle-aged chemistry professor from Long Island, and Joe Gillin, an aerospace engineer from the D.C. area who was one of Charles Miller's original 1995 MarchStorm foot soldiers and who serves as ProSpace's secretary. Slightly graying and bespectacled, with a high, round forehead and shy smile, Joe was a calm counterpoint to Howard's hyperkinetic enthusiasm—a man who, one senses, would be perfectly happy sitting all night on a hillside by himself, staring silently up at the stars.

During our first briefing, I was handed the lead, since the House member we were visiting represented California, my state of residence (another MarchStorm dictum is that "home state" team members open discussions with any representative from their state). True to prediction, we saw nothing of the Republican lawmaker who occupied the office and instead were greeted by a snappily dressed twenty-something legislative assistant who looked as though he'd need another few cups of coffee before he was ready to talk to anyone.

We settled into the legislator's luxurious office; holding forth in the boss's office seemed to be a favorite activity of young staffers. As I bumbled through the first few phrases of an introduction I'd prepared only the night before, the staffer actually seemed to be falling asleep. I had plenty of notes at hand, but reading from notes is discouraged, since it detracts from the all-important personal connection March-Stormers always attempt to make. To quote the old comedian's line, I was dying out there.

"Have you ever dreamed about going into space?" Joe interrupted in his soft, bookish voice. The staffer seemed momentarily puzzled by the question, or perhaps Joe had simply woken him up. After a few beats, he answered, "Well, sure, I guess that would be interesting. Yeah, it would be great. It would be very cool."

"I think it would be great, too. But we're not going to get there from here," Joe said calmly, gazing directly into the staffer's now fully alert green eyes. "We need to develop space the way frontiers have always been developed. There's tremendous potential there, market potential. Once market potential begins to be realized, then there will be opportunities for people, regular people, to experience space. ProSpace has some proposals that we believe can move us in that direction, to everyone's benefit. The cornerstone of our agenda is a tax credit plan that would . . ."

And on it went throughout the day, as Joe and Howard used a deft combination of heartfelt passion and capitalist immediacy to laser their way through thinly veiled boredom, smiling indifference, obvious confusion and, on one occasion, even mild hostility from a veteran staffer whose boss represented a district fat with NASA contracts and who smelled danger in the ProSpace agenda. By the end of the meeting, though, even this tough nut had been cracked, and we left him nodding at Joe's gentle but persuasive assertion that the ProSpace propositions would simply allow NASA to use its resources more effectively.

No such cajoling was necessary during MarchStorm 2003. The reason, of course, was *Columbia*. "You know, space wasn't on our radar screen until February 1," offered one Senate staffer, who was so young that he confessed to barely remembering the *Challenger* crash (although the *Columbia* disaster "just shocked" him). After *Columbia,* the staffer said his senator had "wondered out loud several times what exactly NASA was doing in space, with human missions, and what America was doing there." Another young staffer in another Senate office seemed positively indignant at the *Columbia* disaster. "Bureaucracies should not be allowed to create their own future," he growled, before asking to hear more about the Alternate Access program. "America needs to do better than this, because if we don't, you can be damn sure someone else will."

The briefings I attended all went more or less along the same lines. What was surprising is that, unlike 2001, not a single staffer seemed indifferent to the issues at hand. Something about the loss of *Columbia* had combined with a sense of confusion about the mission of the ISS (a point of commentary by several staffers), frustration at NASA's inability to delineate a clear vision and the uncharted geopolitical waters facing the nation to make the MarchStorm message seem, if not as urgent as ProSpace might like, then at least of greater interest than ever before.

As in previous years, the efforts of MarchStorm 2003 did not escape NASA's attention. The week after MarchStorm's Alternate Access blitz, Scott Pace, NASA's deputy chief of staff, said the agency was taking another look at Alternate Access concepts and would compare them with Russian and European alternatives. Pace stopped well short of opening the door to the program's reinstatement, and his comments contained a familiar qualifier—one that has preceded the demise of many a promising project introduced from outside the NASA kingdom. "Folks have been in to talk about the technologies and capabilities they have developed under Alt Access . . . ," Pace told *Space News*. "The problem many of them face is that their systems do not yet exist."

Such are the pronouncements of a confident monopoly. It is hardly surprising that the Alternate Access vehicles do not yet exist. NASA axed the Alternate Access program well before it had spent any money on demonstration work, effectively keeping Alternate Access concepts confined to virtual, not actual, reality. As for private options to bring Alternate Access competitors to the demonstration stage, what investor in his right mind would spend a dime on such endeavors when the only Alternate Access market, the ISS, is owned by an international government coalition that seems bent on sponsoring (not to mention owning) the competition? Although mildly worded, Pace's refrain was one that has echoed down the years, from the Industrial Space Facility to a commercialized Mir: no one but NASA has what it takes to lead in space.

Whatever the fate of the Alternate Access program, Marc Schlather was pleased that MarchStorm 2003 had at least raised its profile among legislators; he seemed even more pleased that several congressmen had committed to signing a ProSpace letter to Sean O'Keefe directing NASA to reinstate the program. If Alternative Access hadn't been revived by the end of 2003, Schlather said, ProSpace would probably make it the top MarchStorm agenda item in 2004, or for as long as it took for someone to force NASA to reinstate the program, or at least the concept. "We'd love for change to happen more quickly," Schlather said, "but we recognize that change usually happens by accretion, not epiphany."

ProSpace is at something of a crossroads. As its efforts have grown in impact and ambition, more resources have been required to put people like Schlather in front of legislators on a more regular basis in order to keep the ProSpace agenda moving forward between MarchStorm sessions. To rely only on the annual MarchStorm efforts would have little lasting impact: the lobbying equivalent of a single wave lapping against a panoramic beach, and then only once a year. As with all things in Washington, the key to having more impact is having more money, and while ProSpace receives some sponsorship (mostly from entrepreneurial space companies), the organization does not have nearly enough resources to consistently compete head-to-head with NASA and its Big Aerospace partners.

Thus of late, not only have MarchStorm lobbyists had to pay their own way to participate, they've also been charged a fee to offset the cost of materials preparation and staff time. This in turn has created a heated debate within the organization. Should ProSpace charge its volunteers even more to pay for increased recruiting efforts—thereby bolstering its ranks and strengthening its influence—or would charging a higher fee actually reduce the MarchStorm ranks, since participants effectively pay to participate already by donating their own time and expenses? Is high-level corporate sponsorship the right option, or would that necessarily create a more partisan agenda, thus eroding ProSpace's independence? These are tough questions, and ProSpace is still looking for the right answers.

If imitation is the sincerest form of flattery, then ProSpace and MarchStorm have had another, more unexpected impact; they seem to have reawakened the National Space Society. The NSS remains the largest advocacy organization in the world focused primarily on human spaceflight, numbering some thirty-five thousand members scattered across the globe. Yet in recent years, the much smaller ProSpace has had by far the louder, more effective voice on Capitol Hill (the formal lobbying arm of the NSS, Spacecause, folded in the 1990s). However, in April 2003, the NSS inaugurated a new event called Making a Difference in Washington, a push to deliver the NSS message to Congress through small, informally trained teams of NSS members hastily prepared to give quick-and-dirty briefings to legislative staffers—in other words, virtually a carbon copy of the MarchStorm effort.

There is some history between the NSS and ProSpace. When Charles Miller put together the original MarchStorm, he wanted the NSS to lead the lobbying charge by having the organization recruit its members to the MarchStorm effort and by providing logistical and legislative appointment coordination. Had that happened, MarchStorm might now be an NSS program. The NSS leadership first agreed to such participation and then, at the eleventh hour, backed out. The reason, according to Miller, is that the NSS got cold feet over concerns that the lobbying effort would turn into a NASA-bashing fest. "I was obviously pissed off," Miller said.

Perhaps understandably, ProSpace's leadership has mixed feelings about the new NSS lobbying effort, even though it was conceived by a former MarchStormer and uses, with ProSpace's permission, the briefing analysis report employed during MarchStorm. On the one hand, more people getting in front of legislators about space can't be all bad; on the other, NSS is necessarily treading on ProSpace's turf. Yet there is another reason why the new legislative effort by the NSS may be more troubling than encouraging.

The National Space Society says most of the right things about what's needed to realize a more ambitious human space future: that human presence in orbit should expand, that private-sector involvement is critical and that lowering launch costs is essential. Certainly, its vision statement should be music to the ears of space enthusiasts everywhere: "People living and working in thriving communities beyond the Earth."

But one look at the NSS recommendations for America's space trans-

The Russian space station Mir. *(Photo courtesy of NASA)*

Mir was the original destination of space tourist Dennis Tito, pictured here with crewmates Talgat Musabayev and Yuri Baturin just prior to their April 28, 2001, launch from the Baikonur Cosmodrome in Kazakhstan. When Mir was de-orbited in March 2001 under intense NASA pressure, Tito flew instead to the International Space Station. *(Photo by Greg Klerkx)*

Tito's flight infuriated then NASA administrator Dan Goldin, who once questioned Tito's patriotism for flying with the Russians. *(Photo courtesy of NASA)*

John Glenn *(right),* former U.S. senator and the first American to orbit the Earth, giving a press conference with astronaut Curtis Brown during their 1998 mission on *Discovery. (Photo courtesy of NASA)*

Glenn's flight ignited a debate among NASA's leadership—including Alan Ladwig, who was then senior adviser to Dan Goldin—as to how "nonprofessional" astronauts should be selected. Ladwig felt Glenn's slot should have been given to Barbara Morgan, the schoolteacher next in line to fly after Christa McAuliffe. *(Photo courtesy of Alan Ladwig)*

Barbara Morgan *(left)* and Christa McAuliffe in an undated photograph. *(Photo courtesy of NASA)*

Columbia, the first of NASA's space shuttle orbiters ever to fly, at liftoff on January 16, 2003, on what would turn out to be its final mission. This photograph was taken at 10:39 A.M.; approximately eighty-two seconds later, a large piece of insulating foam broke off the shuttle's external fuel tank and struck the leading edge of *Columbia*'s left wing, an event that led to the craft's destruction sixteen days later. *(Photo courtesy of NASA)*

The last crew of *Columbia* poses for an in-flight portrait. *Bottom row, from left:* Kalpana Chawla, Rick Husband, Laurel Clark, and Ilan Ramon. *Top row, from left:* David Brown, William McCool, and Michael Anderson. On February 1, 2003, these seven astronauts were killed when *Columbia* broke up during its landing approach. *(Photo courtesy of NASA)*

The *Columbia* disaster was the first major challenge to new NASA administrator Sean O'Keefe *(center, inspecting* Columbia *debris),* who took over from Dan Goldin a little more than a year before *Columbia*'s destruction. *(Photo courtesy of NASA)*

Physicist Gerard O'Neill *(center)* with several MIT students who built a working prototype of O'Neill's mass driver, one of the tools he had devised for his expansive extraterrestrial agenda. *(Photo courtesy of Space Studies Institute)*

Gerard O'Neill's vision of space as a personal frontier galvanized an "alternative" space movement, with former O'Neill protégé Rick Tumlinson as one of its leading figures. *(Photo courtesy of Rick Tumlinson)*

An artist's rendering of life in an O'Neillian space colony. *(Photo courtesy of Space Studies Institute)*

The Flashline Mars Arctic Research Station is a privately funded prototype of a research facility that might house human explorers on Mars. The "Hab," located on Devon Island in the Canadian Arctic, was completed in 2000. *(Photo courtesy of NASA Haughton-Mars Project 2000/P. Lee)*

The Hab's construction team, including Hab's founders Robert Zubrin *(far left)* and Pascal Lee *(far right)*, poses triumphantly on the day of the station's completion. *(Photo courtesy of NASA Haughton-Mars Project 2000/M. Webb)*

Pascal Lee in an experimental Mars exploration suit built by NASA spacesuit maker Hamilton Sundstrand. *(Photo courtesy of NASA Haughton-Mars Project 2000)*

"Analog" activities like those conducted on Devon Island have a long history in space exploration, as demonstrated here by astronauts Harrison Schmitt *(foreground)* and Eugene Cernan, who are practicing their lunar geology on a Florida beach a few months before their *Apollo 17* mission to the moon in December 1972—the last time humans traveled beyond Earth's orbit. *(Photo courtesy of NASA)*

This photograph, taken in 2001 on Devon Island, offers a real-life glimpse of what human exploration of Mars might look like. Although Mars has long been heralded as the next logical step in human space exploration, NASA currently has no plans to send humans to the red planet. *(Photo courtesy of NASA Haughton-Mars Project 2001/P. Lee)*

portation policy—released shortly after the *Columbia* disaster—sends
another message. At the top of the NSS policy list are recommendations
to boost shuttle funding and accelerate the Orbital Space Plane, goals
that are wholly consonant with NASA's ambitions. In terms of next-
generation space transportation, the NSS recommends investment in
"long-range technologies that could reduce the mass to orbit required
to support crewed space flight," a position that virtually parrots the
goal of "new technologies" work proposed under the downsized Space
Launch Initiative. Where the private sector is concerned, the paper
simply states that "U.S. space transportation policy should provide reliable markets and engage the private sector to the maximum extent
possible"—words that merely echo what is already supposed to be happening under the Commercial Space Act, rather than prescribing how
the government, and NASA, might set about achieving these goals.

The National Space Society is a well-run organization with an impressive array of advisers and directors. It enjoys strong support from
many segments of the space community. It receives substantial Big
Aerospace funding, and its membership is stocked with both NASA
supporters and advocates of a supra-NASA space future. Too often,
though, the NSS has fallen into step with Freeman Dyson's "paper
NASA" paradigm: laying out a dreamy space future while backing
near-term efforts designed primarily to ensure the status quo.

When the NSS hit Capitol Hill in 2003, however, its tone and agenda
were unusually direct—and like ProSpace, much of this directness was
channeled toward restoration of the Alternative Access program. "The
loss of *Columbia* dramatically underscores the urgency to develop a
secondary capability to launch crews to and from ISS, and it is not clear
that this sense of urgency is shared by all of NASA's managers at the
program level," NSS executive director Brian Chase told a Senate committee during the 2003 NSS lobbying push. "The [Alternative Access]
program should get a fresh look from NASA so that, when combined
with the Orbital Space Plane program, we will have both assured crew
and cargo access to the International Space Station."

It is unlikely that the NSS will become as confrontational regarding
NASA's agenda as ProSpace, at least not anytime soon. But it is telling
that an organization that has long been a leading cheerleader for

NASA's von Braunian spin seems now to be voicing some serious second thoughts. If the NSS can stay the course it seems to have begun with its Capitol Hill push in 2003—if it can join the ranks of organizations like ProSpace that hold NASA to account—it may well have a significant impact on how the agency does business.

In and of itself, supporting NASA's human spaceflight program is not a de facto negative: the agency isn't going away, and change (such as it is) will necessarily happen only by working both with the agency's supporters and with its detractors. No one at ProSpace doubts that in most cases, the NASA vision will prevail, at least for now—and as Schlather and other ProSpace leaders advise, NASA-bashing does not go far on Capitol Hill.

All of that being true, cracks in the establishment's armor have been appearing with increasing regularity of late, even before the *Columbia* disaster. The revamping of the Space Launch Initiative, for example, drew some unusually quick and harsh criticism in the space trade press. "Are NASA officials and the Bush administration about to commit the agency to maintaining the status quo for another two decades or more?" a *Space News* editorial asked bluntly. "If they are, how long will it be before the public just loses all interest in the space program?"

Almost immediately after NASA's announcement of the Orbital Space Plane, Congressman Ralph Hall, the ranking Democrat on the House Science Committee, commented that "it seems to me that rather than spending $10 billion or more on an Orbital Space Plane, we should stabilize the [ISS] program by developing a cheaper, dedicated Crew Return Vehicle based on the X-38 design," the latter referring to the experimental vehicle NASA abandoned in 2002. "The balance of the funds could be used to restore space station research capabilities and to deal with other pressing NASA needs," Hall concluded. Republican legislators have voiced similar hard-edged concerns. Senator John McCain, for instance, remarked on the OSP's future funding prospects that "if [NASA had] had the efficiencies for the space station and maintained their budget, they'd have $6 billion to play with."

More jeers were heard when, in 2003, NASA quietly mentioned that the "P" in OSP wasn't necessarily binding. The OSP might be something other than a planelike vehicle, and one NASA sketch suggested that NASA could find itself building a slightly more advanced version of the Apollo capsule it abandoned three decades ago. Even in the most generous spirit, it is hard to call this progress.

However the OSP evolves or whether or not it is even built, the space station is at the heart of every move NASA makes these days. What NASA calls its Integrated Space Transportation Plan—encompassing the shuttle, what's left of SLI and now the OSP—focuses even more of NASA's forward-looking resources on getting the most out of the space station. And the station is the one major NASA project ProSpace has largely steered clear of, although privately many of its members and leaders question its worth. Scientifically dubious, commercially convoluted, ridiculously over budget, the ISS nonetheless remains politically bulletproof.

"Any attempts to kill ISS could never succeed. You're always affecting a significant constituency of someone powerful," Schlather told me. "There are solutions, but they're either unrealistic or unpleasant. One, you shut down the station, which isn't going to happen. Two, you can screw the little guys and only work with the big players, which stifles entrepreneurship and defeats the purpose of expanding competitiveness. Or three, you can reward NASA for incompetence. I don't know what the better answer is."

Besides, Schlather said optimistically, "We're looking at the bigger picture. We're looking beyond the ISS."

Schlather was speaking figuratively, since his goal and ProSpace's goal are to open up near-Earth space to commerce in order to establish the foundation for activities like popular human spaceflight. NASA, however, continues to divert attention from its troubles in orbit by repeatedly pumping up the ISS as a necessary step to zoomier voyages elsewhere. Even though the agency is perpetually short on details, at the end of the day NASA knows that spinning visions of distant exploration is what keeps the public happy, and what keeps the money flowing.

A prime example of this strategy was on display only a month before NASA's announcement of its SLI reconfiguration. In October 2002, the World Space Congress—a decadal gathering of the space community in all its forms—was dazzled by the elaborate future space agenda delivered by the NASA Exploration Team, or 'NeXT.' As a reality-free exercise, the NeXT space future is a fabulous future indeed: it includes research bases on the moon (which has been notably absent from NASA's exploration agenda, despite being the agency's prime point of glory) and an occupied space station at Lagrangian Point 1, which is about 84 percent of the way to the moon and, according to NeXT, a handy place for reaching other stations that might be built at other gravitationally stable Lagrangian points, including an Earth-sun point known as L2 (notably, the NeXT team did not mention L5).

Not surprisingly, the NeXT presentation was a star attraction at the World Space Congress and earned a healthy chunk of media coverage. To be fair, NeXT has a better-than-average record of bringing blue-sky ideas to at least the study stage, as evidenced by ongoing research in new propulsion techniques like nuclear-electric engines that had their origins as NeXT proposals. With a budget of only a few million dollars a year, however, NeXT is a decidedly small effort in the NASA universe— and, of course, its propositions have no hard and fast timelines or cost projections.

However, the biggest giveaway that the NeXT effort is merely another fine example of the "paper NASA" hard at work is its consistent recital of the need for process over vision. Going to the moon, L1 or anywhere else will happen, NeXT contends, only if the technology can be developed to make such things happen safely and efficiently. How soon that will happen is anyone's guess, so setting concrete priorities for human missions is, therefore, a futile endeavor. "There are many exciting places to explore," cheers the NeXT web site, "we want to go where the action is." Where, exactly, NASA believes the action to be is not specifically stated.

In waffling about the details of its future human-spaceflight agenda, NeXT echoes Sean O'Keefe's new NASA, which firmly favors incremental technology development unmoored to any specific goal for the use of such advancements. This strategy flies in the face of what drove NASA to prominence in the first place, and this may be intentional; after all, NASA was quickly abandoned to the budget cutters after reaching its last firm destination, the moon. In this spirit, the current NASA vision statement, pronounced early in O'Keefe's tenure, is vague and tentative and reads a bit like haiku-by-committee: "To improve life here, to extend life to there, to find life beyond."

Among the several problems with such a broad statement of modus operandi is that it makes NASA even more inscrutable to the public, which is about as likely to be excited by advancements in spacecraft propulsion as by progress in building more efficient automobile engines. A more efficient automobile engine would, at least, have some direct relevance to the average person. Like the ISS, nuclear-electric engines would be a marvelous technological achievement. But what would they be for? At this point, NASA is not saying.

NASA, it seems, is waiting for direction and guidance; nowhere is this more evident than in its human-spaceflight program, in which the International Space Station goes round and round and the shuttle goes up and down. Yet in the middle of the proverbial room representing NASA's future, in which dreamy NeXT propositions float toward dusty stacks filled with their "paper NASA" predecessors, sits the elephant no one in NASA's leadership wants to acknowledge. Lurking between the lines of every NeXT proposition, wedged between each proposed improvement to the shuttle, and weighing on every word written about the ISS is the phrase NASA dares not commit itself to: humans to Mars.

More than even O'Neillian space habitats, Mars was always the next stop for humans on the final frontier; in many ways, it is the last unanswered call of the original Space Age. From the standpoint of vision, the possibility of sending humans to Mars gave birth to both the space shuttle and the space station—which returned the favor by being so costly, unwieldy and politically contentious as to make a mission to Mars seem unthinkable.

Mars is where von Braunian and O'Neillian futures begin to converge, since any effort to send human explorers there would almost certainly mean the development of an ambitious infrastructure that could support more expansive human space activities closer to home. Even the most sanguine members of the alternative space community know that Mars is a step too far for the private sector alone, but they are once again trying to lead the way. And once again, NASA is being compelled to follow.

The Emperor of Mars

On a cool day in the late summer of 2001, the acolytes gathered on the campus of Stanford University. In the sparkling morning sunlight they sat on the steps outside Dinkelspiel Auditorium, sipping tepid coffee and chewing doughy bagels. Mostly in their thirties and forties, they themselves had not long ago navigated the Stanfords, MITs and Caltechs of their youth, many earning graduate or doctoral degrees with the highest honors and thereby vaulting into the top ranks of their chosen callings.

They had become doctors or lawyers, engineers or laboratory scientists; they had made a killing in the stock market or bet wrong on the new, new thing. Yet whatever the path, whether rich or poor or somewhere in the middle, they had arrived here unfulfilled, disillusioned, restless. The real world was a faded promise. The acolytes seemed to be waiting for something, or someone, to show them that days of wonder still lay ahead. And then, their leader arrived.

Robert Zubrin is an unlikely rebel. At the Stanford gathering, he was once again wearing his standard uniform: rumpled sport coat, lived-in corduroys, tennis shoes, black dory cap and dingy backpack, the resulting look being that of a slightly unkempt college professor about to go on a long hike. But here as elsewhere, Zubrin—science teacher turned aerospace engineer turned coarchitect of the most realistically ambitious human-space-exploration program since Apollo—had an undeniable magnetism, a clear aura of confidence that radiates from absolute devotion to an idea.

As Zubrin made his way up the steps toward the auditorium, the acolytes sidled up to him with something approaching reverence; some asked questions, while others appeared to want no more than to hover

in his proximity and imbibe the energy of his passion. Zubrin was generous with his time and words, speaking at least for a moment to almost
anyone who sought an audience. Then, politely, he excused himself
and entered the auditorium. His entourage crowded in behind, arriving inside only to discover with dismay that many, having strategically
sacrificed an opportunity for a personal moment with Zubrin, had
already grabbed the choicest seats.

The lights in the auditorium dimmed, and following a brief introduction Zubrin took the stage. He is a small man who works his com-
pact body like a pugilist: head low, eyes darting across the audience in
a dare to all comers, shuffling from side to side on his feet as if about to
throw a punch at an invisible opponent. He spoke in a pinched tenor
that rose and fell with a Brooklyn bounce; a city kid's twang, the delivery halting yet authoritative.

"People often debate in this community whether it's better to pursue
humans to Mars through public resources or to do it privately," Zubrin
declaimed to a rapt audience of about five hundred. "It doesn't matter.
I believe ultimately that Martian society, whether it is started by a
NASA base or by a private consortium, will develop into an independent branch of human civilization. Our task is just to make it happen."

He then looked directly at the audience long and hard, attempting,
so it appeared, to laser his words and their gravitas directly into the
very souls of each individual present. Zubrin wanted to send them
back into the world with the same sense of mission he has, and with at
least a modicum of his zeal and eloquence on the subject that has
become his life's work. He wants Mars to be their life's passion, as it is
his. Zubrin knew instinctively that he was speaking their language.
It is the language of people who have become disheartened because
reality has not, in any measurable way, matched their aspirations. He
knows that language all too well.

Zubrin waited several beats, the room filling with silent anticipation.
"What we're talking about is a second chance for humanity," he then
said quietly. "Mars is that second chance, and a second chance is worth
a lot."

And this is The Message, delivered again and again with the pum-
meling consistency of a politician on the campaign trail, whether on

TV talk shows, through the pages of best-selling books or in private conferences with congressmen, tycoons and scientists. Robert Zubrin is building a movement. His followers are scattered across the planet and number in the thousands, as evidenced by membership in the organization he leads, the Mars Society. What Zubrin proposes is an interplanetary exodus, a journey across millions of miles into a desert far more lethal, far more challenging than any on Earth. If he succeeds in realizing his ambition, though, Zubrin will trump even Moses. Zubrin has put it in all but the biblical terms that so naturally suit his enterprise: Let my people go . . . to Mars.

This is Zubrin's long-range goal, but in the here and now, he'd settle for even the suggestion that NASA, or anyone, is laying plans for a first human visit to the red planet. How long he'll have to wait is very, very hard to say. If you'd put this question to Wernher von Braun back in the 1960s, he'd have said the 1980s, at the latest. In the 1970s, Gerard O'Neill at his most optimistic would certainly have been surprised had the first human set foot on Mars any later than the turn of the millennium. Toward the end of the 1980s, George Bush the Elder was aiming for the first quarter of the twenty-first century—but more on that later.

The bumpy and largely abortive history of humans-to-Mars planning began in the earliest days of the Space Age. NASA's very first long-range strategic plan, released to an eager public in 1959 (only a year after the agency's founding), trumpeted missions to the planets before the end of the century. Always, Mars was at the top of the list. Between 1950 and 2000 more than a thousand Mars mission studies were conducted by NASA, industry, private think tanks or novel combinations thereof. Tens of millions of dollars were spent on these studies; untold days and months of brainstorming, calculating, arguing, planning, writing, presenting and lobbying were expended. In terms of sending humans to Mars, not a single effort amounted to anything at all.

At the advent of a new century, Zubrin and others are weaving a new humans-to-Mars framework that neither puts the space establishment in charge nor excludes it from the game. Significantly, they are achieving a level of success, as measured by funding (much of it private) and public interest, not seen since the Apollo days. They are doing so not by demonstrating that Mars has any immediate economic or even political value, but by taking the lure of Mars back to the people—whether scientists, philanthropists or even political leaders who themselves grew up convinced that humans would one day explore other worlds. They are asking the people to make Mars exploration happen, knowing that unlike missions to the moon (always a political construct) it was with the people that the dream of Mars began.

The first humans to set foot on Mars were gunned down by one of the locals; more precisely, they were murdered by a jealous Martian husband whose wife was flirting (in her dreams, telepathically) with the would-be pioneers as their rocket approached from Earth. "It came from a long way off," wrote Ray Bradbury in this first story of his 1950 classic, *The Martian Chronicles*. "One shot. . . . And then a second shot, precise and cold, and far away."

As Bradbury's epic progresses, more humans come, fleeing an Earth on the brink of nuclear annihilation. The native Martians wipe out one human expedition after another, but like rats leaving a sinking ship the Earthmen keep coming. Eventually, the Martians are felled in their millions by something far deadlier than guns: microbes, chicken pox to be precise, a fate consciously analogous to that of the early Native Americans after their first contact with Spaniards. Continuing the analogy, humans finally arrive in droves, settling the land (Bradbury's fictional Mars conveniently has a breathable atmosphere), and re-creating the American West all over again, complete with whiskey towns and wagon trains.

Like many early science-fiction writers, Bradbury used the American West as a metaphor for human ambitions in space, but he is no friend of Frederick Jackson Turner. On the contrary, like many writers of his day Bradbury viewed western expansion with more cynicism, and in *The Martian Chronicles* he uses it as a cracked mirror to reflect how utopian ambitions invariably lead us back to where we started, which is face-to-face with the problems we were attempting to leave behind in the first place. On Mars, as on any new frontier, tabula rasa quickly becomes terra cognita, yet the travelers rarely recognize their own emotional baggage, a point *The Martian Chronicles* sums up concisely:

> They came because they were afraid or unafraid, because they were happy or unhappy, because they felt like Pilgrims or did not feel like Pilgrims. There was a reason for each man. They were leaving bad wives or bad jobs or bad towns; they were coming to find something or leave something or get something, to dig up something or bury something or leave something alone. They were coming with small dreams or large dreams or none at all.

The Martian Chronicles was a huge success for Bradbury and would over time become one of the most revered works in all of science fiction. Bradbury later expressed surprise that people considered the book to be science fiction at all. He claimed it was nothing more than a set of thought exercises using Mars as a unifying theme, not an attempt at envisioning a realistic future for humans on Mars; in that respect, one could say the red planet was little more than red clay for Bradbury's musings. Be that as it may, *The Martian Chronicles* was one of the first popular works of fiction that tapped into a growing idea of Mars as the next frontier, the next logical place for humans to explore and, possibly, to call home.

Bradbury's book was published at a particularly curious time in the history of humanity's fascination with space travel. In 1950, the public still thrilled to the Buck Rogers romanticism of "outer space," and rockets still possessed an almost innocent, gee-whiz quality as vessels of adventure. Although rockets had been used to deadly effect in World War II, they did not appear on the world stage as weapons of mass destruction until the end of the 1950s, when the Soviets launched the first ICBM and laid down the gauntlet for the Space Race.

At the same time, in the years immediately following Hiroshima and Nagasaki, the world was in an understandable panic about nuclear Armageddon. As a result, it became common for writers like Bradbury to play with apocalyptic scenarios that involved humans fleeing to other worlds: given an extraterrestrial Eden, would humans do things differently, or make the same old mistakes? In one of the most popular films of the era, the 1951 release *When Worlds Collide,* a group of privately funded scientists build nothing less than a rocket-powered Noah's ark, thereby saving the best of Earth ("best" as defined by Hollywood in the '50s: white, American, et cetera) from cataclysmic destruction. Although the escape was not from atomic devastation—instead, a rogue planet was zooming unstoppably toward the Earth—the film's apocalyptic flavor was distinctly postnuclear: panic in the streets, the manipulation of political and economic power and interpersonal relationships for survival, and particularly the portrayal (popular at the time, and still popular today) of scientists as misunderstood and often inscrutable oracles in a universe that was rapidly becoming too complex for mere mortals to interpret for themselves.

In those early days of the Cold War, rockets could still be viewed as humanity's salvation from its self-annihilating instincts, rather than as engines of the apocalypse. Frequently, Mars was used as an extraterrestrial object lesson. The 1950 film *Rocketship X-M* finds five astronauts accidentally landing on Mars (they were trying to reach the moon) only to discover a dying civilization that had nearly obliterated itself and its once-lush planet in an atomic war. The few Martians left are reduced to living in caves. Horrified, the astronauts spend most of the film trying to return to Earth, there to impart the lesson of Mars in the hope of convincing Earth's leaders to avoid the same fate.

In such a context, Mars had much to recommend it as a convincing point of comparison for the Earth—at the least, Mars seemed a better bet to support life than other places in the solar system, even given what little was known about the solar system prior to the robotic explorations of the mid-1960s. Despite being our closest neighbor, anyone with a good backyard telescope could see that Luna didn't amount to much compared with Earth. Venus is the closest planet to Earth (at its closest, about twenty-six million miles away compared with about thirty-five million miles for the nearest proximity of Mars), but its thick cloud cover was soon proved to be the visible layer of an atmosphere so thick it would crush robotic probes like paper cups. Mars, on the other hand, was relatively close, had visible polar caps (possibly containing frozen water), a day only thirty-seven minutes longer than the Earth's and an atmosphere that in the 1950s was thought to be relatively Earth-like, although this would prove to be far from true.

Not surprisingly, it was Wernher von Braun who sketched out the first coherent plan for sending humans to Mars. In keeping with Mars as a symbolic fulcrum for fact and fantasy, von Braun's ideas first appeared in a novel he wrote in the late 1940s while interned with the rest of his V-2 team in New Mexico. The novel was never published (the consensus being that von Braun was a brilliant engineer but a lousy novelist), but the book's humans-to-Mars concepts were supported in the index by painstakingly rendered calculations that were published to great fanfare in 1950 in West Germany as *Das Marsprojekt*.

Das Marsprojekt is positively Wagnerian in scale. Von Braun envisioned an armada of ten four-thousand-ton spaceships assembled in orbit from pieces flown up from Earth, a concept foreshadowing the construction schemes of both Mir and the ISS. Seventy explorers would blast out of Earth orbit and continue in a flotilla to Mars over a period of eight months, there to spend a year surveying the Martian surface before departing on another eight-month return trip to Earth. Once in Mars orbit, a team of astronauts would climb into a space glider carried in the belly of their spaceship—these ships were designed not to land but rather to forever traverse space and circle planets, thereby foreshadowing *Star Trek*'s famous *Enterprise*—and glide to a ski-skid touchdown at one of the polar caps, since von Braun concluded that polar ice would make for a smooth landing.

Once on the Martian surface, the astronauts would drive a pressurized Mars rover nearly four thousand miles to the Martian equator, there to build a landing strip for other gliders. Von Braun thought, correctly, that it would be warmer at the Martian equator and therefore more comfy for the explorers. More gliders would land, and the newly minted Mars explorers would then set up an inflatable habitat and begin their yearlong survey. The gliders would also serve as the explorers' means of returning to their orbiting ships for the trip back to Earth. The astronauts would simply detach a glider's delta wings, prop the glider on its end, reattach the wings as fins and voilà! An instant V-2-style rocket. With their delta wings, ability to travel as both rocket and glider, capacity for both passengers and cargo, and focus on reusability, von Braun's Mars gliders had all the elements that NASA would later incorporate into the space shuttle.

Considering how little was known at the time about space travel or Mars itself (and that this scheme was, after all, created in the service of a novel), it is all the more incredible that von Braun's plan became the humans-to-Mars gospel for both the American and Russian space programs over the next four decades. By the time *Das Marsprojekt* was published in English, it was 1953 and von Braun was a media superstar, thanks largely to the *Collier's* space series and his collaborations with Walt Disney. Ever the master salesman, von Braun used the tail end of the *Collier's* series to promote his *Marsprojekt* ideas to the general public, but with a key addition that would inadvertently set humans-to-Mars planning back by decades. The *Collier's* article titled "Can We Get to Mars?" was published on April 30, 1954, and included the same elaborate mission plan and the same fleet of massive spaceships as in *Das Marsprojekt*. However, instead of being assembled in free-floating space, the huge Mars spaceships would be assembled at an enormous, wheel-shaped space station that von Braun had described in great detail in an early *Collier's* article focused on building moon bases. NASA would later adopt the space station idea and von Braun's concept for the shuttle system to build and service it, and these would ultimately become the primary roadblocks for humans-to-Mars exploration—all the while being justified by NASA as absolutely necessary to send humans to Mars at all.

A series of books by von Braun and his fellow German expatriate Willy Ley would come after the *Collier's* articles and solidify—both for the public and for NASA—the *Marsprojekt* approach to manned Mars expeditions, even if some of the details would change over time. Subsequent versions had fewer ships and smaller crews, different propulsion schemes, and after the discovery that the atmosphere of Mars was less than 1 percent as thick as Earth's (and therefore allowed for very little aerodynamic lift), no gliders.

Heavily influenced by von Braun throughout its formative years, NASA began putting together concepts for Mars expeditions while it was still the National Advisory Committee on Aeronautics. In November 1957, only a month after Sputnik's startling flight, researchers at the NACA Lewis Research Center, near Cleveland (later to become NASA's Glenn Research Center), began sketching out nuclear-thermal and electric propulsion systems for interplanetary flight. In April 1959, less than a year after NASA officially opened for business, the Lewis team was granted funding for a Mars expedition study. This was clearly thinking ahead: Yuri Gagarin would not make his historic flight for another two years.

Like the Americans, Soviet rocket designers were also profoundly influenced by the V-2 team, both because of von Braun's global reputation and because the Soviets had spirited away V-2 personnel and hardware shortly after the war. Clearly, Sergei Korolev had read *Das Marsprojekt* by 1956, the year he directed a team to develop a manned expedition to Mars. The result was *Das Marsprojekt,* Soviet style: the Martian Piloted Complex (MPK). The massive MPK ship would be lifted to Earth orbit by a succession of N-1 rockets (which were devised by Korolev for this purpose and only later redesigned for the moon). The Korolev plan also included landing craft, rovers and a yearlong expedition, although he proposed a much smaller crew than von Braun's.

The Soviets refined their humans-to-Mars plans over subsequent years, but in truth, the Soviets could barely keep pace with NASA once the moon race was on. Korolev's untimely death and the failure of the N-1 kept Soviet humans-to-Mars planning on a more theoretical level than in the United States. At a conference once, I asked the MirCorp consultant Vladimir Syromiatnikov, who had worked closely with Korolev, if he thought cosmonauts might have planted a flag on Mars by now had Korolev lived. "Mars is very hard," the old engineer replied. "Korolev was a genius, but he was not a god."

While Kennedy's mandate may have forced NASA to focus on the moon, it did not in the least blunt the energy or money expended on NASA's planning for human missions to Mars. On the contrary, the historian David Portree calls the period from 1962 to 1965—the height of technology development and testing for Apollo—"the most intense period of piloted Mars mission planning in NASA's history." In the agency's salad years, the Turnernian steps were clear: today the moon, tomorrow Mars.

Appropriately for a superpower with interplanetary conquest on its mind, this aggressive period of humans-to-Mars planning began with a 1962 study called EMPIRE, an acronym yielded by the wonderfully tortured label Early Manned Planetary-Interplanetary Roundtrip Expeditions. EMPIRE was the brainchild of the Marshall Spaceflight Center, which by 1962 had already staked out its ground as the place where rockets were designed and built. EMPIRE was a placeholder for Marshall and for its director, the ever wily von Braun, who anticipated a slump in the rocket-building business after the moon landings unless human exploration of Mars could be set firmly on course. Importantly, von Braun knew that landing humans on Mars would be much harder than landing them on the moon (which would be hard enough), particularly if the intention was to get them back again.

Von Braun needed time to build a firmer technological foundation and weave the political web necessary in Congress and within NASA to make humans-to-Mars happen. But in the meantime, he needed to keep his team together and keep the money flowing; as Portree puts it in *Humans to Mars,* an outstanding history of Mars mission planning, "What Marshall needed was some kind of short-term interim program that answered questions about Mars while still providing scope for new rocket development." In that sense, EMPIRE may have been a prototype for the many "paper NASA" exercises that would eventually befog the agency's mission.

The EMPIRE studies covered most of the primary elements that would be part of nearly all future Mars plans by NASA and its aerospace partners: development of a post-Saturn heavy-lift rocket, the use of artificial gravity to keep astronauts healthy during the long voyage, the use of nuclear propulsion, a combination of crewed and automated ships and, thanks to von Braun and *Collier's,* on-orbit assembly using a space station. However, EMPIRE stopped just short of actually landing humans on Mars, focusing instead on human flyby missions where astronauts would see Mars as passers-by (snapping pictures and deploying probes) before swinging back around for Earth.

Thus EMPIRE was premised on a fatal flaw that would come to haunt all plans for piloted space exploration after Apollo. At the time, few believed that Earth-launched robotic probes would become reliable proxies for human explorers. Such lack of prescience can be forgiven in the case of the EMPIRE study, since at the time robotic probes failed far more often than they worked. (Piloted flyby missions were also proposed by the Soviets in advance of Mars landings; of late, the Chinese have talked about piloted flyby missions to the moon.) The first successful flyby of another planet, the visit of Mariner 2 to Venus, established its success midway through the EMPIRE study and the first of the Ranger probes to the moon didn't succeed until 1964. A successful robotic flyby of Mars, courtesy of Mariner 4, wouldn't happen until 1965.

As Apollo began to take shape, more was learned about spacecraft design, propulsion systems and human survival in space: for example, radiation shielding in manned spacecraft, not considered in *Das Marsprojekt,* was deemed essential after the discovery in 1958 of the Van Allen belts. In the process, Mars mission concepts became more sophisticated and mission planning moved from flyby to landing. Innovative concepts emerged for minimizing mission weight, usually by cutting down on propellant; scientists proposed, for instance, that a spacecraft could dip into a planet's atmosphere to slow it down for landing or to achieve orbit, a maneuver called aerobraking. This technique would be used first in 1993 by the Magellan spacecraft to adjust its orbit around Venus; the first use of aerobraking on a Mars spacecraft was in 1997 when Mars Global Surveyor aerobraked to achieve orbit.

Studies continued and plans were put forward at breakneck pace. By November 1964 NASA had, in a mere seventeen months, shelled out $3.5 million (about $20 million in current dollars) on twenty-nine piloted planetary mission studies, each adding something new to the overall body of knowledge in the field and gradually building consensus on how a human expedition to Mars should be conducted.

Over the next seventeen months, however, NASA would commission only four humans-to-Mars planning studies at a total cost of less than half a million dollars—studies that would, for the most part, represent increasingly desperate attempts to salvage a cut-rate human mission to Mars from some of the most ambitious planning in NASA's history. In retrospect, though, even those efforts were doomed to failure. By the end of 1964, nearly five years before the landing of *Apollo 11,* the prospects for human missions to Mars were already as good as dead.

By late 1964, amid the mounting cost of America's escalating military involvement in Vietnam and growing social unrest at home, the Johnson administration was waging an ongoing battle with Congress simply to keep Kennedy's moon mandate funded. This was not lost on NASA leaders, who feared that the agency's first-ever cutback might occur in fiscal 1966. As a preemptive strike of sorts, in November 1964 NASA presented a plan to Johnson that proposed the use of existing Apollo technology for future activities, rather than developing next-generation spacecraft: this resulted in George Mueller's Apollo Applications program. While it did produce Skylab, Apollo Applications was a major blow to the humans-to-Mars planners, who were counting on new heavy-lift rockets in their planning. To make matters worse, NASA's concession didn't have the desired effect: as feared, Congress dealt the agency its first cutback in fiscal 1966. That cutback would be the first of many to follow, and plans to develop a new heavy-lift rocket would never be resurrected.

In October 1968, Thomas Paine was named acting NASA administrator, taking over from the retiring James Webb. By the spring of 1969, with the launch of *Apollo 11* only months away, Paine had become convinced that an aggressive humans-to-Mars program—boldly presented to the Nixon administration, which he hoped would embrace it and then boldly present it to Congress—was the endeavor that could ensure NASA's continuing robustness. Working with George Mueller, Paine crafted a daring twenty-year plan that included several space stations, a lunar base and a new reusable spacecraft called a space shuttle.

Built atop the plan's other components was an ambitious humans-to-Mars scheme developed by none other than Wernher von Braun. Although a decade had passed since his apex as a public icon, von Braun still held substantial sway among scientists and certain members of the American political class. On August 4, 1969—only weeks after the successful landing of *Apollo 11* had sent NASA's stock soaring in political and public circles—Paine and von Braun pitched their plan to a special advisory group set up by President Nixon to give him direction on what to do next in space.

Von Braun's Mars plan was typically ambitious, yet tailored to the emerging new reality of post-Apollo space exploration. Targeting a

1982 departure from Earth orbit to Mars, von Braun proposed to use Saturn rockets to boost the components of a Mars spacecraft into orbit. The mission itself would consist of two spacecraft, each with six crew members, that would each be composed of three space shuttles powered by nuclear engines (von Braun called them nuclear shuttles). The shuttles would be ganged together around a "mission module" consisting of a crew habitat, a piloted landing craft and a variety of automated probes that would land and assess whether any hazards existed to prevent a human from landing.

If things were OK, a three-person crew would land and conduct a one- to two-month survey of the planet. The nuclear shuttles would then return the expedition crew to Earth (von Braun pointed out that the nuclear shuttles, like other space shuttles, could be refurbished and reused). The price tag for von Braun's plan—by far the most sweeping and carefully considered of any yet presented—was roughly twice what the Apollo program had cost.

The pitch landed with a thud. About a month later, a Nixon-appointed space advisory committee issued its own report, including much of NASA's visionary language and acknowledging the potential of space to enhance America's future. However, there was little advocacy for an aggressive humans-to-Mars program and, more critically, no call for the funding necessary to turn the plan into reality. More nails in the humans-to-Mars coffin would follow in short order. The 1970 NASA budget was the lowest yet, and that January, Paine announced that the Saturn production line (suspended toward the end of James Webb's tenure) would be shut down permanently. Soon afterward, Apollo, too, was dead. Under heavy pressure from the Nixon administration, Paine canceled a planned tenth lunar landing in favor of launching Skylab, thus leaving the *Apollo 17* crew to conduct NASA's final lunar mission in the cold twilight of an all-too-brief era of unprecedented human exploration.

Although the Paine–Mueller–von Braun Mars plan hadn't been embraced by the Nixon administration, it also hadn't been eliminated outright. Taking that as a positive sign, NASA's Mars planners quickly set about further refining the plan in the context of the new post-Apollo reality for space exploration. Much of their hope rested with a nuclear propulsion system called NERVA on which development had continued apace since the late 1950s with ever-increasing success, including a faultless sixty-minute test firing in 1967. Nuclear engines held the promise of greater efficiency and lower fuel-mass requirements, both highly desirable for long-haul human spaceflight. However, NERVA was also destined for the congressional chopping block: the last NERVA test was conducted in July 1972, and the program was terminated soon afterward.

Frustrated with the lack of top-level support for space exploration and the calcifying bureaucracy that was enveloping NASA, Wernher von Braun retired from the agency on June 10, 1972, to work for a private engineering firm. (He died in 1977, still faintly hopeful of a von Braunian space future.) Stripped of funding and advocacy, if not hope and morale, NASA stopped planning, in any meaningful way, to send humans to Mars.

And so it would remain for nearly two decades, as first the space shuttle and then the space station would turn NASA from a pioneer of new frontiers to a government jobs project, a personality shift not lost on many of its scientists—some of whom, like Dennis Tito, decided to jump ship. Humans-to-Mars planning would continue among small pockets of scientists and engineers both at NASA and at its aerospace contractors, but most of these efforts were either labors of love conducted by scientists and engineers in their spare time or underfunded exercises in organizational morale building. Mars would remain firmly on the proverbial back burner until 1989, when President George H. W. Bush dropped a bombshell in the midst of his remarks commemorating the twentieth anniversary of the *Apollo 11* landing.

When the dust cleared, the movement to send humans to Mars would be owned, with an iron grip—and for better or worse—by Robert Zubrin.

If you were to be lifted straight from the comfort of your living room and dropped onto the surface of Mars, the only upside is that you probably wouldn't remain conscious long enough to fully register your fatal predicament. Unless you were sitting in your living room wearing

Antarctic survival gear, you would most likely go into thermal shock almost immediately: the average temperature on Mars is about minus 82 degrees Fahrenheit. Even if you could stay warm enough, you wouldn't be able to breathe. Unlike Earth's atmosphere (a comfy mix of about 78 percent nitrogen, 20 percent oxygen and a smattering of other gases) the Martian mix is pure poison for humans: about 95 percent carbon dioxide with a bit of nitrogen and trace elements thrown in.

There's also the problem of atmospheric pressure: Mars doesn't have any, at least not compared to the Earth. The thickness of the Martian atmosphere is less than 1 percent of Earth's, which means that being on the surface of Mars without a pressurized space suit or spacecraft is only marginally better than being naked in the vacuum of space. Humans have been built to order for Earth's atmospheric pressure: too much pressure, and our body's all-important cavities and passageways (which allow us to respire and circulate blood and other fluids) get compacted. Too little pressure and everything expands, which is why creatures adapted to life in the crushing pressure of the deep ocean literally explode when brought to the surface. An unprotected human body on Mars would face a situation somewhat less dramatic than that of a relocated deep-sea creature, but you might find yourself wishing for such a quick and explosive demise, since the end result would be similar—it would just take a little longer.

So here's the summary: dropped unprepared onto the surface of Mars, you'd go into thermal shock, inhale lungfuls of poison, begin to feel extreme unpleasantness in your insides (that would be your blood beginning to boil in the super-low atmospheric pressure) and die within a few minutes. And have no illusion that you would be found in a well-preserved state sometime in the future by better-prepared Martian explorers. Adding insult to injury, the ultraviolet radiation hitting the surface of Mars (unimpeded by any ozone layer, as it is on Earth) would quickly char your corpse black. Soon afterward, your body would mummify as its liquid was sucked away by the near-vacuum Martian atmosphere.

The bottom line is that while it may be the most hospitable place in the solar system other than Earth, Mars is not exactly welcoming. Still, for all of this, Robert Zubrin wants to send you there, permanently. And soon.

"In the twenty-first century, Earth's population growth will make real estate here ever more expensive," Zubrin wrote in his 1996 book *The Case for Mars.* "At the same time, the ongoing bureaucratization of daily life will make it ever harder for strong spirits to find adequate means for expressing their creative drive and initiative on Earth. A confined world will limit opportunity for all and seek to enforce behavioral and cultural norms that will be unacceptable to many.

"A planet of refuge will be needed, and Mars will be there."

One way or another, Mars has always been part of Robert Zubrin's life since he was growing up in Brooklyn, where he was born in 1952. Zubrin's father was the marketing manager for Topps Chewing Gum— the company that brought the world the trading-card series "Mars Attacks!" which was later made into a zany movie of the same name. Zubrin told me that "Mars Attacks!" was his idea: "They were looking for a new kind of trading card, and I said, 'Why not invaders from Mars?'"

By the time the trading-card series appeared, Zubrin was already well on his way to being a full-fledged space enthusiast. "I remember I read this book *Rusty's Spaceship,* about some kids who were building a toy ship, out of wood. Fortunately, an alien lands nearby and they take the wooden ship and put it on the flying saucer and it becomes a real spaceship and they visit the planets and so on. That's just one example of the kind of children's science fiction I was reading by the time I was about five.

"And then Sputnik flew, and then Sputnik Two, which had Laika [the world's first space traveler, a dog] in it. I can remember looking at the picture of Laika in the *Weekly Reader,* which was a children's magazine. I was looking at it and looking at it. And to the adult world, while Sputnik was a terrifying event, I remember being exhilarated by it, because what it said is that this whole world of space and science fiction I had been reading was real, that this was the future, that this was happening, that there was going to be this grand adventure of expansion into space. So I wanted to be part of it."

Much like Homer Hickam, the coal miner's son turned NASA engineer who wrote the book *Rocket Boys* (which in turn became the much-lauded 1999 film *October Sky*), Zubrin was inspired by Sputnik to begin building and launching rockets himself. In urban Brooklyn—unlike

the wide-open spaces of rural West Virginia where Hickam experimented—this was no small feat: "Let me put it this way, I never set the
house on fire and no one was injured," he told me with an impish grin.
Prompted by the Jacques Cousteau adventures popular on American
television at the time, Zubrin and his friends built a diving bell in
which he stayed underwater for over an hour in nearby Little Neck
Bay. Zubrin's teenage engineering team also, he says, started working
on a submarine that they never finished.

At the same time, Zubrin was gobbling up every math and science
course his high school could throw at him, and eventually he converted his predilection for invention into a bachelor's degree in engineering from the University of Rochester. It was at the university, in
the early 1970s, that Zubrin was first "alienated from the government
enterprise" as he got caught up in the anti-Vietnam protests of that
period. He also tasted his first bit of cynicism about space exploration.

"Somehow, from someone, I got the message that real people are not
astronauts, real people are not planetary mission designers, those are
the people you see on TV," Zubrin said. "They're like movie stars; they
exist in this other reality on the other side of the TV screen. Real people
are teachers, doctors, lawyers, accountants: this is what you can be."

Taking such discouragement to heart, Zubrin detoured from a career
that might, one way or another, have gotten him into space. Instead, for
the next eight years Zubrin taught science and mathematics at both the
elementary and high school level, eventually finding himself back in
Brooklyn teaching students at the prestigious Brooklyn Boy's Chorus
School. Those students redirected the course of Zubrin's life back
toward Mars.

"I always tried to convince the kids, 'Look, if you do something
important in science, that's something that will be important five hundred years from now, as opposed to, you know, being a basketball star.'
And then the kids would say, 'Well, how come you're not doing it?'
And I had a ready response, which was 'I'm teaching you and that's
part of the overall process of science, and so on.' That's a logical argument, but part of me was thinking, 'You know, they're right. I want to
be a scientist. I shouldn't be here.' "

As he watched his boyhood dream of expansion into space wither after the Apollo era, Zubrin began to be haunted by the feeling that something more directly connected to space awaited him. In 1983, at the age of thirty-one, Zubrin left teaching and enrolled at the University of Washington, which had a strong nuclear engineering program. At that point, he was convinced that the most important invention of the remaining years of the twentieth century would be controlled nuclear fusion: "I thought fusion power—and I still think this—was not only an energy source for Earth but it would make possible interstellar travel and it's the only source that's likely to do so." Believing he could make his contribution to space exploration through this avenue, Zubrin finished his master's degree and began work on a doctorate, but by that point—the mid-1980s—the American nuclear-fusion program was in contraction. Around the same time, Zubrin discovered the Case for Mars.

Before it was the title of the book that would help propel Zubrin to fame, the Case for Mars was a conference series. Before that, it was part of the title of a 1978 paper by Benton Clark, an engineer with Martin Marietta, the prime contractor for the Viking probe that sent back the first images from the Martian surface. Viking was a turning point for interest in Mars—the first spacecraft to successfully land on Mars, take pictures and, critically, conduct experiments to determine whether life existed there. Although it failed to discover evidence of life, Viking did tell scientists much about resources in the Martian atmosphere and soil that could, possibly, be exploited by human explorers through a process called In-Situ Resource Utilization (ISRU)—in other words, living off the land.

In "The Viking Results—the Case for Man on Mars," Clark noted the rich elemental resources Viking had discovered and posited that Martian explorers could manufacture rocket fuel and water using machines that, say, broke down carbon dioxide into water and methane via interaction with hydrogen. Further processing would yield liquid methane and liquid oxygen, an excellent combination for rocket fuel. The ability to manufacture fuel and water on Mars could significantly cut down on mass loads for Mars expeditions, a major technological hurdle in Mars mission planning. Clark's paper also included a bit of humans-to-Mars salesmanship that would have made Wernher von Braun proud:

"The ultimate scientific study of Mars will be realized only with the coming of man . . . who, with his inimitable capacity for application of scientific insight and methodology, can pursue the quest for indigenous life forms and perhaps discover the fossilized remains of an earlier biosphere."

Although NASA had first explored ISRU during the Apollo program, Clark's paper became a clarion call to the Mars faithful. It revived ideas of establishing bases on Mars by showing how such bases might maintain themselves without expensive launches of water and fuel from Earth. It also sparked the first serious and widespread consideration of an idea that had previously never made the leap from science fiction to actual science, an idea as grand and wild as they come. With enough extant raw materials and sufficiently developed ISRU, it might be possible for future residents of Mars to warm up the planet a bit, thicken its atmosphere to the point where humans could walk around without pressure suits (although they'd still need scuba-style breathing devices) and create, or reinstate, liquid water on the Martian surface—transforming Mars, in other words, into a place resembling, at least superficially, Earth. The idea had been given a name some years before: terraforming.

The term "terraforming" has its origins in the pages of fiction, courtesy of an American pulp sci-fi writer named Jack Williamson, who coined the term in a 1942 novella called *Collision Orbit*. While the folks at NASA dreamed of being pioneers on a new frontier, Williamson had actually done it: born in 1908 in what was then still the Arizona Territory, he was seven years old when he arrived with his family in New Mexico, courtesy of a covered wagon.

For Williamson, terraforming was little more than a convenient plot device, and for years the idea remained in the same category of sci-fi fancy as *Star Trek*'s matter-scrambling transporters and crystal-powered warp engines. But by the end of the 1970s, humans had sent probes to Mars and had built large orbital structures, and would soon launch the space shuttle. The combination of the shuttle's promise and the findings of Viking sparked an unprecedented round of thinking within the Mars-interested science community in which crazy ideas like terraforming were back on the agenda of possibility.

Two weeks after *Columbia*'s maiden voyage on April 12, 1981, two University of Colorado doctoral students, Christopher McKay and Carol Stoker, convened a public conference in Boulder to brainstorm concepts supporting what they believed was a new era of possibility for the human exploration of Mars. Launch vehicles, habitation structures, ISRU, rovers and timelines for establishing and expanding bases were all on the table. Terraforming was also on the agenda, and conference-goers were thinking very big indeed: Could giant orbiting mirrors reflect enough light onto the Martian poles to melt its ice, thereby creating gushing rivers and sparkling lakes? Would terraforming take a hundred years or a hundred thousand years? Would the Martian population eventually eclipse the Earth's and if so, what would that mean?

The first conference attracted a crowd of three hundred, a turnout whose size surprised the organizers (they had expected perhaps a few dozen). Along with the anticipated engineers and scientists, a large number of attendees were nonscientific space enthusiasts who, for one reason or another, had never gotten Mars entirely out of their heads. The second conference in 1984 drew the former NASA administrator Thomas Paine and produced a Mars mission design so detailed that the Case for Mars leadership was invited to deliver it to NASA. Not unexpectedly, Paine was excited by the reinvigorated thinking about humans-to-Mars: when tapped the next year by Ronald Reagan to head the National Commission on Space (which also included Gerard O'Neill), Paine coaxed that group to conclude that America should have a human outpost on Mars by the early twenty-first century.

In 1987, Zubrin got wind of the third Case for Mars conference (which featured a keynote address by Carl Sagan) and connected with its organizers. The conference changed Zubrin's life. "I was exhilarated," he recalled. "There were so many different ideas being discussed: life support, nuclear propulsion, antimatter propulsion, artificial gravity. It was incredible. I was like a warhorse responding to a trumpet." The Case for Mars conference inspired Zubrin to pursue a second master's degree, this one in aeronautical engineering, which he completed in 1988. Through his new Case for Mars contacts, Zubrin landed an engineering job at Martin Marietta and joined a team doing contract studies for NASA to jump-start its humans-to-Mars agenda, which had been dealt yet another blow when the *Challenger* tragedy once again altered the agency's political landscape.

Shortly after Zubrin joined Martin Marietta, the newly elected president, George H. W. Bush, set up a National Space Council and put Vice President Dan Quayle in charge of coming up with a set of twenty-first-century goals for America in space. The result was the Space Exploration Initiative, which was announced by Bush while flanked by then NASA administrator Richard Truly and *Apollo 11* astronauts Neil Armstrong, Buzz Aldrin and Michael Collins on the steps of the Smithsonian Institution on July 20, 1989, two decades to the day since the first humans had walked on another world. To those disillusioned by the directionless nature of the space program, Bush's words were music:

> We must commit ourselves to a future where Americans and citizens of all nations will live and work in space. We have an opportunity. To seize this opportunity, I'm not proposing a ten-year plan like Apollo. I'm proposing a long-range, continuing commitment.

Bush then proceeded to step through what were, by then, familiar phases of a future plan for exploring space: a space station, advanced development of the shuttle and the establishment of a lunar base—"Back to the moon," Bush hurrahed, "and this time, back to stay." Finally, Bush took it to the level everyone was hoping for: "And then, a journey into tomorrow, a journey to another planet—a manned mission to Mars." Ten months later, during a May 11, 1990, commencement speech at Texas A&M University, Bush did something no president had ever done; he publicly set a specific target for landing humans on Mars, the year 2019, the half-century anniversary of the *Apollo 11* landing.

This was a bold public commitment from a surprising source. As vice president, Bush had demonstrated no measurable interest in space or space policy, reflecting the relative disinterest of his boss, Ronald Reagan. True, Reagan had approved the construction of the space station and kept the shuttle alive after *Challenger,* but those were largely economic decisions rooted firmly on the Earth. Both programs, after all, represented thousands of jobs in dozens of important congressional districts. Bush may also have been looking to use space to stimulate economic growth, or perhaps his Space Exploration Initiative was an attempt to step out from the shadow of one of the most popular presidents in American history by championing a powerful vision of America's place on the high frontier. But instead of making Bush appear Kennedyesque (or even Reaganesque), the Space Exploration Initiative turned into a major embarrassment; sweeping and visionary, certainly, but desperately ill-conceived and fatally flawed from the start.

The Quayle team had consulted with NASA prior to Bush's announcement, but despite the thousands of pages of previous study on sending humans to Mars, the agency didn't have much to offer Quayle in the way of detail. Following Bush's announcement, Quayle directed NASA to conduct a more detailed study, in essence asking NASA to tell the Bush administration, after the Space Exploration Initiative had been announced, what it would take to realize the initiative's goals. For an agency whose glory days had been shaped by a very simple political and economic equation (presidential mandate = big funding), Quayle's directive seemed like an invitation to write a number onto a blank check.

The report NASA sent back to Quayle, the now-infamous 90-Day Study, concluded to no one's amazement that the Space Exploration Initiative was feasible, at least from a technological standpoint. All it would take was an expanded space station (even though after five years, the program hadn't settled on a design or produced a single piece of hardware), a new class of heavy-lift rockets (not yet designed by a team not in existence), a beefed-up space shuttle (this only three years after *Challenger* cast doubt on the entire program), and a lunar base that would mine oxygen from the lunar regolith by a process that even its supporters said would be extremely challenging. As for Mars, the plan was von Braunian in the extreme: build ships at the space station, fuel them on the moon, launch them to Mars, stay for a month or so and then come back.

And it could all be done on schedule and on budget (making it an anomaly in NASA's history) for a mere $400 to $500 *billion* of taxpayer funds.

Beyond the sticker shock, what's stunning is how much of the 90-Day Study's budget was consumed by a cost cushion: as much as 55

percent of the entire budget estimate. This seemed to indicate that
NASA was unsure how to accomplish the Space Exploration Initiative—or how much it would ultimately cost. As word of the 90-Day
Study leaked into the science community, other organizations—some
outside NASA, some within NASA centers—scrambled to offer alternatives for accomplishing SEI. Some of the alternative studies dis-
pensed with returning to the moon and focused on getting humans
straight to Mars. Others focused more heavily on ISRU, and most proposed using existing expendable rockets (forgoing the expense of a
new heavy-lift program) like the Titan or Delta. Significantly, each of
the plans carried budget estimates far lower than NASA's, with most in
the tens (not hundreds) of billions of dollars.

But when all was said, space was NASA's domain, and the official
NASA response to Quayle's directive was the 90-Day Study and its
enormous price tag. The 90-Day Study, therefore, set the baseline for
the Space Exploration Initiative, which was then sent to its death in
Congress: no funding was allocated to the initiative in fiscal years 1991,
1992 or 1993. By fiscal year 1994, with a new president, Bill Clinton, in
the White House, the Space Exploration Initiative had disappeared
altogether.

There are some who believe NASA intentionally killed the Space
Exploration Initiative with the 90-Day Report. At the time Bush announced SEI, NASA was fighting a long, hard battle to fund a replacement for the lost *Challenger;* only the applied muscle of Big Aerospace
and the military made that happen, amid calls for the shuttle program's
termination and a desire to move on to something new. Having just
won a reprieve in a bread-and-butter program, NASA may well have
viewed SEI as a threat to business as usual, just as normalcy was being
reestablished.

From her vantage point at the time as president of the National Space
Society, Lori Garver agreed that SEI seemed to be well received until
the 90-Day Report was returned. "It was only NASA's response that
caused that program to be dead on arrival in Congress," Garver said.
American University professor Howard McCurdy, who has written several books on NASA, has noted that the first Bush administration
clearly felt stung by NASA in the SEI debacle, so much so that the 90-

Day Study was almost certainly a motivating factor in the White
House's call for Richard Truly's resignation in 1992. Garver said that
Dan Goldin was appointed to succeed the ousted Truly in part "to make
SEI a reality." But such a feat was beyond even Goldin's considerable
powers of persuasion; Goldin let SEI die a quiet death and moved on to
the "real NASA" problem of getting the space station off the ground.

That the Space Exploration Initiative would ultimately be killed was
not really in question once the 90-Day Study hit the streets, but the
process of killing it was, like many politically sensitive things, both
subtle and costly. In what amounted to little more than political dumb
show designed to gracefully extract the Bush administration from the
SEI debacle, NASA solicited ideas from the science community and
from its contractors on how to refine SEI, a process called the SEI Outreach Program. One of the companies that participated was Martin
Marietta, and on the company's SEI response team was Robert Zubrin.

"In the aerospace world," Zubrin recalled, "people were saying,
'This is crazy, why don't they just let us do it another way?' But at the
same time, the aerospace companies don't want to be saying to NASA,
'Look, you're stupid' because they want to be responsive, especially
if the RFP [Request for Proposal] says these are the requirements."
Because it had been the prime contractor for Viking, there was a fond-
ness for Mars that permeated Martin Marietta from its staff engineers
to its top brass, so when Zubrin and other colleagues asked Martin's
management for some latitude to craft their own humans-to-Mars proposal, free from any of the SEI requirements, they were granted their
request. The result was Mars Direct.

Mars Direct is less an original idea than an original packaging of
existing ideas. Virtually all of its key concepts can be traced to the
flurry of humans-to-Mars mission plans produced in the 1960s and
1970s, including extensive use of ISRU, a "split-mission" profile using
robotic ships in advance of crewed ships, and the use of artificial gravity to keep astronauts healthy on the way there and back (since long
exposure to microgravity severely degrades muscle and bone). Zubrin
and his Mars Direct collaborator, David Baker, did propose a new
heavy-lift booster called the Ares (the Greek name for Mars), but in
reality the Ares was little more than a reconfiguration of existing hard-

ware: a modified shuttle external tank with four space shuttle main
engines mounted on it plus two solid-fuel rocket boosters.

However, in a significant departure from most previous plans, rather
than rely on spacecraft assembly in orbit or refueling at the moon,
Mars Direct launched its mission straight from Florida to Mars. Three
Ares rockets would launch in carefully timed succession over a period
of several years to take advantage of the relative positions of the Earth
and Mars and therefore the shortest route between the two planets,
which is important because the difference between Mars's closest pass
to Earth and its most distant is more than two hundred million miles.

The first Ares would deposit onto Mars an earth-return vehicle with
an empty fuel tank, an ISRU "propellant factory" and a small nuclear
reactor mounted on a robot truck. In a year, the propellant factory
would manufacture enough fuel for the earth-return vehicle. Three
years after the first Ares launch, two more Ares rockets would lift off:
one would carry a cargo manifest identical to that of the first Ares (as
backup) and the second would carry a four-person crew in a two-story
drum-shaped "habitat" that would land on Mars and serve as home for
a surface expedition lasting more than a year. All of this was a radical
departure from *Das Marsprojekt:* the only major element of von Braun's
original concept to survive in Mars Direct was the pressurized rover.

Mars Direct's other radical departure was its estimated cost, a mere
$40 billion: in current dollars, about half what the Apollo program
cost, and a fraction of the 90-Day Study estimate. True, Mars Direct did
not include a moon base and it bypassed the space station altogether,
but everyone agreed that the ultimate goal of SEI was Mars, which
Mars Direct convincingly proposed to deliver at a reasonable cost.

Zubrin and Baker went public with Mars Direct at a National Space
Society conference on April 20, 1990, in Anaheim, California. Shortly
thereafter, the *Boston Globe* ran a front-page story on Mars Direct that
was syndicated around the world, giving Zubrin his first major media
exposure. The Martin Marietta brass was enthusiastic enough about
Mars Direct to send Zubrin and Baker on a promotional tour of NASA.

In 1991, Zubrin was called in to present Mars Direct to a new Bush-
appointed advisory body called the SEI Synthesis Group, which was
charged with pulling together and summarizing the various inputs

(more than two thousand in all) derived through the SEI Outreach Program. The chief innovation of the resulting report, ominously titled *America at the Threshold,* was the elimination of the space station as a necessary step in sending humans to Mars (although it did not recommend the elimination of the space station altogether). Otherwise, the Synthesis Group more or less stuck to the 90-Day Report script: lunar base and lunar resources, heavy-lift vehicles, hundreds of billions of dollars.

Zubrin's last shot at making Mars Direct the SEI Mars architecture came in the summer of 1992, when he was asked to advise yet another NASA group that was charged with further refining the recommendations of the SEI Synthesis Group: in retrospect, the last gasp of SEI. The resulting plan was known as the Mars Design Reference Mission and was heavily influenced by Mars Direct (Zubrin derisively calls it Mars Semi-Direct). The plan recommended less reliance on ISRU, a six-person crew rather than Mars Direct's four, and splitting the function of Mars Direct's earth-return vehicle across two vehicles: one that would ascend from the Martian surface to dock with a second vehicle in orbit, which would make the trip home. (Mars Direct had the Mars explorers blasting from the Martian surface to Earth in one vehicle.) At about $55 billion, the Design Reference Mission cost more than Mars Direct, but even at that price it was a vast improvement over the 90-Day Study.

NASA accepted the Design Reference Mission as its official plan for sending humans to Mars, then proceeded to dismantle the internal NASA infrastructure needed to make it happen. The Exploration Office at NASA headquarters, set up by Truly in 1987 to give some heft to moon and Mars planning, was closed at the end of 1992 and its responsibilities were shuffled down to the Johnson Space Center's Planetary Projects Office (which was closed in 1994). By the early 1990s, that great sucking sound heard throughout NASA was Space Station Freedom, vacuuming funding from every corner of the agency.

The Design Reference Mission itself survived and was dusted off several times in subsequent years whenever NASA felt confident enough to talk about humans-to-Mars without any perceived threat to the station-shuttle conglomerate: the excitement generated by the Pathfinder/Sojourner mission in 1997 was one such occasion. However, once snug within the space station and shuttle programs, NASA was not about to encourage anything so grand as a mission to Mars.

Zubrin, once in the media spotlight, proved to be adept at staying there. He was an engaging public speaker, whether talking in excruciating technical detail to groups of engineers, or sermonizing to general audiences on the promise of Mars exploration: a talent that is perhaps a legacy of his early teaching days. Whatever the reason, reporters began to call Zubrin more frequently, both to talk about Mars Direct and for general commentary on space-related issues. After the death of SEI, Zubrin's Mars Direct coauthor, David Baker, retreated back into his lab, then left Martin Marietta to start a private consulting firm. At that point, Mars Direct became Zubrin's baby.

Eventually, Zubrin found his roles as high-profile humans-to-Mars advocate and Big Aerospace engineer to be incompatible, particularly as the former role involved increasing public criticism of NASA, a primary customer of his company. In the mid-1990s, Zubrin followed Baker's suit and left Martin Marietta to form his own company, Pioneer Astronautics, which develops and promotes technologies to advance human exploration of space. One of his first efforts was collaboration with Mitchell Burnside Clapp on the Pioneer Rocketplane concept (Burnside Clapp told me they parted ways "more or less" amicably).

In most subsequent media coverage of Mars Direct, Baker was forgotten: it was all Zubrin. With his engineering pedigree and easy eloquence, Zubrin was anointed by the alternative space community as the new champion of humans-to-Mars, perhaps even the natural heir to von Braun: an ironic twist to be sure, since Zubrin was known to frequently ridicule *Das Marsprojekt* and its numerous descendants as the "Battlestar Galactica" approach to sending humans to Mars.

Zubrin also began to speak more broadly about what Mars might mean to the future of humanity. The vehicle in which Zubrin's views and ideas finally found a focused voice, his 1996 book *The Case for Mars* (cowritten with Richard Wagner), bears more than a passing resemblance to O'Neill's *The High Frontier*. Like that book, *The Case for Mars* had something for everyone: science-y detail for the nerds, historical perspective sympathetic to the NASA faithful (while also critical of its leadership), and depictions of Space Age dreams painted with enough humor and eloquence to appeal both to those enthusiastic about space exploration and those skeptical of its value yet interested enough to pick up the book in the first place.

Amid the step-by-step space engineering lessons and historical perspective, as *The Case for Mars* builds up a head of steam, a curious thing happens: readers find themselves in the midst of a call to arms— a call, in essence, to invade Mars. Using Mars Direct as the mission platform, Zubrin envisions at first a small cluster of successive landing craft with a rotating complement of explorers shuttling in and out much like Antarctic researchers at their remote science bases. Within about a decade of the first human landing on Mars, Zubrin envisions a veritable town of habitats clustered on the barren Martian surface. In-situ manufacturing of materials, powered by solar and nuclear energy, allows other structures and tools to be built. "As more people steadily arrive and stay longer before they leave, the population of the town will grow. In the course of things children will be born, and families raised on Mars—the first true colonists of a new branch of human civilization."

For Zubrin, though, this is only the beginning. Mars colonization, he writes, will be "the keystone to human expansion throughout our planetary system." Terraforming is the ultimate means of building humanity's new home, with Zubrin saying it might take as little as fifty years to warm Mars up enough to give it a thicker, more human-friendly atmosphere, albeit still a poisonous one. That timeline, too, could become shorter as yet-to-be-developed engineering tricks accelerate the increase of oxygen and nitrogen in the Martian atmosphere. In his gripping *Red Mars* trilogy, for instance, writer Kim Stanley Robinson turns Near Earth Object nightmares on their head by having his Mars colonists intentionally redirect water-rich comets to smash into Mars, albeit in uninhabited areas. Some scientists have proposed the artificial generation of greenhouse gases—something we try to avoid on Earth—to warm things up on Mars and help block deadly radiation.

The true potential of such schemes is hard to determine precisely, because even after all this time, there's a lot we don't know about Mars. Robinson makes life easier for his *Red Mars* colonists by having them discover aquifers practically bursting with liquid water, but this is inference at best where science is concerned. As to whether an artificially generated greenhouse effect might improve a planet's livability, the results of such activities on Earth have not exactly caused people to cheer for more of the same.

When it comes to concepts like terraforming, though, the knowledge gap also gives Zubrin plenty of room to wax expansive about Mars's potential as a vast new frontier. While he nods here and there to the rest of the world, Zubrin's case for Mars is clearly pitched at America: "Looking around the world today, it is not difficult to pick out dozens of small nations in Asia, Africa, the Middle East, the former Soviet Union, and Europe that adjoin larger nations that now or in the past have demonstrated the desire to subjugate their neighbors. Again, there will be wars, and losers, and millions of emigrants willing to take on the hard challenges of making a new life on the frontier rather than accept subjugation."

In offering such heady (not to mention inflammatory) stuff, Zubrin makes no secret that he completely accepts Frederick Jackson Turner's frontier hypothesis, which he calls "a brilliant insight into the basis of American society and the American character." At some points in *The Case for Mars,* Zubrin seems to be writing nothing less than an extraterrestrial sequel to Turner's 1893 treatise. Consider the following from Turner:

> The existence of an area of free land, its continuous recession, and the advance of American settlement westward, explain American development. Behind institutions, behind constitutional forms and modifications, lie the vital forces that call these organs into life and shape them to meet changing conditions.

In *The Case for Mars,* Zubrin picks up almost exactly where Turner left off:

> The true value of America was as the future home for a new branch of human civilization, one that combined its humanistic antecedents and its frontier conditions to develop into the most powerful engine

for human progress and economic growth the world had ever seen. Every feature of frontier American life that acted to create a practical can-do culture of innovative people will apply to Mars one hundred-fold.

Zubrin showed his Turnerian bent again in a 1999 follow-up to *The Case for Mars* called *Entering Space: Creating a Spacefaring Civilization.* Yet Zubrin clearly understands that there is a time for evangelical sermonizing and a time for more measured political savvy. In our conversations, Zubrin sounded less like Moses than like Colin Powell: he discussed technology, talked about how to generate the "political will" for exploration and urged building coalitions of private and public organizations to achieve Martian exploration goals. Like any good politician, Zubrin tailors the message to the audience. In a May 2001 speech to a U.S. Senate committee (stumping for more Mars funding in NASA's 2002 budget) Zubrin's most extreme bit of Turnerism was to state the "need to rekindle the national optimism that made the United States the greatest country on Earth. . . . A strong sense of national optimism is the best vehicle for continued prosperity."

Zubrin's rhetoric has consistently found keen audiences, often in the oddest of places. In the early 1990s, Newt Gingrich summoned Zubrin to Capitol Hill for the first of what turned into a series of meetings about how to advance Zubrin's Mars agenda: the meetings occurred around the same time that Gingrich was considering advancing Charles Miller's lunar-settlement proposition. Gingrich was impressed with Mars Direct and wanted to support Mars explanation with legislation, and he directed his conservative Progress and Freedom Foundation to work with Zubrin to that end. The result was the Mars Prize bill: the U.S. government would post a $20 billion award to the first private organization to successfully land a crew on Mars and return them to Earth. Zubrin claimed that minus NASA requirements and NASA-style management, the *real* cost of a manned mission to Mars was more like $4 to $6 billion; thus, the Mars Prize winner stood to reap a handsome profit.

The Mars Prize was, again, distinctly American-focused. Conditions included that a majority of the crew be Americans and that at least 51 percent of the cash value of all hardware used to win any of the prizes must be manufactured in the United States. The winner would also be required, at the government's option, to sell up to three additional copies of the winning flight system to the U.S. government, at a cost per copy of no more than 20 percent of the prize. In return, the U.S. government would support all competing missions with NASA's space tracking network and ground and mission-control support from the Kennedy Space Center and other launch sites, at cost.

With Gingrich's fall from grace during the second Clinton administration, the Mars Prize went nowhere except into the pages of fiction, where a multibillion-dollar private incentive serves as the basis for the first human mission to Mars in the 1999 novel *The Martian Race* by Gregory Benford. Instead of a government endeavor, though, the money in Benford's story is put up by a mega-billionaire whose "Mars guru" is none other than Robert Zubrin: older and grayer (since the first mission doesn't launch until 2016) but still, as Benford describes him, "hot-eyed" with passion for sending humans to the red planet.

By 2016, Zubrin will be sixty-four years old: not an unreasonable age to realize one's lifelong dream. But fictional timelines are notoriously difficult to live by, a lesson the space community has learned the hard way. By the time Zubrin left Martin Marietta in the mid-1990s, SEI was dead, the Mars Prize was taking up shelf space in Congress and NASA's humans-to-Mars planning was once again dormant. In short order, Zubrin's chief challenge was to find something new, something tangible, to keep people engaged with the idea of sending humans to Mars until NASA or someone else could be convinced to make it happen.

Without some demonstrable proof that a human mission to Mars was possible and of interest to the broader public, Zubrin knew that he might well find himself celebrating his sixty-fourth birthday and still wondering why humans hadn't yet set foot on Mars. Then in 1998, he met Pascal Lee.

Mars on Earth

The day has ended badly for Pascal Lee. Sitting on a plastic crate on the lower level of the spacecraft he commands—a two-story cylindrical tub perched on struts at the rim of an enormous crater—Lee has just told me that the medical officer for his new crew will not be coming. There is no last-minute emergency, no change of heart, no schedule conflict and no logistics problem. There is, however, a serious political problem emanating from a familiar source: NASA.

We are not, of course, on Mars proper, and Lee's spacecraft will never go into space: it is, as Lee describes it, "a lo-fi simulation," although a simulation modeled closely on what scientists hope will ultimately be the real thing. Dubbed the Habitat, or just the Hab, it is a semifunctional replica of the landing craft proposed in both NASA's Mars Design Reference Mission and Zubrin and Baker's Mars Direct. The Hab's physical dimensions (about twenty-seven feet in diameter and twenty-six feet high) are such that it could fit comfortably inside the upper stage of a Saturn rocket, or Zubrin and Baker's Ares, or perhaps even a Russian Energiya. Whatever the vehicle, the Hab is designed for launch from Earth to Mars in one piece—no shuttle-intensive orbital assembly required.

But the Hab's design is secondary in terms of why it has captured the attention of the world. Far from being a sleepy research project in a lonely NASA back lot, the Hab sits smack in the middle of a pockmarked, windswept desert called Devon Island, probably the most Martian-like place on Earth. Devon Island is a 54,100-square-kilometer dogleg of land only somewhat smaller than West Virginia, and it is the world's largest landmass completely devoid of regular human habitation.

Lee had invited me to visit Devon Island in August 2001, and I planned to spend eight days shadowing Lee and his colleagues as they went about their work. Lee and I had known each other through the SETI Institute, which cosponsored Lee's work on Devon Island. "There is no place you will ever visit that is like Devon," he'd said to me only a few weeks before my visit. Then he added, with the slightest of knowing smirks, "Unless you go to Mars."

The truth of those words took hold immediately. As Lee and I talk in the Hab, it is about 10:00 P.M. on a late summer evening, yet there is no sign of twilight, nor would there be; at that time of year and only about nine hundred miles from the north pole, daylight reigns for twenty-four hours a day. Outside, a light rain drizzles across the cold boulder-strewn landscape; the clouds crowd low to the ground, trailing off in tendrils of mist and fog. It is starkly beautiful.

Being so high in the Canadian Arctic, it is also a logistical nightmare. Access to Devon Island is by helicopter or, more frequently, via rugged twin-prop airplanes flown by seasoned Arctic bush pilots who land here without control towers, runway lights or runways. Custard-thick fog often prevents anything but seals or polar bears from getting on or off the island for days.

Given this, part of what makes everyone comfortable is having medical expertise on hand at all times. Herein lies Lee's current problem, which now has priority over dozens of other headaches: ensuring that there is enough kerosene for the generators, fixing damaged all-terrain vehicles, writing up research to meet publishing deadlines (which in turn helps keep research dollars flowing) and, above all, preventing several dozen scientific egos from crashing into one another with too much force.

Every week or so, researchers fly in and out of Devon Island in batches: only a few, including Lee, stay for the entire ten-week period during which the Arctic summer makes it reasonably hospitable for human habitation. Leaving on the next plane is the current medical officer, who happens to be Lee's younger brother Marco, an Oxford-educated physician. Ongoing bad weather has kept Marco here for longer than he'd have liked, but this does little to raise his brother's spirits about the situation, the underlying circumstances of which are potentially more far-reaching than a shortage of on-site medical help.

The medical officer slated to replace Marco is Jeffrey Jones, who is based at the Johnson Space Center and who also happens to be NASA's chief flight surgeon for the International Space Station. Since Lee's first expedition to Devon Island in 1998 through the current year, 2001, the Johnson Space Center had provided low-key logistics support for Lee's project. The JSC brass, Lee said, approved of the project since it was costing them little and their scientists were begging to participate; perhaps more than any other NASA center, JSC has never gotten humans-to-Mars out of its system.

In recent years, scientists at JSC and throughout NASA had seen glimmers of hope for the revival of serious planning for sending humans to Mars. Following the devastating back-to-back losses of the Mars Polar Lander and Mars Climate Orbiter in 1999, NASA's robotic Mars exploration program was back on track in 2001 with Mars Odyssey, which successfully attained orbit that October (the probe is expected to return data through August 2004). Among its payloads was an experiment that was explicitly focused on future manned missions, a small device called the Martian Radiation Environment Experiment, or MARIE. Developed at JSC, MARIE was intended to "characterize the Martian near-space radiation environment as related to radiation-induced risk to human explorers," according to the Mars Odyssey web site. As MARIE was being prepared for Mars Odyssey, NASA quietly issued, in February 2001, a Request For Proposals offering small grants for research focusing on human exploration, including habitats and ISRU fuel generation. All of that smelled like early preparatory work for a human mission to Mars.

A few months after the Mars RFP was issued, NASA announced the jaw-dropping $4.8 billion ISS cost overrun. With no money forthcoming to cover the shortfall, NASA headquarters began eyeing projects for termination. Centers began scrambling to demonstrate that everything they were doing was somehow vital to the station: the tail was once again wagging the dog. While Mars Odyssey was already on its way (several million miles into space being safely beyond the reach of the budget cutters), the Mars technology RFP was quietly retracted, as were broader hopes that NASA might finally make a bold commitment to human-centric Martian exploration.

In the bigger picture, NASA brass had anticipated the impact of the

ISS cost announcement the previous year. In October 2000, announcing NASA's Mars plans for the next decade, Associate Administrator Ed
Weiler talked loudly about the agency's 100 percent robotic approach
and was particularly blunt in underscoring that the program "is not
driven by human exploration." The ripple effect from the ISS cost over-
run reached Pascal Lee just before the start of the 2001 Devon Island
field season, when JSC canceled its official participation: they were
preparing to run a mock "mission control" operation as outlined in
NASA's Mars Design Reference Mission.

Once the field season began in late June, though, NASA researchers
were still showing up; the would-be medical officer, Jones, a veteran of
the 2000 field season, insisted that he, too, could still make it. Then
came the word. "It's come from the highest level at JSC," Lee said
dejectedly as we sat in the Hab. "Even if he [Jones] uses his own time
and money they're saying he can't come. It's the publicity. The project
has become high-profile and JSC has suddenly realized they're not supposed to be supporting human exploration of Mars."

The doctor debacle was symbolic of Lee's prime dilemma: how to
conduct an ambitious humans-to-Mars research project using NASA
resources and NASA researchers at the same time that NASA, officially,
didn't want anything to do with it. No one disputes that good science is
happening on Devon Island. What's more, since it operates under a
NASA contract, the project is one of the better examples of research
done faster, better and definitely cheaper: the whole shebang on Devon
Island costs about $500,000 a year, barely an accounting error in
NASA's annual budget.

Lee created the Devon Island project almost entirely on trust: people had provided spare equipment, redirected grant money (including
NASA dollars), quietly allowed their graduate students to disappear in
Lee's care for weeks on end, and generally accepted that Pascal Lee
would make it all work, which he did. By 2001 interest in the project had
grown so fast that Lee had to put a cap on the number of participants—
and still, for the second year in a row, Devon Island would host more
than a hundred people during the June-to-August field season.

An ever-growing number of visitors were media representatives who
wanted to tell the world about what was happening on Devon Island.
In Lee's opinion, that was good because the world needed to get excited

about sending humans to Mars if it was ever going to happen: few scientists were counting on an Apollo-like funding miracle. But greater publicity meant that more NASA scientists might find themselves in Jones's position, possibly including Lee himself.

Oddly enough, the "spacecraft" we were sitting in was a major source of Lee's troubles, since technically it wasn't part of Lee's NASA project at all. As we talked, most scientists on Devon Island were esconced in a small tent city downridge from the Hab, quietly going about their research. Only a few scientists were actually based in the Hab, yet it was the Hab that had drawn most of the publicity that had excited the world and awakened NASA's spin doctors—the Hab and its titular owner, Robert Zubrin, whose drive to bend every research effort on Devon Island to his own ends had driven Lee and many of his colleagues to distraction, if not outright anger. The greater irony is that Lee knew he had no one but himself to blame: not only did Lee help Zubrin get onto Devon Island, but Lee proposed the whole Hab concept in the first place.

Enmeshed in a situation of Jacobean complexity, Lee confessed that every now and then he needed to visualize for himself something beyond the political and logistical details that seemed so often to consume his life; he needed to remember the larger goal. As we sat in the perpetual twilight of that Arctic evening, I watched a moment of calm clarity come over Lee, almost as if bestowed upon him from beyond. We sat and talked about the JSC headache for a few minutes, and then Lee's gaze wandered to one of the Hab's round portholes; his eyes narrowed slightly and in an instant, his mind seemed to be flying across the rocky landscape, perhaps already on a trajectory for Mars itself.

"Once we have been to Mars and explored Mars thoroughly, I don't want to settle there, to build a house and live there," he said, as if from a great distance. "I want to move on to the next planet and the next. I feel like we are so close sometimes. I feel we could really get there in my lifetime."

One can only wonder why Devon Island ended up without human habitation of any kind while nearby Cornwallis Island became home to the tiny horseshoe of buildings collectively known as Resolute, the second most northern municipality in the world and the staging base for the research being conducted on Devon. Cornwallis Island has a long history of being unloved: in 1854, the crew of HMS *Assistance,* which got stuck in the area while looking for the missing British explorer John Franklin, wrote in a report that Cornwallis Island "is perhaps one of the most dreary and desolate spots that can well be conceived." The crew may have been especially bitter since the *Assistance* had to be abandoned offshore, one of countless ships that fell victim to the crushing polar ice during the nineteenth century as Britain probed the barren, frigid north (none of Franklin's crew was ever found alive; evidence of cannibalism was later discovered). Save for the prefabricated buildings and all-terrain vehicles, one imagines that Cornwallis Island looks today much as the crew of the *Assistance* saw it a century and a half ago: a mud-brown lump of land layered with rubble and shards of jagged limestone, its glacier-scrubbed surface nearly devoid of life.

Viewed from the air beneath a dome of hulking gray-black clouds, Devon Island is a far more intriguing place. Laced with pocket-sized canyons and deep craggy gullies shot through with melt-fed creeks, the island's landscape is painted from a palette of lush chocolate browns and creamy shades of taupe and tan, here and there washed with whorls of golden yellow or pale, gingery orange. Although not usually visible from the air, life is relatively abundant here, including small yet persistent populations of musk oxen, lemmings and Arctic foxes, as well as sizable seal rookeries and one of the Arctic's largest polar bear populations.

Even with its austere charms, few would mistake Devon Island for Shangri-la. But this wasn't always so. More than twenty-three million years ago, when humans were barely a gleam in Mother Nature's eye, Devon Island was something of a minor-league paradise. Geologists believe that freshwater lakes glistened here and rivers coursed across the island's gentle plains, and in those warmer days of the Miocene epoch, fauna on Devon Island were straight out of *Alice's Adventures in Wonderland:* giant rabbits bounded across rolling hills and miniature rhinoceroses grazed amid piney forests.

Things went bad in a big way. The asteroid or comet that came screaming in from space straight toward Devon Island measured about a mile in diameter: even slowed by the atmosphere, the object was traveling at an estimated forty-three thousand miles per hour when it hit terra firma. If you're wondering how fast that is, consider that the speediest fighter jets lumber along at just over two thousand miles per hour at best.

The object slammed into Devon Island with a force roughly equivalent to the explosion of one hundred thousand *million* tons of TNT (by comparison, the atomic bomb that leveled Hiroshima packed a force equal to about fifteen thousand tons of TNT). The impact carved a hole in the island more than seven miles wide and more than a mile deep, blasting into the sky untold hundreds of thousands of tons of rock, dirt, water and biology. In a finger snap, an entire ecosystem was wiped out: plants and animals that had enjoyed the sun only moments before the impact were instantaneously entombed or vaporized altogether. In addition to denuding Devon Island, the impact probably obliterated all life within several hundred miles of the impact point.

But the party wasn't over yet. The steaming mile-high walls of Devon Island's new crater began to collapse inward in a domino effect of rumbling landslides that buckled the earth for miles, ultimately doubling the crater's width. In a matter of seconds, the collapsing earth piled into the initial impact hole in a backfilling job that a thousand bulldozers couldn't achieve if they worked around the clock for a hundred years. Minutes after the impact, Devon Island was a scorched, lifeless wasteland.

To all appearances, Haughton Crater today is a wide, lumpy valley that resembles what most people think of as a crater only in satellite photos. Because of the sheer scale of the impact and the massive backfilling that followed, the fourteen-mile-wide crater is simply too big and too shallow to see in its entirety from ground level: the drop from highest rim to deepest floor is at most a few hundred feet. But what it lacks in dramatic topography (the much smaller Meteor Crater in Arizona, where the Apollo moon walkers trained, is far more telegenic), Haughton Crater more than compensates for in the sheer immensity of its scale, and the unearthliness of its landscape. The terrain is a confusion of sedimentary muck, flash-baked limestone, stony outcroppings, and ridges of hard, grayish stuff called breccia—molten earth baked and blasted from the initial impact.

The wild landscape often surprises even veteran scientists working here. The plains are unpredictably squishy with Arctic quicksand that is likely to suck in all-terrain vehicles to their wheel tops, while scattered fields of car-sized boulders and knife-edged rocks are, in places, nearly impassable. Rocky and wind-blasted, the land a pastiche of reds, browns and pasty yellows, and with not a shred of plant life in sight, Haughton Crater and its environs look stunningly, eerily, like Mars.

Pascal Lee first learned of Devon Island in the early 1990s when he was a graduate student at Cornell University. When he saw the island for the first time, it appeared, as he later described it, to be "Mars on Earth." Although only in his early thirties at the time, Lee may already have been the sole person in the world with the right combination of experience, scientific knowledge and sheer will to build an expeditionary base there, particularly one focused on realizing his lifelong dream of exploring Mars.

To watch the Lee brothers together is to observe the result of an unintentional social experiment. Marco, younger by five years than Pascal, is charming, ebullient and openly emotional: a few days after the September 2001 terrorist attacks in America, Marco sent an e-mail to his friends, including his Devon Island companions, describing his thoughts on what had happened. It was beautifully written and intimate, clearly a note from his heart and soul. Pascal plays his emotions closer to the vest; when one gets too close to a deeply held feeling, he will offer an embarrassed smile, but still you'll only rarely learn what's on his mind. Marco is playfully rakish and naturally sociable; Pascal is cheerful yet possessed of the distant dreaminess of a lifelong loner.

Perhaps the most striking difference between the Lee brothers is their accents. Marco speaks pure Oxford English, clear and precise. Pascal's English is riddled with Americanisms, all wrapped in a confident lilt that is unmistakably Parisian French. For their differences as well as their obvious closeness, the brothers jokingly blame their parents: raised in Hong Kong by a French mother and a Chinese father, Pascal and Marco were always different from everyone else they knew. "It was still very rare in the 1960s to have a Eurasian family in Hong Kong," Marco recalled. "So we went to schools which were 99 percent Chinese and we were obviously Eurasian, and in that sense I guess you could say we were more active. We got into a lot of trouble, playing pranks, not doing homework, that sort of thing."

Being more active had a negative impact on the brothers' performance in school, and so when he was eight years old, Pascal, being the elder, was sent to live with his grandparents in France. It was the defining event of his life. "All of a sudden, I found myself alone at eight years old in boarding school, in a foreign country whose language I did not speak, with completely new customs and new ways of doing things," Pascal told me. "I missed home very much. All of a sudden, I straightened out my act and wanted to do well." (Courtesy of some family friends, Marco would go to England a few years later for his overseas education.)

From being near the bottom of his class in Hong Kong, Lee went to the top of his boarding school class in France. By the time he was accepted at the University of Paris, he had already set his sights on Mars. He had long been a science-fiction fan, but two nonfiction books—the French translation of Carl Sagan's *The Cosmic Connection* and the French scientist Albert Ducrocq's *À la recherche d'une vie sur Mars* (which described the findings of the Viking missions)—converted Lee's fascination with fantasy into a thirst for the real thing. "There was a time I was interested in becoming an engineer and building rockets," Lee said. "But after reading these two books, I made the decision that I was more interested in planets and exploration and what was out there—not so much how to get there."

As a teenager, Lee joined a space club founded by Ducrocq and eventually became its leader. He wrote articles and organized a lecture series in Paris on space exploration. All the while, he was thinking about how he could best position himself to be one of the chosen when NASA or ESA actually sent human explorers to Mars. At the age of nineteen, Lee put the question to Ducrocq, who by that point had become a mentor. "His advice was, 'In order for anyone at this stage to have a hope to go to Mars, you just have to be the best person they have to send,'" Lee remembered. "That was it. It wasn't very directive, but it's true at many levels. Your goal is to be the obvious choice, the guy who just has to be on board that ship for that mission."

And so Lee set about trying to give himself the background of someone who would be an obvious choice when it was time to send humans to Mars. First came his education: after considering engineering and physics, Lee settled on geology, concluding that one of the first activities for humans on Mars would be assessing its physical environment. Having met several astronauts at Paris air shows, he knew that a disproportionate number of space travelers were also pilots. So he got licensed to fly both airplanes and helicopters, and even became an instructor. Being a geologist and pilot was a good start, but Lee was convinced he needed to be, as he put it, "more exceptional."

After graduating, Lee secured a national service posting (required of French citizens) as a geologist in Antarctica, the place where his career as an explorer would begin in earnest. Lee arrived in December 1987 at the French Antarctic station at Dumont d'Urville, where his official tasks were to analyze fluctuations in the Earth's magnetic field and monitor seismic disturbances (including any generated by nuclear weapons testing). Dumont d'Urville's location on a rocky coast gave Lee ample opportunities to learn how to conduct geology in a polar desert—the same climate that exists on Mars. Lee was stationed at Dumont d'Urville for 402 days straight "with thirty-one other men, no women, the strict code of the French military, and nowhere else to go." While some of his colleagues grew depressive, reclusive or even hostile, Lee fell in love with the isolated, harsh polar climate. "It was the experience of my life," he says with an unmistakable tone of fond nostalgia.

While serving his Antarctic tour, Lee was notified that he had been admitted to Cornell University for graduate work, another step in his master plan: "I knew I had to go to America at some point, because that was where space was at." The choice of Cornell was not accidental: it was also home base for much of the imaging work being done for interplanetary probes, including those sent to Mars. Once at Cornell, Lee soon found himself working full-time to help select imaging targets for the next NASA probe, the Mars Observer.

On its final approach to Mars in August 1993, NASA lost contact with Mars Observer, never to regain it—another dreadful blow to the agency's Mars exploration program. Lee was devastated but found some solace by returning to Antarctica in 1995 as part of an American meteorite-hunting expedition. Because of his previous experience, Lee was assigned to a small team charged with scouting out new sites. The team was led by one of the world's most experienced Antarctic researchers, John Schutt, and the expedition would profoundly influence Lee's own planning for Mars research in the Arctic.

"I learned a lot of things from John," Lee said. "One thing I learned in particular is that you don't improvise polar expeditions. The secret is in the planning. If you improvise, if you do things driven by your ego, you end up in a glorious death. If you do it right, you're successful and you actually get things done."

Lee returned from his second Antarctic stint in early 1996 and set about the next phase of his mission to Mars. By 1996, Chris McKay—cofounder of the Case for Mars conferences and of the "Mars Underground" that attracted Robert Zubrin—had become one of NASA's most respected and vocal proponents of sending humans to Mars. Lee says he felt "a kinship" with McKay, who not only was a fellow Mars enthusiast but was himself a veteran of Antarctic research. McKay was based at NASA's Ames Research Center, and Lee told McKay that he wanted to come to California to work with him.

And Lee had an idea. "He said, 'I want to go to Haughton Crater and study it as a Mars analog,' and he wouldn't take no as an answer," McKay recalled. "Pascal didn't just bring the project to me, he sort of jammed it into me. There were some bureaucratic problems with [NASA] headquarters but he stuck with it and pushed it through, and he made it happen."

With McKay's sponsorship, Lee moved to California and secured a National Research Council grant to mount a four-person, three-and-a-half-week expedition to Devon Island in July 1997. While Haughton Crater had been studied sporadically over the years, no scientists had explored the site looking at it as a Martian analog. With their "Mars eyes," as Lee described it, the research team noted valleys and canyons on Devon Island remarkably similar to those seen by robotic probes looking at Mars, which has a surface sculptured in part by just the sort of impact that produced Devon Island's present-day landscape. In short, Lee and his team found far more than they could study in a few weeks. Even before the team left Devon Island, they were planning future, more elaborate operations. They also gave their work a name: the Haughton-Mars Project, or HMP.

In the following months, Lee worked with a team from Carnegie-

Mellon University (CMU) testing robots as meteorite hunting tools, work that resulted in Lee's third Antarctic trip; there would later be a fourth trip as well. Lee invited the CMU team to Devon Island for the second HMP expedition to see how their machines, including a robotic helicopter, might perform as assistants to human explorers on Mars. Thus did the HMP team begin to take on a more multidisciplinary approach, adding specialists in robotics, information technology, biology and, eventually, space-suit and spacecraft design.

HMP '98 lasted for more than a month and involved more than two-dozen researchers. The project's participants were decidedly international, with Americans, Canadians and Europeans all working together in pursuit of their common goal of learning how to explore Mars by analogy on Earth. Multinationalism thus became a signature theme of the work on Devon Island, often in complex permutations: indeed, from the beginning, HMP was led by a French-Chinese scientist employed by the American space agency in an Inuit-governed Canadian province.

Another HMP hallmark was introduced in 1998 when the project received its first private funding: $15,000 from the National Geographic Society to pay for aerial surveys of the crater, something that couldn't be easily justified in any NASA or NRC grant request. Private funding further distinguished HMP from other NASA-led projects. In this, Lee also drew lessons from Antarctic research: early explorers like Shackleton, Amundsen and Mawson had all been backed by private funds.

In addition to the funding, National Geographic also sent a writer and photographer to Devon Island and ran a feature story on the project in its magazine. Thus began the extensive, continuous media exposure that would prove to be both boon and thorn for Lee and the research being done on Devon Island. It would also be the last time that media coverage would focus solely on the science being done on Devon Island. The following year, Robert Zubrin would arrive, and everything would change.

A few months before the start of HMP '98, Zubrin met at the NASA Ames Research Center with McKay and other leaders of the Mars Underground. Still looking for the big idea that would keep his humans-to-Mars agenda in the spotlight, Zubrin planned to form an international organization of Mars exploration advocates that would demonstrate to NASA, private industry and politicians the popularity and viability of sending humans to Mars. Zubrin wanted the Mars Underground on board. At the last minute, McKay invited Lee to the meeting, where he met Zubrin for the first time.

"I knew of him," Lee said. "He wrote really cool books on Mars and I was looking forward to meeting with him. When I met him, he sort of fit the picture. I thought he was very intense. That wasn't a bad thing from my standpoint."

While Zubrin and the others discussed piggyback missions on NASA rockets and other ideas to put the new Mars Society on the map, Lee was formulating a simpler idea: a Mars mission simulator. Simulators have a long and distinguished history in the space program, from centrifuges that replicate the physical effects of liftoff and reentry, to parabolic flights that mimic weightlessness, to high-tech computer-enhanced simulated space shuttle cockpits that even incorporate the vibration produced by the shuttle's engines. Apollo astronauts spent extensive time in simulators, and Lee said that the helicopterlike contraption used to practice lunar landings is what motivated him to go for his helicopter license.

Lee is quick to note that the idea of a Mars simulator is not original, nor is even the idea of a simulator in a polar environment. "It was an idea that had been floating around for some years although very specifically in the context of Antarctica," Lee said. "But Antarctica is really a difficult place to do that, because intrinsically it's about an order of magnitude more expensive than the Arctic and there's a lot of red tape, so it didn't really go anywhere."

When the meeting was over, Lee was in the Ames parking lot and saw Zubrin walking to his car. Like his astronaut heroes, Lee at the time drove a black Corvette, in which he pulled up next to Zubrin. "I said to him, 'I think there's a better thing the Mars Society can do than putting piggyback payloads onto NASA missions. I think we should build a habitat in the Arctic.' "

"It was just the thing to do," Zubrin recalled of the suggestion. "I knew that was the way to begin to make things really happen."

From that point forward, Lee and Zubrin would agree on little else.

During HMP '99, Zubrin landed on Devon Island to discuss Hab site selection with Lee and others. "Basically, Robert turns up in camp and anything we'd say, even if it's something he really should be agreeing with, something basic, he immediately takes the opposite view," recalled Steven Braham, a British physicist who became chief engineer for both the HMP and the Hab. "For instance, we picked the present site of the Hab on Haynes Ridge, and Robert spent a whole week arguing why this was a bad site. He didn't have an alternative; he just didn't want someone else to pick the site. He finally said OK, and then he immediately told the press how he had selected the site."

Surviving the site-selection process was only the beginning of what would become an ongoing battle of wills and egos. After the HMP '99 field season, Zubrin and the Mars Society went to work raising funds to build the Hab, tapping the closet space geeks in the red-hot tech sector (only a year before it went cold) and securing six-figure contributions from the dot-com company Flashline and the private foundation of the Internet entrepreneur Steve Kirsch. Eventually, Zubrin and others secured more than $1 million to cover the Hab's construction and five years of operation, all from private sources.

The Hab was fabricated in parts and scheduled for arrival on Devon Island in the summer of 2000 via airdrop by the U.S. Marine Corps, who were already supporting HMP courtesy of Lee's connections at the Johnson Space Center. The Marines air-dropped seven pallets of Hab equipment onto Devon; the first six landed smoothly. The seventh created the island's second impact crater when the pallet separated from its parachute a thousand feet above the island and then smashed into a ridge at several hundred miles per hour. The Hab's fiberglass floors, a trailer to move the Hab parts to the construction site and the crane needed to assemble the structure were destroyed. The construction team hired by the Mars Society concluded that the Hab couldn't be built and walked off the job.

By all accounts, Zubrin immediately turned his energy to begging, sermonizing and haranguing Mars Society volunteers and even a few stunned journalists to help build the Hab anyway. Lee conscripted some Inuit workers from Resolute, and the group salvaged what they could from the shattered pallet, then jury-rigged construction equipment and even a new floor for the Hab. After a week of sweating, swearing and freezing, the Hab was standing and functional. Zubrin inaugurated the faux spacecraft by smashing a bottle of champagne on its main hatch.

At one point or another, several people told me, almost everyone doubted whether or not they should proceed in building the Hab with the materials and expertise at hand. The only person who didn't have any doubts was Zubrin. "After the construction crew walked off the job, the journalists came up to me and said, 'So, Dr. Zubrin, how would you compare your failure here to that of Mars Polar Lander?' " Zubrin told me later with a triumphant grin. "And what I said was, 'Yeah, there are similarities: we both hit a rock. But the difference is that we have a human team on the scene and we're going to find a way out of the situation.' And that's what we did."

Not everyone shares Zubrin's perspective. "After the fact it looks OK because it got done and no one got hurt, or at least seriously hurt," Steven Braham said, noting that there were numerous cuts, bruises and even sprains from lifting and manipulating the heavy Hab parts, some of which weighed hundreds of pounds. "But there were some situations where people nearly got killed, like pallets dropping and missing people by a few inches. Anything like that could have spelled the end of the project, maybe the end of all the work on Devon."

This would be the beginning of a clash that would play out more fully during the 2001 field season: maintaining a precarious balance between the need for political and financial support and finding the degree to which vision and truth can, or should, be bent to those ends. Those on Devon Island would, once again, struggle to find the balance between spectacle and science: this is the fundamental dispute that has been at the heart of every major effort in the history of space exploration.

The Haughton-Mars Project occupies two modest plateaus looking out over the northwestern rim of Haughton Crater, one for the base camp and another for the colorful quilt of small, hard-weather tents that house an ever-changing group of researchers. The base camp has a stun-

ning view of von Braun Planitia, a vast and undulating plain stretching
as far as the eye can see. The ridges around the base camp are covered
with fragments of petrified ocean bottom, blasted to the light of day by
the Haughton impact practically in one massive, if fractured, piece.
Haughton Crater and its environs are a fossil-hunter's nirvana: walk
virtually anywhere and you are likely to find yourself padding across
huge slabs of flash-dried coral or oblong bricks of ossified sponge.

Geologists have dated many Haughton-area fossils to the Ordovician
period, a time well before the continents had moved into their current
configuration: it's not unusual, therefore, to pick up a rock-embedded
trilobite petrified half a billion years ago. Geologists have studied
Haughton Crater for several decades, both due to its enormous size and
because it is the only exposed crater on Earth located in a polar desert,
which among other benefits means that its geology is relatively well
preserved. While HMP geologists rave about the crater's intrinsic geo-
logic value, the aim of every scientist approved for detail to the
Haughton-Mars Project is to inform the study of Mars. And researchers
get in line to participate: HMP project leaders always receive more
requests for field work than they can accommodate, and there is a waiting list, as the camp can handle only about twenty-five people comfortably at any given time.

The waiting list certainly isn't due to the lavish living conditions,
yet the relative comfort of life with HMP is remarkable considering that
the island lacks even a shrub to blunt winds that routinely kick past
fifty miles per hour—or that snow squalls are not uncommon in July
and August. In the mess tent, there is an extensive kitchen staffed by a
full-time cook and there is running water pumped up from a nearby
meltwater stream playfully dubbed Lowell Canal after the nineteenth-
century astronomer Percival Lowell, who reported to the world that
he'd discovered artificially constructed canals crisscrossing the Martian
surface.

Next to the mess tent is a wooden shack bifurcated into a washroom
(two taps and a plastic basin) and a shower using water heated on the
mess-tent stove. Gerald "Oz" Osinski, a British geologist from the University of New Brunswick who in 2001 was spending his third field
season on Devon, told me that the shower had arrived during the 2000

field season. Before that, the only way to bathe was to bring plenty of Handi Wipes or walk down to Lowell Canal and splash in the icy snowmelt while shivering in the Arctic air: "You definitely came away refreshed," Oz remarked dryly. Toilets at base camp remain a somewhat primitive affair, the source of continuous jokes and occasional exasperation on the part of newcomers and HMP denizens alike: suffice it to say the process involves a tent, a tin can, some plastic bags and a folding chair with a hole cut into the seat.

On a cold, cloud-shrouded morning, I went out for a daylong "traverse" with Oz and Graham Lawton, a journalist with *New Scientist* magazine. We ate breakfast and then packed lunch and a spare set of clothes, including an emergency blanket in case we were stranded overnight, either by bad weather or because our ATVs broke down (a common occurrence in the harsh polar environment). Oz's equipment included a shotgun, slung across his back; a harmless blast into the air would probably be enough to scare away any polar bears that came too close.

Tall and lean, with piercing blue eyes and rock-star good looks, Oz was in his twenties, making him one of the younger scientists with the HMP but by no means alone in his demographic: the under-forty crowd included Lee, Braham, HMP chief biologist Charles Cockell and HMP science adviser and engineer Kelly Snook. In the otherwise-dry HMP base camp, Oz was famous for generously sharing his stash of aperitifs. But when it came to his geology traverses across the island, Oz was all business: an accomplished mountaineer and survivalist, like Lee he placed great value in methodical preparation. As we got ready, Graham and I joked about polar bears and the guilty pleasures of recreational four-wheeling over virgin tundra; Oz simply smiled and prepared silently.

We roared out of camp on our ATVs, the thick, nubbly wheels spitting mud as we raced down into the crater at full throttle. A light rain stung our faces, a rarity in these typically dry months. Liquid water on Devon, whether trickling in small streams or pooling in shallow lakes, is almost exclusively from snowmelt; the island gets about an inch of rainfall each year, a reminder of why Devon is classified as a polar desert. Snow blankets much of the island all year: even at the height of the Arctic summer, broad bluish-white slabs of sun-hardened snow were draped across gravelly hills and stacked in rocky crevasses, dribbling liquid onto the land in countless thousands of busy rivulets.

Ten minutes into our ride, we were at the bottom of the crater, roaring to a stop at a gurgling meltwater creek. The base camp was out of sight, but in the morning haze we could just see the Hab, looking like a small white mushroom growing on the crater's edge. Oz waded into the ankle-deep creek and picked up a flat black rock, which he handed to me. "This is shatter cone," he said. He ran a finger down a web of hair-thin striations stretching from one end of the rock to the other, the lines radiating from a single point in a fan shape; my first thought was that this was an exquisitely beautiful work of nature's art. Oz saw the shatter cone more pragmatically: to him the rock spoke volumes about what happened here twenty-three million years ago, holding clues about the exact speed and trajectory of the impact and the nature of the object that transformed Devon Island.

Thinking of the modern reliance on technology to conduct scientific research, I asked Oz what computer modeling offered to the process of learning what happened on Devon. "Not much," he said. "The problem with most people who rely on computer modeling for impact studies is that they've never been to an actual crater. You have to come here to learn what happened. That's what it'll be like on Mars, too. Robots and computers will be able to tell us only so much."

Leaving the creek behind, we gunned the ATVs across a broad muddy plain and rode past Trinity Lake, which looked more like a pond. Just as many of the land formations near Haughton Crater are named for scientists or explorers, many of the bodies of water are named for institutions of higher learning: in addition to Trinity, there are Oxford, Cambridge and Stanford, among others. Oz said naming the landmarks was Pascal Lee's idea: "It's helpful, so we know where we're going and so others know too, instead of just saying, 'I'm going over to the little lake on the other side of the crater.' "

Through a narrow canyon and across a boulder-encrusted ridge, and we were on the edge of a vast, desolate plain that was wet and spongy from the summer snowmelt; this was a sedimentary lake bed where geologists have found fossil bones of toy rhinoceroses and giant proto-rabbits, along with bark and pinecones from trees blasted to smithereens by the impact. This is the largest and best preserved Miocene site in Canada and possibly, Oz said, the world. Oz led Graham and me over to a small chimney of rust-covered rock, which was actually a dried-up hydrothermal vent.

"After the impact here life almost certainly recolonized the crater, and it would be interesting if these vents were the first place life did recolonize," Oz said. "One of the important things about when we go to Mars is that we're going to have to find very specific targets to look for life, so what we learn at the crater about vents, where they are and what they're like, could help us a lot in determining where to go on Mars." HMP chief biologist Charles Cockell later told me that impact events are "magnificent" at creating life as well as destroying it. "This has relevance to Mars because Mars is an impact process planet," said Cockell, who is based at Cambridge University and is, like Lee, an Antarctic veteran. "So if you're going to look for life on Mars at all, you're going to have to look for the right habitats to try and find life."

We motored across the lake bed and into a bowl-shaped gulch simply known as the Canyons, where massive spires of rock jutting out at every possible angle gave the impression that we'd just driven into an enormous snaggletoothed mouth. Oz spent a few minutes here taking notes and photographs, then we threaded our way out the other side and onto a particularly mucky plain where the ground was so gooey that all three of our ATVs got stuck; one by one, we revved the ATVs hard and spun our way out, spraying showers of mud in every direction. Free of the bog, we moved on, kicking the ATVs into high gear and thundering across the land as if we owned it, not an animal or plant in sight for mile upon mile.

Finally, we roared onto a plateau above a massive, heart-shaped valley at the center of which sat the first proper lake I'd seen on Devon. I asked Oz the name of the lake and he shrugged and said, "It's just a lake." We were on the outer edge of the typical zone of HMP research; with some access restricted by the Inuit-led government in the Canadian Arctic, the team can study much, but not all, of the crater.

This area was where Oz spent most of his time this field season, moving from spot to spot and, in essence, mapping this section of the crater much as human explorers might work on Mars. Oz climbed an embankment and went to work for about half an hour, leaving Graham and me to wander with our thoughts. Standing here, listening to the wind whistle and scream, it seemed as though it could be A.D. 2001 or twenty-three million years ago, just weeks or even days after the impact. It was hard to believe that anything could live here, yet even on Devon's barest plateaus life had taken hold: turning over round stones or peeking into flinty nooks, one could invariably spy green-blue algae, lichen and occasionally a proper scrubby plant, albeit nothing bigger than a thumbnail—the most basic forms of life, true, but life nonetheless. Such finds gave Cockell and others encouragement for what they might find on Mars.

Soaked, stinking and exhausted, we arrived back at base camp just as dinner was being served. After several days on Devon Island, I'd also come to expect dinner conversation to lead inevitably to Robert Zubrin and the Hab. While the HMP is technically a separate endeavor from the Hab, the neighborhood is small and many scientists with HMP assignments—eager to extend their time on Devon Island—also secure research stints in the Hab. The Hab also relied heavily on the HMP for key logistics support, including communications support, equipment repair and flight logistics for personnel and supplies. Necessarily, the activities of one project affected the other on a daily basis.

The nature of Zubrin-themed discussions fell into two categories: grumbling about his messianic vision of interplanetary colonization and griping about actually having served with him in the Hab while he was "commander." Steve Braham complained loudly to me several times on the latter theme: "I spent three weeks in that Hab with Robert doing nothing because he was so dedicated to getting EVAs [extravehicular activities] out to look good on TV, without even asking us what we need for the science. That just basically blew a big chunk of our time." In 2001, Braham was the single person on Devon Island assigned to live in the Hab for the entire two-month field season: during my eight-day visit, however, he didn't stay in the Hab once. Lee admitted that after serving time with Zubrin, Braham had been "let out" of the Hab, ostensibly to fix some systems at base camp but more to air his nerves a bit.

Regardless of whether or not they'd spent time in the Hab, Zubrin seemed to be on the minds of many of the scientists on Devon Island. Was he really helping the humans-to-Mars effort? By extension, was the Hab? Both were certainly gaining popular attention for the idea of sending humans to Mars—and largely, it was agreed, because of Zubrin's tireless efforts to engage the media. But was Zubrin, who frequently inserted anti-NASA barbs into his dialogue with reporters, doing more harm than good? Since many in the room were themselves NASA scientists, it wasn't an unimportant question.

Listening to the various gripes about Zubrin's megalomania was even stranger knowing that he isn't particularly interested in going to the red planet himself. "It's not about me getting to Mars," Zubrin told me just prior to the 2001 field season. "Even as a kid, that wasn't at the center of things and now I'm forty-nine years old and, you know, one can recognize what the chances are of me actually going to Mars." But he added quickly, "If I was less than fully involved, I'd be less than fully living. I have to take responsibility to actually make this happen."

The following evening, I climbed onto an ATV and bounced across a half mile of gently rolling tundra to see for myself how the other half lived on Devon Island. Up close, the Hab looks like a pair of gigantic tuna cans stacked one atop the other, rigged up with various antennae and power lines stretching to nearby generators and satellite dishes. Although I had been invited to spend some time in the Hab, I was a bit nervous walking up to the main hatch: the crew was supposed to be in simulation, and on Mars it's unlikely that a neighbor would simply drop by for an after-supper chat. On Zubrin's watch, only the Discovery Channel—which had an entire film crew planted on Devon for the 2001 field season (and had contributed a reported $70,000 for that privilege)—was allowed into the Hab, and then only at definitely prescribed times. There was no doorbell, so I banged on the hatch; from deep inside, I heard a muffled "Come on in, it's open."

Inside, the Hab is a bit like an eccentric country cabin. The submarinelike main hatch and "airlock" (complete with massive metal turn-wheel and loudly clanking bolt lock, although lacking any actual compression equipment) lead into an "EVA prep area" where I was greeted by Pascal Lee, who had left the HMP operation in the hands of his Antarctic mentor John Schutt in order to lead the current Hab crew. The EVA prep area is a pie-wedge room with space suits hanging on coat pegs along a wall; helmets were lined up neatly on a shelf over each suit. The suits were handmade, the work of Mars Society volunteers, and they weigh about twenty-five pounds each. Like the Hab itself, the suits are lo-fi.

A small doorway leads from the prep area into the laboratory and engineering area, the largest section of the Hab's lower level, with two oval windows facing out onto the crater. There is a small sample-return hatch built into the crater-facing wall, used by space-suited scientists conducting EVAs to transfer samples from Devon's surface into the lab. "On Mars, of course, we would not be able to simply open the hatch," Lee explained with a wry smile. "But it helps to get used to the process."

To the right of the sample hatch is a long counter that follows the curve of the Hab's hull and is crammed with a variety of scientific equipment, including an oven for "cooking" rock samples and a sparkling new epifluorescent microscope (courtesy of Olympus, one of more than a dozen private companies sponsoring work on Devon Island) used for close examination of microbial specimens. On the other side of the room are a few tables piled high with an assortment of robots. These were the charge of John Blitch, a cheerful, beefy Marine Corps colonel and DARPA engineer, and Arnis Mangolds, a robotics specialist with the company Foster-Miller.

Both Blitch and Mangolds have designed and built robots used by the U.S. military in combat zones and for search-and-rescue operations. Blitch showed me a small robot nicknamed Stumpy, which is about the size of a child's wagon and looks like a miniature, riderless ATV. "On Mars, little robots like this could be dropped off all over by a bigger robot and do all sorts of exploring while the humans sleep," Blitch said with the glee of a tinkering teenager. "Then the mother rover picks them all up, and humans would spend time analyzing the results rather than doing first coverage of that ground ourselves. They'd then point us to the really interesting places to explore firsthand."

Mangolds was busily dismantling a robot whose belly pan had been torn out during a test drive over some unusually rocky terrain earlier in the day. "That usually doesn't happen," he said with a mixture of annoyance and admiration. "This is tough territory." Like many Hab participants from private industry, Mangolds also had a NASA pedigree; Mangolds had worked on a number of NASA projects as a JPL engineer before joining Foster-Miller.

Although its robots are among the most sophisticated of their kind, the company doesn't work with NASA; most of Foster-Miller's customers are military. "We don't play with NASA anymore," Mangolds said matter-of-factly before launching into another lament on the brother-in-law problem. "We've been, many times, in situations where someone at NASA likes our proposal, thinks the budget makes sense, and then says, 'Let me take this to my boss.' The boss then sees another proposal that's maybe not as good and more expensive, but from a tried and true friend like a Boeing, who also has other projects with them. He's going to pick Boeing every time. We're just not going to do that anymore. What's the point?"

Mangolds had come to Devon Island on his own dime, and for one reason: "The spirit has been missing from the space program for years," he said. "What is there for people to get excited about anymore? This is bringing the excitement back."

For the most part, Hab participants understand that a Mars simulation under such low-budget circumstances can be analogous to the real thing only to a very limited extent. During a subsequent visit of mine, also while the Hab was in "simulation" mode, the Hab lights dimmed then punked out altogether: one of many power failures due to the tough Arctic environment. Wearing only a T-shirt and jeans, Blitch popped open the rear hatch (leaving it swinging open) and sprinted through the freezing drizzle to fix the generator. It goes without saying that such a move on Mars would have resulted not only in an agonizing death for Blitch, but also in the depressurization of the entire Hab.

But the imperative for most Hab crew members was meaningful research, not vigorous adherence to pseudo-Martian living conditions. The exception, again, seemed to be Zubrin, whose Hab diaries underscore his relentless pursuit of verisimilitude, including the following entry from July 9, 2001, in which he criticized some of the practices in the first Hab crew rotation (which was also under Lee's command):

"The crew has been quite liberal in taking hot showers, which in my view is not realistic. . . . Conserving water is essential to reduce the mass and cost of human Mars missions, so I intend to be much more strict in this area. Instead of daily showers, my crews will be held to sponge baths every other day."

Lee didn't see the need for such strictness because, in truth, the degree to which the Hab is an accurate representation of what will happen on Mars is limited not only by funding but by the obvious fact that Devon Island isn't Mars. "It's a big confusion to call confinement a simulation," Lee said. "A simulation, to me, is make-believe that is driven by the desire to learn a lesson. That means there are some parameters that we decide to leave untouched, and we focus on areas where research needs to be pushed forward."

For Lee, that also means not artificially restricting the crew's grooming and cleanliness or, as another example, requiring them to don simulated space suits every time a repair needs to be done outside the Hab. At one point, Braham told me, Zubrin wanted him to go outside in a space suit to set up antennae, fix electronics gear and otherwise learn about the difficulties of working in a space suit. Lee said this was ridiculous, since the bolts, screwdrivers and other tools used on Mars would be designed with space suits in mind, which isn't the case on Devon. "You're basically kidding yourself if you're saying you have to stay in the suit at all costs," Lee said. "Meanwhile, all you're doing is delaying how quickly you can get things repaired."

Touring the Hab easily leads one to conclude that while it may be the most accurate Mars simulator around, it is still very much a work in progress. The upper portion of the Hab is reached via a simple household ladder canted steeply through a semicircular cutaway in the wooden ceiling. In the center of the upper level is a cramped live/work area with simple folding tables arranged in an L shape and typically cluttered with laptops and papers filled with scientific notations. On the left, two porthole windows give a view onto the crater; straight ahead, a counter curving halfway around the room holds cooking utensils and dishes, a hot plate, a small refrigerator and a microwave. On the right are six narrow white doors, set evenly apart, leading to tiny "crew" berths with a simple bunk and a tiny space for clothes and personal belongings.

Each crew member has put his or her own personal stamp on the Hab's live/work area: posters and stickers with space or science themes are everywhere, giving the place the look of a nerdy college dorm room. There is a bookshelf replete with, among other items, all the books Zubrin has published. The other books on the shelf bespeak the stew of ambitions and influences on the Hab's residents: a biography of the American frontiersman Jedadiah Smith, Jules Verne's seminal *From the Earth to the Moon,* Carl Sagan's *Pale Blue Dot,* and both the *Iliad* and the *Odyssey.* The films on the shelf are somewhat more enigmatic in their meaning to the crew: the selection includes the 1978 sci-fi thriller *Capricorn One* (in which the government fakes a mission to Mars) and also *Being There,* the 1979 film starring the late Peter Sellers as a charming idiot who stumbles into Washington's most rarefied corridors of political power.

The University of Arizona researcher Peter Smith has draped a string of Tibetan prayer flags above the crew berths. "It's great, it's intense," he said excitedly of his Hab experience. "We really pack it in, research-wise. I was up until three this morning, but of course it's pretty hard to tell because it's always light out."

Smith is no stranger to the ups and downs of Mars research: he played critical roles in devising the imaging systems for both the triumphant Mars Pathfinder and the ill-fated Mars Polar Lander. On Devon, he was conducting research on a prototype instrument to be incorporated into the lander of ESA's Mars Express spacecraft, which was scheduled to arrive at Mars in late 2003. The British-sponsored lander is called Beagle 2 after Darwin's ship of exploration, and not only because of its scientific legacy: just as Darwin went hat in hand to both public and private sources to fund his expeditions, so did the British space agency. Beagle 2 will cost about $40 million, a relative bargain: fund-raising is ongoing, and even the pop band Blur got into the act by donating the proceeds from one of its songs to the lander's development. (Launched on a Soyuz from Baikonur, Mars Express also brought Russia back into the Mars exploration business.)

When I visited, Smith was analyzing a sample retrieved the previous day by one of Blitch's and Mangolds's robots. He and three other members of the crew were also preparing for an EVA the next day to follow up on what the robots found. Smith's enthusiasm for the EVA was clearly tepid. "A space suit, at least our space suit, is not comfortable," he said mournfully. "It's cramped, it's hot, you've got a microphone [for communication with other team members] jammed in your face, you can barely move. I can't even bend over."

Despite Smith's complaining, he would indeed suit up the following day and set out across Devon Island with his crewmates. Like it or not (and most scientists didn't), EVAs—humans in space suits, simulating exploration of a Martian surface—are the star attraction on Devon Island. Even more than the Hab itself, these modest rehearsals for human planetary exploration are what have grabbed dozens of feature pages in publications as diverse as *Popular Science* and *GQ,* not to mention prime coverage on television programs and web sites around the world.

Uncomfortable and irritating though they may be to researchers like Smith, to a large extent the EVAs make Robert Zubrin's case. After all, if the world is this captivated by a lo-fi rehearsal for sending humans to Mars, how might it respond to the real thing?

The Marsnauts were scheduled to leave the Hab at 2:00 P.M. after running through system checks and donning their suits, which have only two mechanical systems: a radio headset and a small cooling fan, located in the suit backpacks. The checks done, the Marsnauts spent thirty minutes in the airlock to simulate the prep time needed to adjust to Mars's lower atmospheric pressure, as they would if this were Mars. Lee later showed me a picture he took of this particular EVA team in the airlock; a close-up of all four Marsnauts smiling in a "Go team!" huddle, their four heads filling the frame inside four fishbowl helmets and looking strangely like beheaded lab specimens inside cylindrical jars. In addition to Lee (who will lead this EVA) and Peter Smith, today's explorers included the NASA Ames roboticist Larry Lemke and Carol Stoker, who has long worked at Ames with her Mars Underground cofounder, Chris McKay.

The 2:00 P.M. start time was supposedly a sure thing, but like star performers building anticipation in a giddy audience, the Marsnauts were fashionably late: it was past four o'clock by the time they popped out of the airlock hatch, four bubble-headed figures in cream-white suits, looking as if they were stepping onto the set of the 1950 sci-fi classic *Destination Moon*. This was cinema verité in reverse, enhanced by two members of the Discovery Channel film crew who thrust a video camera and boom mike at the somewhat startled explorers as the hatch opened. From inside her plastic helmet, Stoker flashed a look at the Discovery team that bespoke a deep desire to have a ray gun at her disposal; one of the Discovery cameramen would later grumble that she intentionally ran over his foot with her ATV. Nevertheless, for the most part Stoker and the other Marsnauts gamely attempted to ignore the cameras: a photographer from London's *Daily Mail* and an independent French photojournalist were also firing away from very close range, while *New Scientist*'s Graham Lawton and I stood to the side.

Lee led the team through a final radio check as the Marsnauts climbed onto their ATVs, which this day as on other EVA days were playing the part of pressurized Mars rovers. Mounted side by side on their rovers, the Marsnauts resembled a lineup of futuristic race horses, an impression enhanced by huge numbers stamped on their space-suit packs so the Marsnauts could identify one another if the comm systems failed or if there was trouble: Lemke was 3, Stoker was 4, Lee was 5 and Smith was 6.

Their final checks finished, the Marsnauts kicked their ATVs to life, the engines rumbling as they rolled away from the Hab onto the rubbly ground: a veritable flotilla of media followed behind on their own Kawasakis and Hondas. Today's mission was scheduled to last no more than two and a half hours, start to finish, a time frame corresponding to expected EVAs on Mars after factoring in the likely robustness of space suits and rovers, and the harshness of the Martian environment (dust, temperature, radiation exposure). The crew's assignment was to motor over to a network of valleys known as Site 6. As Lee described it, "The main goal of the EVA was to determine the origin of the small valleys. Were they of tectonic origin, the result of glacial meltwater discharge, the result of massive outflows, the result of glacial carving, or perhaps combinations of several of these or of other processes? We'd have to find out."

The Marsnauts drove single file around the rim of the crater, main-

taining a distance of about fifteen feet between one another and intentionally keeping their ATVs at an excruciatingly slow speed intended
to mimic the likely cruising speed of a real Mars rover. About twenty
minutes after leaving the Hab, the team halted abruptly, nearly causing
Devon Island's first fender-bender as the media ATVs, following the
Marsnauts one another a bit too closely, slammed on their brakes and
lurched to a nerve-racking stop. Lee stood tall on the metal stirrups of
his ATV, alternately photographing the terrain and consulting a map.
He gestured and said something to the other crew members, and the
Marsnauts began to chatter back and forth: the communications were
faint and crackly, but it sounded like they were confirming directions.
Their ATVs revved, and we all roared forward.

After cutting to the northwest and bouncing tediously across some
exceptionally boulder-heavy landscape, we rolled down a gentle slop-
ing hill onto a flat saddle of mud through which Lowell Canal, am-
plified by a freakishly heavy rainfall the night before, was flowing
with respectable force. Lee eyed the stream carefully, searching for the
shallowest point, then led his team across; we followed, gunning our
ATVs to push through the bracing water, which all but swamped the
machines and soaked us to the knees. On the other side now, the Marsnauts filed away from the stream and into a claustrophobic gulch that
looked so much like something out of a spaghetti Western that I practically expected a posse to be waiting behind the steep pile of boulders
on our right. The team stopped and switched off their ATVs. This was
the place.

After dismounting from their rovers, the Marsnauts meandered
around the gulch for a few moments, pulling sampling tools from bags
and scouting the site to reacquaint themselves with what the robots
had shown them the day before. Such ad hoc exploration was apparently not enough for the Discovery Channel producer, who pulled the
Marsnauts into a close huddle and, like a coach pumping up his team
before a game, asked for some words of pre-exploration enthusiasm
for the camera. The team members exchanged a quick glance of quiet
exasperation, then cleared their throats and did their best to ad-lib a
response.

"Ah, well, I feel a profound sense of isolation here, a real peacefulness," Smith said uncertainly, trying his best to sound like an Apollo-era astronaut.

"I know what you mean," Lemke chimed in, desperately cheerful, then apparently deciding to go for full sci-fi effect. "No human has ever set foot here before."

There was a moment of somewhat uncomfortable silence: was that enough? Then Lee concluded firmly, an unmistakable wishfulness to his words. "It's just the four of us, alone on this planet."

The team marched over to the boulders jutting into the gulch: no Western posse here, just a sheer wall of gray-black rock that appeared to have little to recommend it. On closer inspection, though, the boulders were discolored with patches of greenish-brown scum and faint, yellowish pads of fuzz. The Marsnauts grew visibly excited.

"Could that be a microbial mat we're looking at edgewise?" Lemke asked.

"What do you make of that, Carol?" Smith added. Stoker wasn't sure: "Let's take a sample," she directed. Lemke and Lee pulled out rock hammers and chipped away at the wall, breaking off several small pieces and dropping them into sample bags. The team then moved to the side of the outcropping and crowded around a faint orange patch on the rock, and the Marsnauts let loose an almost unison yelp of excitement; here, apparently, was something the robots didn't see. "Well, that looks like lichen to me, pretty interesting," Lemke said. The group studied the orange patch for a moment, then Lee began chipping vigorously at the lichen, breaking off a number of chunks and stuffing them into sample bags. The Discovery team crowded in, the camera almost touching Lee's arm as he whacked at the rock.

Finished with his rock chipping, Lee turned to the broader group of us, breaking his Marsnaut "character" almost apologetically, as if to underscore that whatever appearances may be, the group was doing its best to simulate a realistic Mars research expedition. "We want to see if the robot missed anything and it did, it didn't see the lichen," he said evenly. "That's why we sent humans here."

Having gotten what they needed from the rock wall, the Marsnauts fanned out again to study the gulch in more detail for other things of interest. The French photojournalist stopped the group to get a shot

and out of the corner of my eye I noticed the Discovery cameraman
throw up his hands in exasperation; the Frenchman, it seemed, was
cramping his style.

At this moment, the Discovery producer decided to take things in
hand, striding purposefully to the center of the gulch; the Marsnauts,
cameramen and journalists were scattered around the perimeter of the
site like bit players waiting for center stage. "OK, can we stop a minute
here?" the producer shouted, his voice echoing off the canyon walls.
"We need to bring some order to this. We have to decide when the still
photographers shoot and when we shoot. I mean, we need to get it
together. This is the end of our story."

There was silence for a moment, everyone seemingly frozen in their
tracks by the producer's harangue; then, mild chaos ensued as at once
everyone scrambled around the muddy gulch trying to do their various
jobs (chipping, filming, writing) at an even more frenetic pace. The
Daily Mail photographer marched up a slope to get a downward shot at
the lichen wall as the Marsnauts studied it, but this appeared to be
exactly where the Discovery Channel was going as well. The Discovery
producer trotted up to the photographer and asked him in no uncertain
terms to move so his team can work unobstructed: "Look," the producer argued, "the Discovery Channel paid a lot of money for this so
we have the right to make something happen." The photographer, who
towered over the producer, looked at him for a moment, then smiled
and shook his head. Lighting a cigarette, he walked back down the
slope and out of the Discovery shot.

Watching the media bickering, the Marsnauts seemed uncertain
about what to do: they'd stopped in their tracks as they were preparing
to go on about their work. The Discovery producer gathered them up
and practically pushed them as a group over to the base of the slope;
his camera team was poised in front of them, ready to walk backward
up the slope and film them moving on to another sample site, although
up the slope was not where the Marsnauts had intended to go. The producer clumped the Marsnauts together and faced them up the hill
anyhow, posing them like models during a fashion shoot. "OK, we're
rolling."

The Marsnauts looked at one another, their faces registering a mixture of resignation and irritation even through their bubble helmets. Then they trudged together up the hill, the Discovery team walking backward up the slope to record their advance. The Marsnauts sampled and reconnoitered for another half an hour, then the team and its entourage climbed back onto the ATVs and rode back through the gulch, across the stream, through a labyrinth of pocket canyons, then up along the crater ridge, the Hab now growing larger in the haze-shrouded distance.

Back at the Hab, the Discovery Channel's sound man reached into the backpacks of the Marsnauts and removed equipment used to record their "mission" conversations. The photographers continued to shoot the Marsnauts as they wound down the expedition.

Graham Lawton watched the scene and shook his head. "Do you know what we are here?" he said to me. "We're Martian paparazzi!" But by the end of the day, even Graham couldn't help himself. Handing his digital camera to me, he walked over to the exhausted Marsnauts and posed with them for a picture.

Possibly the biggest news about Mars to date—bigger, even, than the announcement in 1996 that meteorite AHL-84001 contained fossilized Martian life (although many scientists subsequently concluded that it probably didn't)—came in mid-2002, when the Mars Odyssey probe returned data strongly suggesting that Mars was heavy with water ice just beneath its surface, in the form of something resembling semi-frozen mud. This was joyful confirmation of work Lee and his colleagues had pursued on Devon Island for several field seasons, as they sought to perfect methods for detecting subsurface water ice using both ground-penetrating radar and human-tended drilling techniques.

As expected, NASA made hay with Odyssey's data, but still not a word was spoken about planning for human exploration. On the contrary, only weeks before, Sean O'Keefe made it a point to say that NASA's future should not be "destination-based," and in sketching out NASA's vision through 2030, nothing was mentioned at all about human travel beyond Earth orbit. Within NASA, it seems, only the Johnson Space Center continues to keep the humans-to-Mars flame alive with any vigor, and then only through an educational exercise: its annual Mars Settlement Design Competition for Houston-area high school students, which has run for several years. Even this is tentative, since the scenario asks students to plan for a human mission to Mars not for NASA, but rather for "the Foundation Society, a private organization that promotes human exploration and colonization of space."

While a privately financed Mars mission seems unlikely, both Pascal Lee and Robert Zubrin continue, in their very different ways, to pursue the steps that might one day lead humans to Mars. Zubrin's Mars Society scaled back its presence on Devon Island in order to devote more energy to building additional Hab prototypes: a second Hab, sponsored in part by two labor unions, opened for business in the Utah desert in late 2001 and another was set to open in 2003, in Iceland. Not surprisingly, the extra Habs meant an even greater emphasis on media coverage (the *New York Times,* the *Boston Globe* and CNN all covered the launch of the Utah Hab) and made Zubrin's humans-to-Mars agenda more visible than ever.

It has not been all roses for Zubrin, however. In the fall of 2001, Mars Society infighting led to the resignation of nearly half of the board of directors. Among those who left was Elon Musk, someone whose presence, and money, had held great promise for Zubrin. Musk, an Internet entrepreneur in his thirties, had grown up on *Star Trek,* dreaming of space travel. Hooked into a Silicon Valley fund-raiser for the Mars Society in early 2001, he made a six-figure donation on the spot and was quickly recruited by Zubrin for the Mars Society board. Only a few months after that, I heard Musk speak at the Stanford Mars Society convention, where he talked excitedly about personally putting up "big funding" to help advance the Mars Society's agenda.

On the money front, there was little question that Musk could certainly deliver. When the company he founded, PayPal, went public in February 2002 (in one of the Internet industry's first real recoveries after the dot-com meltdown), Musk's estimated net worth shot up by a reported $143 million. Made rich by PayPal, Musk and his brother set up a private foundation that supports clean energy, education and space exploration, all of which the Musk brothers view as being interrelated. On the last topic, the foundation's purpose is quintessentially

O'Neillian, with a Zubrinesque Martian twist: "Given that humanity faces certain extinction if it only inhabits Earth," the foundation's web site intones, citing nuclear Armageddon and meteor impacts as likely couriers of the apocalypse, "we need to take the next step beyond the Apollo program and establish a permanent, self-sustaining civilization on another planet."

Yet by the time the Stanford convention was under way, ties between Musk and Zubrin were already fraying. Zubrin wanted Musk to bankroll a mission to send a small group of mice into Earth orbit to see if, while spinning in a specially designed capsule to simulate Mars's one-third gravity, they could successfully reproduce. Musk wanted to explore other ideas, but Zubrin insistently, and publicly, pushed what CNN dubbed "mice-tronauts." Musk eventually walked away.

Musk is now working solo on a different yet still ambitious mission: a robotic lander that would attempt to grow samples of food crops in treated Martian soil (an idea long advocated by NASA's Chris McKay). Musk has dubbed the mission "Mars Oasis" and hopes to launch it in 2005 or 2006, although neither launch date nor mission hardware has yet been finalized. Frustrated by attempts to secure a deal to launch with the Russians, Musk joined the ranks of tech-world rocketeers and started his own launch vehicle company, SpaceX. Zubrin continues to press ahead with the mice-tronaut idea, although without anything near the estimated $20 million needed to make it happen.

In early 2003, Pascal Lee proudly told me that in addition to the usual infusions of donated cash and equipment, he'd managed to secure a specially outfitted Humvee for use as a prototype pressurized rover on Devon Island; such a vehicle is a key element in NASA's reference mission plan for human exploration of Mars. The space media company SpaceRef contributed funds for the Arthur C. Clarke Mars Greenhouse to experiment with food production in a harsh, Mars-like environment (Clarke said in a press release, "Look out, Mars—here we come!"). The greenhouse made its public debut at a NASA conference in the spring of 2002 and crops were grown in it on Devon Island in 2003.

Tentative though it may be about humans-to-Mars, top-level NASA has become somewhat more comfortable with the work on Devon Island, if still schizophrenic in its endorsement. At the NASA conference where the greenhouse debuted, Jim Garvin, NASA's chief scientist for Mars exploration, praised Lee's "sheer entrepreneurial spirit and cleverness" in developing the Haughton-Mars Project. "Haughton crater," Garvin told *Science* magazine, "may provide an important test site for future Mars technology because of the infrastructure in place there." Even so, only a few weeks earlier in the *Boston Globe*—where he also praised Zubrin's desert research simulation—Garvin said that NASA had no plans to send humans to Mars.

Through it all, Lee remains optimistic. At the 2002 World Space Congress, Lee announced the formation of a new nonprofit organization to provide more focus to his work: the Mars Institute, which will work with NASA, the SETI Institute and other longtime partners to advance Lee's humans-to-Mars agenda. With Lee as its chairman, the institute is filled with familiar faces from Devon Island: Charles Cockell is its vice president, and its advisers include Steve Braham, Carol Stoker and Peter Smith. Also in advisory roles are the *Red Mars* author Kim Stanley Robinson and the man NASA kept from joining the Devon Island project in 2001, ISS flight surgeon Jeffrey Jones.

When Lee was still in his early twenties, he and some friends in his French space club made bets about when humans would first set foot on Mars. He told me, without bitterness, that some of his friends gave dates that have already passed; others gave dates that remain far in the future. The date Lee gave was 2017, and he is still standing by it, if only for personal reasons; among other benefits, Lee would be fifty-three, a prime age both physically and mentally for human space travel. "It will be a good time to go to Mars," he said with cautious enthusiasm. "The planets will be close together, which will shorten the journey meaningfully. It will also be a time of moderate solar activity, and you're less likely to have solar flares, which is important, since radiation on the trip is a big concern. Third, it's on a timescale that still seems reasonable."

Lee shifted in his seat at this point in this conversation. He admitted that he is not particularly good at making speeches, but every now and then, he cautiously climbs onto the soapbox where Robert Zubrin is so much more comfortable.

"For us to go to the next step, to Mars, we need to be working on it right now," he said with a flicker of a smile. "You see, what puts things in perspective is that I hear that cities are already working to compete for the 2014 Olympic Games. You know, if a city has to worry now about building a stadium for 2014, if we want to get to Mars by 2017, we better start kicking it into gear."

Lighting Out for the Territory

As the work of the *Columbia* Accident Inquiry Board drew to a close in the spring of 2003, two things had become clear to most observers. The first was that the problems that had resulted in *Columbia*'s destruction were not new. The second was that for NASA and its shuttle program, business as usual was about to commence again.

The inquiry had revealed startling lapses in communication within NASA and between the agency and its contractors that had buried concerns about shuttle safety, sometimes willfully. Pages of testimony were taken describing the many, many reasons why the remaining shuttle fleet was far from robust. NASA's "culture" needed to change, the board would eventually conclude, even while offering few ideas about how to make that happen.

For all the red flags, there was little doubt that the important questions about NASA's human spaceflight program—the ones that sought to explore its purpose and meaning in the modern world—clearly took a back seat to the more concrete goal of getting the shuttle flying again. NASA promised new safety procedures; Congress promised more money. Even before the board's recommendations were released in August 2003, the hard-driving head of the inquiry, retired admiral Harold W. Gehman, Jr., said that he saw "no showstoppers" that would prevent the shuttles from returning to flight in 2004—a time frame consistent with what NASA had been planning well before the inquiry process had gotten into gear.

Most agreed that Gehman had done yeoman's work as head of the inquiry board. He had dealt evenhandedly with an organization so hostile to outside scrutiny that in the midst of the *Columbia* inquiry one lawmaker actually introduced legislation to ensure that future inquiries about NASA would be independent of NASA control.

Toward the end of the inquiry process, Gehman, for all of his relative equanimity, seemed to have become a bit nonplussed by his time inside the belly of the NASA beast. During a hearing on May 14, 2003, at which Gehman and Sean O'Keefe were updating the Senate Committee on Science, Commerce and Technology on the inquiry's progress, a rather strange exchange took place. O'Keefe told the committee that he found it "infuriating" that senior NASA managers ignored offers from the National Imagery and Mapping Agency to image *Columbia* for possible damage caused by the falling foam. Rather than suggesting that O'Keefe perhaps hold those managers directly to account for their bad decision making—not an unreasonable proposition, considering the charge of the inquiry board Gehman chaired—Gehman instead blamed NASA's system, adding that there was "not one person responsible" for what had happened. Senator John McCain—acerbic and direct throughout the inquiry process—shot back that it was "equally infuriating that no one is responsible. Those decisions aren't made by machines. Someone is responsible."

At the end of the day, however, Gehman may have come to understand NASA better than McCain. The agency is not the sum of its parts; it is the sum of its history, and of its "culture." People do not run the agency. The agency—stuck in a routine of never-ending PR and bearing pork-barrel politics and the dead weight of unfilled expectations—runs the people.

In his extensive history of what he calls "the first Space Age," *This New Ocean,* William Burrows described the last days of the National Advisory Council on Aeronautics—NACA, the organization from which NASA was created—in terms that have a chilling similarity to the status quo NASA has been stuck in since Apollo ended the agency's true mission. NACA was originally deemed able to take on the Eisenhower administration's new space mandate alone, without the help of any other agency. However, in 1958, on the eve of its transformation to NASA, there was a perception that NACA "had withered into a timid bureaucracy . . . too narrowly focused and too conservative to function in the seemingly boundless world of space."

The story of the first Space Age was largely that of NASA and its Cold War nemesis, the Soviet space program. It was a gripping tale; a battle not only of engineers who sought to conquer gravity but also of two vastly different societies determined to prove the merits of their ideologies on the world stage. The triumph of *Apollo 11* signaled the beginning of the end of the cosmic Cold War well before the Earth-bound conflict itself drew to a close, but the result has been the same. With the Cold War over, both the world and its space community are run by a single superpower.

Founded as a trailblazer and innovator, NASA has become a barrier. Sanguinely, the *Columbia* board opined that "good leadership can direct [NASA's] culture to adapt to new realities." Alas, NASA's post-Apollo track record indicates otherwise. NASA has evolved a cultural and operational ethos that is so infinitely tentacular as to strangle the influence, skill and good intentions of any single leader. All that even the most energetic and clever NASA administrators can do is to make minor adjustments to an organizational character that has calcified beyond redemption. They have by and large labored simply to keep NASA alive to fight another day, even though the "fight" NASA was born for is unlikely ever to come again. All of the agency's strategies for the future, large or small, are thus based on ensuring its own incremental survival.

More often than not, history's most successful pioneers and explorers of the modern age were not government employees like Neil Armstrong but privately sponsored entrepreneurs, like Roald Amundsen and Charles Lindbergh. In some cases, their exploration was fueled by government interest and received government support, but that in turn served as an incentive for private investment with potential gain—be it knowledge, glory or gold—for everyone involved. It is a strange historical anomaly that most space entrepreneurs have thus far been halted at the brink of success (or failure) because the government has not seen fit to give them real opportunities to stretch the limits of technology and imagination.

NASA need not be disbanded outright to advance human space-flight, as has been proposed by several organizations, most notably the National Taxpayers Union Foundation. The foundation has put forward a policy document calling for the federal government to "put NASA on the auction block" and sell all of its assets to the private sector, which would then sell services back to private customers and to relevant government agencies like the National Science Foundation and the National Oceanic and Atmospheric Administration. "Everything, including the shuttle project, International Space Station, and nonprofit scientific exploration, would be vested in private hands," proposed the Taxpayers Union.

Whatever the appeal of such a draconian proposition, I have yet to come across anyone who seriously believes this is likely to happen, at least not in the fashion of federal deregulation that broke up telecommunications dynasties like AT&T in the 1980s or budding commercial airline monopolies in the 1930s. Unlike corporate monopolies in America, which can be dismantled by brute legal force, there is no straightforward means of "deregulating" a government agency. Despite the Taxpayers Union's statement, NASA should continue to pursue science-focused robotic exploration and fundamental research—the work that is closest to its original NACA core competency, and what NASA actually does best. Interestingly, NASA has been far more receptive to entrepreneurial work on its robotic projects than it has in the human spaceflight arena.

NASA should also continue its cutting-edge research into critical spaceflight technologies. One of the brightest developments in Sean O'Keefe's NASA was the announcement in late 2002 of Project Prometheus, a billion-dollar effort to develop a nuclear propulsion system that could, one day, replace the relatively inefficient chemical propulsion systems now used. Project Prometheus is just the kind of big-thinking, big-spending work NASA should be doing more of, although not solely in service of NASA missions. An ideal structure would have NASA turning over development and "marketization" of critical technologies to the private sector, based on a clearly delineated framework that would prevent the kind of anticompetitive behavior that frequently channels the majority of the agency's work to a handful of contractors. In this scenario, NASA's role would be much as NACA's was in conducting the essential research that advanced commercial aviation.

Unlike fundamental research, robotic exploration and satellite monitoring, NASA's human spaceflight work requires a far more radical overhaul. For the near term, if NASA plans to stay the course with the

Orbital Space Plane, its crew capacity should be no fewer than the six people needed to optimize the value of the ISS. Alternate Access for ISS resupply should become a NASA priority at least equal to the Orbital Space Plane, with a commitment to flight testing, not just paper studies. Ideally, NASA and its political allies would leverage the *Columbia* disaster for more funding in order to fast-track both programs—the better to accelerate the retirement of the space shuttle, the continued operation of which threatens, with every mission, to bring NASA's entire human spaceflight enterprise crashing down once and for all.

In the *Columbia* inquiry report, recommendations on the shuttle's future were mixed, if not contradictory. The report acknowledged that the shuttle is, so to speak, on its last legs:

> Because of the risks inherent in the original design of the Space Shuttle, because that design was based in many aspects on now-obsolete technologies, and because the Shuttle is now an aging system but still developmental in character, it is in the nation's best interests to replace the Shuttle as soon as possible as the primary means for transporting humans to and from Earth orbit.

After curtly summarizing the many abortive attempts to replace the shuttle (mentioning the Reagan-era National Aerospace Plane and Lockheed Martin's X-33 by name) as a "failure of national leadership," the board not only recommended that NASA accelerate the development of the Orbital Space Plane, it stated that "whatever design NASA chooses should become the primary means for taking people to and from the International Space Station, not just a complement to the Space Shuttle." The board also suggested stripping the shuttle of its ISS cargo missions: "When cargo can be carried to the space station or other destinations by an expendable launch vehicle, it should be." This point was cheered by those in the entrepreneurial space sector who hoped it would fuel their efforts to restore the Alternate Access program.

In one of its most damning statements, the board said that it had "no confidence" that the shuttle could be operated for more than a few years based solely on more safety checks, more upgrades and what it

called "renewed post-accident vigilance." Yet for all of this strong language, one of the board's more concrete long-range recommendations
was for NASA to "develop and conduct a recertification of the shuttle
program" to qualify it for operating beyond 2010. NASA leadership
wholeheartedly endorsed this recommendation, as one would expect
them to. By putting both the creation and execution of such a "recertification" program in the hands of the organization that stands to gain
the most from the shuttle's continued operation (or so it believes), the
agency would be given permission to keep the shuttle around for quite
a bit longer.

It remains to be seen how much impact the *Columbia* report (CAIB)
will have on NASA and its human spaceflight program. Some longtime
critics, like *NASA Watch* editor Keith Cowing, called the report
unprecedented: "This is not your everyday accident report, for it
reaches into the very soul of NASA," he wrote in his own analysis of
the CAIB report. Others were more cynical. Former NASA chief historian Roger Launius, who because of his profession has a particularly
long view of the agency, pointed out that most of the big-picture findings of the *Columbia* report—bad management, clumsy safety procedures, insular organizational culture—could also be found in the
Rogers Commission report on the *Challenger* accident. In some cases,
you could simply substitute the name of one shuttle orbiter for the
other. "They weren't fixed then," Launius noted ruefully of NASA's
problems, "so what should lead us to believe they'll get fixed now?"

Whatever the ultimate consequences from the *Columbia* report, it is
clear that we must soon reckon with our own Space Age history, and
decide what we want from any Space Age that is yet to come. Like any
endeavor, human spaceflight has its hardened core of true believers
who cannot envision a future that does not involve human expansion
into the cosmos. It is probably true that from the most pragmatic perspective, political interests will keep NASA alive and its human spaceflight program on the funding equivalent of life support for at least
another decade, come what may.

But in time, political and economic power will change hands. With
or without another shuttle disaster, the next generation will ask the
science community what resources should be committed to human

spaceflight. The science community will say then, as it does now, that human spaceflight serves only minimal scientific purposes. Future leaders will also canvass private industry, which will confirm that satellites, not crewed spacecraft, are its bread and butter. Leaders will query the aerospace industry and the military, which will ask for more money to build better expendable rockets, since the one space vehicle that isn't expendable is far too costly and risky for their purposes.

And finally, the question will be put to the public: should human spaceflight continue? That public, too, will reflect a new generation—one whose human spaceflight perceptions have been shaped by *Challenger, Columbia* and the stop-start International Space Station. With nothing approaching Apollo's inspirational scale to generate enthusiasm, budget-friendly robotic missions will look more and more appealing even to those who are, to use a demographic expression, space-interested. Lacking money, vision, purpose and popular connection, it is hard to see how NASA's human spaceflight program will survive.

There are, of course, other national space programs that might take up the gauntlet should NASA's human spaceflight activities end. On October 15, 2003, China became the third nation capable of sending humans into space when its Shenzhou spacecraft carried thirty-eight-year-old Yang Limei, a veteran military pilot, on a fourteen-orbit journey around the Earth. Over the last two decades China has developed space-faring technology at an impressive pace, one that is appropriate to the society that invented rockets in the first place.

Yet China does not seem a likely candidate to pick up the torch of the world's space-faring ambitions. In human spaceflight, China appears to be bent on embarking on a one-nation Space Race—the one that was run, not to say won, more than thirty years ago, and with essentially the same technology (China's Shenzhou spacecraft is essentially an upgraded Soyuz; its Long March launch vehicle is a souped-up ICBM). To say the least, China's ultimate space aims are unclear, but the fiercely nationalist rhetoric surrounding its space program makes the American versus Soviet Space Race seem like a love-in. If the Chinese do send humans to the moon—a goal that is alternately proclaimed then retracted by Chinese leaders—such a feat would be one small step for China, and probably little more.

The European Space Agency is pushing a multidecade plan of exploration called Aurora that would, in its most ambitious incarnation, put humans on Mars in 2025. But ESA's annual budget is less than a fifth of NASA's, and the agency has no independent means of putting humans in space: without a lot more cash and commitment, it is hard to see how ESA could pull off a humans-to-Mars mission, or anything else significant in human spaceflight. As for other national space programs of note, namely those of Japan and India, they, too, lack the capability to send humans into space independent of either America or Russia. For their part, no doubt the Russians would dearly like to reclaim a place of leadership in human space exploration, but their economic fortunes would have to improve considerably for that to be anything more than wishful thinking.

By the same token, enthusiasm alone will not bring success to the space entrepreneurs. It may be true, as George Bernard Shaw once wrote, that "all progress depends on the unreasonable man"—and in the light of that generous spirit, space entrepreneurs are certainly unreasonable, and admirable. But the alternative space community needs to do much more to convince politicians and investors that their ideas have as much "reasonable" merit as those produced by NASA and its Big Aerospace partners.

MarchStorm and the new National Space Society lobbying effort are a start, but unless such efforts can be linked to real economic and political impact, they will likely remain confined to chipping very small chunks off a very large iceberg. The obvious course of action would be for ProSpace and NSS to find a way to pool their experience and resources in a concerted push to effect significant changes in space policy that would open the door wider for the private sector, much as ProSpace founder Charles Miller had originally hoped for. There are ideological differences to be overcome in the pursuit of such a goal, but they are not overwhelming, just challenging.

The alternative space community must also generate more focus on the idea that the average man or woman's desire to travel into space is no less noble than that of any NASA astronaut. Perhaps NASA did learn something from the Dennis Tito experience, since its revival of the Teacher in Space program followed closely on the heels of the Tito and Shuttleworth flights (and their positive press). Indeed, the moment Barbara Morgan does fly to the International Space Station will be a step in the right direction for NASA, particularly in light of the *Columbia* tragedy.

However, even the Teacher in Space program is very much cosseted by NASA tradition. The agency is training educators as professional astronauts, which is very different from sending a message that space travelers do not need to be space professionals any more than airline passengers need to know how to fly a 777. NASA continues to require Herculean standards for spaceflight, both mentally and physically, when even some of its own astronauts dispute how much preparation is actually required. Speaking at a space tourism conference in 2001, the three-time shuttle astronaut Rich Clifford offered the following training regime for nonprofessional space travelers: "I'd suggest that potential space tourists put on an orange pumpkin suit [the pressurized space suits worn by shuttle astronauts in flight], then be shut in a Volkswagen bus," Clifford said. "Then, run the bus for thirty minutes. If you survive, you've passed the physical."

Obviously, NASA does not see it quite this way, particularly after *Columbia*'s loss, and NASA may never view popular human space travel as anything other than crass and hollow. Short of a NASA epiphany, then, the alternative space community must embark upon its own Apollo program for popular space travel if it hopes to see any real progress toward making a convincing case for this embryonic industry. It must commit to proving out near-Earth passenger space travel with an unwavering timeline to produce a working vehicle, and then fly it.

The message to the alternative space community is this: gather every engineer, tinkerer, millionaire, NASA refugee and corporate friend you can find, create a coalescent force for change (a trade association, an *über*-entrepreneurial venture, et cetera), settle on a unified approach for achieving cheap, reliable, safe, reusable near-Earth passenger spaceflight, and do it yourself—NASA be damned. The X Prize is a positive beginning to such an agenda, but loosely combined individualism will not crack this problem, nor will it outshine NASA's discouraging glare in popular human spaceflight; there is not yet enough fertile soil for a thousand flowers to bloom. The alternative space community must find the most fruitful patch available and then apply every ounce of its collective effort to ensuring near-Earth passenger space travel.

The foundation for such an approach may have been laid at a private meeting held in Los Angeles six weeks after the *Columbia* disaster. Convened by the unlikely trio of Rick Tumlinson, Dennis Tito and Buzz Aldrin, the meeting pulled together groups in the alternative space community who had heretofore rarely, if ever, met—between some of them, like the Space Frontier Foundation and the Mars Society, there has long been considerable animosity. In addition to these groups, leaders from the National Space Society, the X Prize Foundation, the Space Studies Institute and the Mars Institute all were present and accounted for, as were the founders of several entrepreneurial space firms.

All in attendance agreed to set aside individual agendas to focus on those goals everyone sought in an ideal space future. At the end of the day, it boiled down to three ideas: to make use of resources in space to enhance life on Earth, to fulfill the innate human drive for exploration and discovery and, perhaps most controversially, to "ensure the survival of human civilization and the biosphere." If the leaders of these groups have thus far stopped short of merging their objectives into a more focused coalition of common interest, they did agree in principle to pool their efforts and influence toward the achievement of a space future they all, at some level, are desperate to realize. As they say, it is a start.

Were NASA to embrace the efforts of the entrepreneurial space sector and actively support broad-based, participatory interest in human spaceflight, the agency could well inspire its own renaissance. One of NASA's greatest laments in recent years has been its inability to draw the crème de la crème of the nation's technical and managerial talent even as its longtime employees continue to retire. NASA itself openly complained about this situation in the wake of the *Columbia* disaster, although if the agency was attempting to link this scenario to *Columbia*'s loss no one bought it.

By embracing rather than shunning the alternative space sector, NASA and Big Aerospace—which has lost potential talent to the computer and Internet industries—could help reverse this trend. Indeed, it is telling that so many of the newer space entrepreneurs, like Elon Musk, hail from the high-tech sector.

To take the step that would naturally follow renewed popular interest in human spaceflight—a human mission to Mars—will require a bolder paradigm shift. Oddly enough, NASA has already laid the foundation of such change with the International Space Station (and this may be the single most compelling case for the station's continued existence). Politically, technologically, economically and culturally, a human mission to Mars must, as its Devon Island rehearsal implies, be an international undertaking: it must be a united effort of the space programs of America, Europe and Russia, and emerging programs like China's. It must tap the experience of Big Aerospace as well as look to the entrepreneurial sector for innovation. The human exploration of Mars would then be a true exploration for humankind.

As for the ISS itself, Rick Tumlinson has championed one of the more tenable propositions for the station's future in describing the creation of a nonprofit, multinational "port authority" to provide a "clear and impartial management structure" to regulate the station's activity. Freed of NASA ownership, private options to give the station more bang for the buck, like the Spacehab/Energia Enterprise module, would get precedence over government options created from expensive whole cloth; autonomous orbital facilities like MirCorp's Mini Station could be built to directly complement—but not necessarily supplant— ISS research and commercial capabilities.

In this scenario, NASA would give up all pretense of being the entity that "commercializes" the station. In fact, the port authority should take a position directly counter to NASA's current ISS commercial policy and offer seed funding specifically to kick-start promising ISS-oriented entrepreneurial ventures. Eventually, all of the station's operations—from transportation to science to maintenance and even to crew—would become private enterprises with the port authority ensuring their work met satisfactory cost, reliability and safety standards.

Shedding the shuttle and operational responsibility for the ISS would free up billions of designated space dollars: in such a situation, the first order of business on the part of NASA and its allies would be to keep this extra cash earmarked for space. The government, perhaps through the Department of Transportation instead of NASA, could use all or part of this money to create an incentive program along the lines of the Kelly Mail Act in which cash is given to any company—Big Aerospace or entrepreneurial—able to send payloads into orbit or to the ISS at a capped cost of $1,000 or even $500 per pound.

The program could also match any investment dollars put forward in pursuit of vehicles to achieve such a goal. Government subsidies like this are tricky, since they can easily morph from incentives to hand-outs. But particularly in America they have often been the deciding factor in whether a particular industry lives or dies. Properly conceived and managed, such a program could jury-rig a robust entrepreneurial space-launch sector into existence.

Just such a conclusion was reached in 2002 by the Commission on the Future of the United States Aerospace Industry, a congressionally created effort to recommend ways to ensure the vitality of American aerospace enterprises. Acknowledging low space-launch rates and expensive launch costs, the report said that subsidies "not unlike what was done in the early rail and airline industries" could be essential to keeping the space industry alive. Ironically for NASA, the report noted that the one bright spot in terms of future markets was space tourism, which the commission said "may be the key to help fund the launch industry through the current market slump by providing increased launch demand and thus helping to drive launch costs down."

Unlike the National Taxpayers Union Foundation and other non-profit think tanks, the aerospace commission has political teeth; whether or not its bolder conclusions have any impact remains to be seen. The same must be said of the *Columbia* inquiry board recommendations. Likewise, whether or not those in power decide to continue periodic yet ultimately cosmetic face-lifts for NASA, or whether there will come a real commitment to major and undoubtedly painful organizational surgery, may well determine whether or not the future of space will look like the present, or whether it will be something we can only begin to imagine.

Anyone with even a passing familiarity with the *Star Trek* universe knows that the United Federation of Planets is not a purely research-oriented entity. It was plainly modeled after America, and its exploratory ambition was backed with firepower. Whatever its incarnation, not only does the *Enterprise* itself have a plainly military command structure, it spends as much time blasting hostile aliens to smithereens as it does conducting scientific surveys.

The military aspect of the *Enterprise*'s mission was actually toned down by Gene Roddenberry, even though he had been a World War II combat pilot who clearly saw a place for the military in space. According to Roddenberry's biographer David Alexander, one of the earliest versions of the show's now-famous opening monologue that begins "Space, the final frontier . . ." called upon the *Enterprise* to contact alien life and "enforce galactic law" (a slightly more benign version said that the *Enterprise* "regulates commerce"). In any event, the *Enterprise* was less Darwin's *Beagle* than a high-tech gunboat whose main job was interplanetary security.

In its seamless joining of the civilian and the military, *Star Trek* both reflected and presaged what actually came to pass. NASA, of course, has long had quasi-military overtones; the shuttle came to life with military backing (and remains a product of military-industrial giants like Boeing and Lockheed). Some of the most promising post-Apollo space projects, like the DC-X, had distinctly military origins.

And thus was it ever. When the Space Race began to heat up, an early U.S. Air Force proposal would have sent a hydrogen bomb as America's first ambassador to the moon; the bomb would then have been detonated on the lunar surface in a cosmic display of American military superiority. By the mid-1960s, both the Army and the Air Force had drawn up detailed plans for military moon bases whose assignments would have included lobbing nuclear devices at the Earth (this turned out to be a less efficient delivery strategy than using ICBMs). The Soviets weren't exactly cosmic pacifists: the Salyut-3 space station, for example, was mounted with a modified jet-fighter cannon that was actually test-fired in orbit.

Warfare is a booming business in space these days, a trend unlikely to diminish any time soon. This is true particularly in the United States, which is already well ahead of the rest of the world in military satellites—the backbone of what's brusquely dubbed "milspace" in defense circles—both in technological sophistication and in sheer numbers.

The Stockholm International Peace Research Institute reported that, as of the end of 2001, the United States had 110 operational military satellites, accounting for two-thirds of the military satellites in orbit; Russia had 40, and the rest of the world combined claimed 20.

Well before the September 11 terrorist attacks, the American military was beating the drum for stepped-up capability in space-based weaponry and space transportation systems. The possibilities bandied about with serious money in mind included *Star Wars* stuff like laser weapons and self-maneuvering mines, as well as pre–Space Age concepts like antipodal bombers: piloted or unpiloted craft that would travel partly through space (much like ICBMs) to achieve rapid global strike capability.

In the summer of 2001, the U.S. Army graduated the first class from its Space Operations Officers Qualification Course in Colorado Springs. The fourteen officers in this inaugural group were to be the vanguard of a growing force trained to focus on those elements of warfare dependent on space-based assets. "We in the Space business in the U.S. Army, and the Army itself, understand the criticality of Space to its future," Lt. Gen. Joseph Cosumano, Jr., head of the Army's Space Command, told the graduates (in this verbatim extract from an Army press release). "And so it is an uncharted path, you graduates will go down that path and mark the trail for those who will follow."

After September 11, the "trail" got crowded fast. The Pentagon began framing what it called a National Aerospace Initiative to push the development of new propulsion technologies and better, faster access to space. Under former navy secretary Sean O'Keefe's leadership, NASA quickly abandoned any pretext of keeping military concerns at arm's length and happily joined the accelerated military space planning efforts. "I think it's imperative that we have a more direct association" with Department of Defense space efforts, O'Keefe told a media gathering in 2002, adding that recent developments in space-relevant technology had "taken us to a point where you really can't differentiate . . . between that which is purely military in application and those capabilities which are civil and commercial in nature."

It is unclear how military interests will ultimately be blended into NASA's future. One area in which such cozy relations might affect

NASA is on the budget side. If military needs are integrated more thoroughly into NASA's work in propulsion and launch-vehicle development, the DOD's massive annual budget ($355 billion in 2003) could provide a healthy shot in the arm for NASA, giving it more fiscal room to salvage the wobbly space-station program.

Where space warfare is concerned, of course, Big Aerospace stands to gain considerably. "The whole area of integrated battle space—we see that as a big growth market," Boeing's president for space systems, James Albaugh, told *Aviation Week* shortly after Boeing merged its civilian and military space operations into one unit. In the same article, Lockheed Martin's Albert Smith, executive vice president for Space Systems (and a former CIA engineer), noted that Lockheed, too, was more than ready to step up to the challenge of converting as much government funding as possible into space-based military assets, describing such work as "a very stable, good-margin business."

Entrepreneurs have not been shy about grabbing pieces of the new milspace, too, since for most space entrepreneurs, military space has long been just another possible path to the future. In 2002, Pioneer Rocketplane was among a half-dozen companies awarded between $1 million and $2 million each from DARPA to develop competing designs for a small-satellite launch system with a reusable, jet-powered first stage and rocket-powered second stage. Pioneer was not the final winner, but another entrepreneurial company was: Space Launch Corporation of Irvine, California, which received about $10 million to fund a flight demonstration in 2006 from Vandenberg Air Force Base.

According to the Stockholm Institute, in 2001 the governments of Earth spent about $772 billion on military activities, of which America's share was a staggering one-third. However one may feel about this, it is only realistic to say that the military is not going away anytime soon, nor is its presence in space likely to diminish. If anything, it is increasing. For those of us who are simultaneously opposed to the global war machine and believers in a human space future based on peaceful cooperation, this is a particularly challenging reality.

Yet these positions do not have to be mutually exclusive. It may well be that one of the best, and most optimistically subversive, uses of military spending is to pursue better, cheaper and more reliable spacecraft. After all, the $60 million the military spent on the DC-X not only went further than the $1 billion NASA spent on the X-33, it also kept at least $60 million from being spent on bombs. There is also reason to believe that any military advances in space-launch technology could make their way into broader use faster than those developed under NASA. Again, the aviation industry offers precedent: the jet engine was developed under wartime pressure in distinctly military circumstances, yet commercial jetliners using what was, essentially, modified military technology were beginning to crisscross the globe within a decade after the war's end.

Along with its quasi-military zeal, *Star Trek* also conveyed a sense that any human space future would necessarily require the collective effort of many nations and many cultures. Unlike most previous science fiction, the original *Enterprise* had a decidedly multiethnic, multinational crew: there was no doubt that Americans ran things, but there were Russians and even Chinese in significant positions of authority and control (although "taikonaut" Mr. Sulu was portrayed by a Japanese-American, the actor George Takei). This flew in the face of the political ideology of the 1960s, which outlined space-faring prowess as a clearly nationalist endeavor.

Future installments of *Star Trek* reflected loosening social strictures in Western society, the softening of Cold War rhetoric and the opening of political and cultural borders. Jean-Luc Picard, the *Enterprise*'s captain in *Star Trek: The Next Generation,* was a European whose long-suffering second-in-command was American; *Deep Space Nine*'s commander, Benjamin Sisko, was black, and *Voyager*'s captain, Kathryn Janeway, was the first female commander to lead a *Star Trek* series.

As with its exploratory zeal and fusion of civilian and military exploration, in its presentation of space as a personal frontier regardless of race, culture or nationality, *Star Trek* had some real-life ramifications. The astronaut Mae Jemison, who in 1992 became the first black woman in space, said repeatedly that one of her chief role models was Lieutenant Uhura, the black female communications officer from the original *Star Trek* series played by Nichelle Nichols. Jemison initiated each of her shuttle crew shifts with Uhura's famous line, "Hailing frequencies open"; she later had a cameo appearance in a *Next Generation* episode. Worldwide, some version of *Star Trek* is shown on television hundreds of times daily, translated into many languages and shown in dozens of countries, thereby cutting across cultural barriers in reality just as it sought to in fiction.

There is a nonfictional place where multinational, multicultural and multiagenda space interests converge for their common advancement, much as they do at *Star Trek*'s fictional Starfleet Academy. In the course of its fifteen-year history, the International Space University, typically known by its abbreviation, ISU, has become perhaps the only truly neutral ground for the worldwide space community: a place where bottom lines, political maneuvering and international rivalries are left outside of Strasbourg, France (the university's home since 1994), in the interest of trying to find a common future for humans in space.

The university's board of advisers includes Sean O'Keefe, who heads NASA, and Yuri Koptev, who heads the Russian Space Agency, while its lecturers and panelists have included Jeff Manber, Walter Anderson (also an early financial backer), Rick Tumlinson and Robert Zubrin. ISU's alumni include managers and engineers from Lockheed and Boeing (who are among the university's biggest private funders), NASA, ESA, and the Russian, Canadian, Japanese and Chinese space agencies, with some astronauts also on the alumni rolls. Many of the scientists leading work on Devon Island are also ISU grads; so are staff members working for high-profile entrepreneurial companies like Kistler Aerospace and Space Adventures.

Considering all of the NASA and Big Aerospace support for ISU, it is all the more unexpected that the university was created specifically to mold future space professionals whose space-faring philosophy was in sync with that of Gerard O'Neill. The university's cofounders were among the earliest O'Neillian disciples: X Prize and Zero Gravity Corporation cofounder Peter Diamandis, aerospace entrepreneur Robert Richards and the late Todd Hawley, the university's first chief executive and one of the most active members of the alternative space community before his untimely death in 1995. To date, Hawley is the only one of his ISU cofounders to have made it into space: in 1997, seven grams of his cremated remains were launched into orbit along with those of O'Neill himself. They were on the first flight sponsored by Celestis, which to date has performed such services for more than one hundred people, including Gene Roddenberry.

ISU may be the only place on Earth where anyone with a vested interest in space can actually say what's on their mind and know that someone will listen. By the standards of most universities, ISU is a small operation. Its budget is a few million euros a year, and it may never rival the world's great educational institutions in those qualities that are held as standard measures of first-class academia. But where space is concerned, ISU may prove to be something of a new beginning.

In an early promotional brochure supporting ISU, Arthur C. Clarke wrote, "The first universities helped to bring mankind out of the Dark Ages and into the Renaissance. In our day, there are few institutions which satisfy any higher individual aspirations or greater interests of humanity." Clarke was obviously painting with broad brush strokes, but he hit the essential mark about the appeal of space travel. It endures as a dream to which people will give their lives—sometimes literally, like the crews of *Challenger* and *Columbia*—because it is, if nothing else, a pursuit they believe to be greater than themselves.

In so many ways, "space"—whether you call it a place, a resource, an ideal or an idea—is still in the Dark Ages. It remains strange and inscrutable to most of the world, and access to its mysteries and potential remains open to only a few; knowledge is still hoarded by the monks, so to speak, and they are loath to relinquish it.

As of this book's publication, nearly thirty-five years have passed since humans first left Earth—or as Konstantin Tsiolkovsky once described it, "humanity's cradle"—and set foot on another world. Tsiolkovsky famously followed his description with another remark: "You cannot remain in the cradle forever." Yet unless the agencies, the industries and the entrepreneurs can make a better case for the importance of human spaceflight to the future of humankind, we may well find ourselves three decades from now—after the ISS has followed Mir into the ocean, after the last shuttle has finally given up the ghost— with no greater support, no further reach and no better sense of what humans might achieve in space. Making the "case" effectively depends, as all relevant things do, on inclusiveness. For that to happen, everyone must be invited along for the ride.

In the fall of 2002, the International Space University got a brand-new, twenty-million-euro home courtesy of various forms of French governmental and private support. A few months before it opened for business, I toured the still incomplete building, which was engulfed in a frantic buzz of workmen and littered with sawhorses, coiled wiring and great slabs of freshly hewn floorboards. Stepping over bales of insulation and around rows of plastic-wrapped desks, I found myself smiling, thrilled to see the place in such a prenatal state: before the inaugural parties, before the first students blinked nervously in the light of their gleaming new classrooms—before the building became, as all buildings become, settled and marked by the casual wear of its users.

It was already a pretty place, even for all the clutter. The lecture halls were spacious and fitted with the latest state-of-the-art gear to facilitate maximum connection between lecturer and student. There is a "high bay" for examining and even building satellites and other space hardware, and seemingly the entire east wing of the building is devoted to a library and multimedia center, the better for developing the compelling marketing and public relations presentations needed to excite investors, politicians and the public.

Everything was new. Everything was possible. But my favorite part of the campus looked to the casual observer like a piece of incomplete workmanship. Spanning the entire front width of the new building is a wide concrete pathway overlooking terraces of lawn on which future generations of space dreamers may huddle together and, perhaps, devise the breakthrough that will do away with all the desperate space futurism by creating realities more tantalizing, more exciting than any-thing that might have been imagined before.

The pathway is neatly squared off by a metal rail at one end of the building, offering a view of thick woodlands beyond. At the other end, where it arcs over a narrow finger of a large pond, the pathway simply ends; the concrete stops abruptly and wild, weedy brush begins, loom-ing for about a hundred yards before being tamed by the main thor-oughfare that winds through the industrial park and that leads, in essence, to the rest of the world. I asked the ISU representative who was showing me the building when the pathway would be completed. "We left it this way intentionally," she said. "We want students to create their own path."

It is heartening to envision the leaders of a new Space Age striding down a path they have literally made themselves: scientists, engineers, inventors and entrepreneurs whose ideas and drive will rival and even surpass any who came before them. From here, they'll harness creativity, ambition and the human desire for advancement in order to build a future in space that everyone, everywhere can have a stake in.

It may take time; there will certainly be resistance. The walls that now protect old ideas and old ways of doing things are sturdy, and they won't fall easily.

But I have to believe that change is inevitable. Many might argue for a more dour perspective, but I'll gladly hang on to mine. There is something positively joyful in thinking that as a species, there's nowhere for us to go but up.

Reference Notes

Prologue: Lost in Space

15 **In October 2002:** "Futron Releases New Space Tourism Publications," press release, Futron Corporation, October 7, 2002.

The Price of "Peace"

29 **In a 1996 report to Congress:** Congressional Research Service Issue Brief for Congress, 93017: Space Stations, updated December 12, 1996, Marcia S. Smith, Science Policy Research Division.

29 **Between 1957 and 1998:** "Worldwide Space Launchings Which Attained Earth Orbit or Beyond, Calendar Years 1957–1998," NASA Aeronautics and Space report to the president, 1999, as accessed at http://www.aia-aerospace.org/stats/facts_figures/ff_99_00/Ff99p063.pdf.

30 **Jeffrey Manber:** Author interviews with Jeffrey Manber in July and September 2001 and various e-mails and telephone calls in 2002 and 2003.

33 **On February 21, 1988:** "American Company and Soviets Agree on Space Venture," William Broad, *New York Times,* February 21, 1988.

34 **"You've got a guy":** Author interview with Rick Tumlinson in August 2001 and various conversations, e-mails and telephone calls, 2001–2003.

36 **Testifying before the House:** "Building Alpha Town: The International Space Station as a Precursor to First City in Space," Rick Tumlinson, testimony before the House Space and Aeronautics Subcommittee, Washington, D.C., April 9, 1997.

37 **Doing ISS training duty:** "Navigating on Mir," keynote address by Michael Foale at the 10th Anniversary Mathematica Conference, *Mathematica Journal,* June 19, 1998.

39 **In a February 1999 Associated Press article:** "MIR Mission Could Be Last," Associated Press, February 13, 1999.

39 **Semenov quickly agreed:** Fax from Yuri Semenov to Rick Tumlinson, March 15, 1999.

40 **When confronted:** "Russians Drop Plan for British Businessman's Flight on Mir," Vladimir Isachenkov, Associated Press, May 27, 1999; "U.S. Space Buffs See Otherworldly Profit in Russia's Rickety Mir," Neal E. Boudette, *Wall Street Journal,* June 16, 2000.

41 **With that loose agenda:** Protocol of the meeting between FINDS, RSC Energia and Gold & Apel [Gold & Appel], September 29–October 1, 1999, and signed Memorandum of Understanding.

44 **"If there is a possibility":** "Cosmonauts Blast Off for Mir Station," Simon Saradzhyan, *Moscow Times,* April 4, 2000.

45 **The Russians:** "GOP Rep Reams NASA over Russian Relationships," Paul Hoversten, SPACE.com, February 16, 2000; "Goldin: Russia Not Only Cause of ISS Delays," *SpaceViews,* February 17, 2000.

45 **Well before MirCorp:** Details of NASA's work on *Armageddon* are from *NASA/TREK,* Constance Penley, Verso, New York, 1997.

46 **Lori Garver:** Author conversation with Lori Garver in July 2001.

47 **NASA also took:** "NASA Kills the Eye in the Sky," CBS News.com, June 4, 2000.

48 **In a February 2000:** "GOP Rep Reams NASA over Russian Relationship," SPACE.com, February 16, 2000.

49 **MirCorp's leadership:** "Even if MirCorp Finds Funds, Will It Be Too Late?" Yuri Karash, SPACE.com, October 19, 2000.

49 **Covering the first:** "Name That Thing," Marcia Dunn, Associated Press, October 30, 2000.

51 **That same day:** "Mir to Be Dumped in February," Associated Press, October 23, 2000.

51 **Manber responded:** "MirCorp president Jeffrey Manber argued in a letter to Russian President Putin that he must help save the Mir space station," SPACE.com, October 27, 2000.

52 **In 1993:** The tug-of-war over the ISS orbit can be found in Congres-sional Research Service Issue Brief for Congress, 93017: Space Stations, updated December 12, 1996, Marcia S. Smith, Science Policy Research Division.

One Giant Leap

59 **Clay tablets:** The story of Etana's flight is from *Science Fiction and Space Futures, Past and Present,* edited by Eugene M. Emme, American Astronautical Society, San Diego, Calif., 1982.

64 **In the early 1950s:** Wernher von Braun's fame in *Collier's* is detailed in *Space and the American Imagination,* Howard McCurdy, Smithsonian Institution Press, Washington and London, 1997.

65 **"The Cosmos is all that is":** *Cosmos,* Carl Sagan, Random House, New York, 1980.

67 **Humanitarianism:** *World Population Prospects: The 2000 Revision,* report of the United Nations Population Fund, found at www.unfpa.org/swp/2001.

68 **The wealth of the Earth's:** *The High Frontier* (Third Edition), Gerard K. O'Neill, Apogee Books and the Space Studies Institute, Burlington, Ontario, Canada, 2000.

68 **In his 1977 book:** *Colonies in Space,* T. A. Heppenheimer, Stackpole Books, Harrisburg, Pa., 1977.

70 **Despite being published:** Michaud quote is from *Space and the American Imagination.*

71 **Even O'Neill's "mass driver":** Referenced in *The High Frontier* (Third Edition), Gerard K. O'Neill, Apogee Books and the Space Studies Institute, Burlington, Ontario, Canada, 2000.

72 **Even Carl Sagan:** *Carl Sagan: A Life,* Keay Davidson, John Wiley & Sons, Inc., New York, 1999.

76 **The popular Apollo astronaut:** Russell Schweikart's comments on Gerard O'Neill are excerpted from a NASA Ames Research Center web site discussion of O'Neill's space colonization ideas, found at http://lifesci3.arc .nasa.gov/SpaceSettlement/CoEvolutionBook/.

76 **O'Neill had a similar effect:** The proposed $750 million increase in NASA's budget for O'Neillian colonies is described in *Colonies in Space,* T. A. Heppenheimer, Stackpole Books, Harrisburg, Pa., 1977.

76 **NASA then tasked O'Neill:** The NASA study headed by Gerard O'Neill is *Space Resources and Space Settlements,* edited by John Billingham, William Gilbreath and Brian O'Leary, National Aeronautics and Space Administration, Washington, D.C., 1979.

77 **When O'Neill's supporters:** O'Neillian versus NASA cost estimates for a proto-space city are from *Space and the American Imagination.*

78 **The commission's report:** "Pioneering the Space Frontier: An Exciting Vision of Our Next Fifty Years in Space," the Report of the National Commission on Space, NASA, 1986, as accessed at http://history.nasa.gov/ painerep/begin.html and also published by Bantam Books, New York.

79 **He and a colleague were hired:** Author interview with Rick Tumlinson in August 2001 and various conversations, e-mails and telephone calls, 2001–2003.

81 **Among the early corporate pioneers:** Information on the McDonnell-Douglas electrophoresis program and other early space-manufacturing propositions is from *NASA and the Space Industry,* Joan Lisa Bromberg, Johns Hopkins University Press, Baltimore, Md., 1999.

83 **Very quickly, NASA employed:** "NASA Announces Long Range Plan for Industrial Space Facility," NASA News, Release No. 84-29, February 28, 1984.

85 **NASA began its formal assault:** *Report on the Potential NASA Utilization of the Space Industries Partnership Industrial Space Facility (ISF),* December 1987.

85 **An internal NASA white paper:** *Industrial Space Facility (ISF) White Paper,* NASA Office of Commercial Programs—Assessment and Recommendations, December 8, 1987.

86 **To Jamie Whitten:** Letter from James C. Fletcher to Jamie Whitten, January 6, 1988.

86 **To the influential:** Letter from James C. Fletcher to James A. Baker III, December 31, 1987.

87 **This was disingenuous:** "Outsourcing controversy blasts NASA; PA chief is latest to resign," Carolyn Myles, *Government Executive,* Aug. 6, 2001.

87   **In a January 15 editorial:** "A Rival for the Space Palace," editorial, *New York Times,* January 15, 1988.

88   **The institute's most notable success:** Lunar Prospector details and Alan Binder's comments are from Binder's presentation at the 2001 Lunar Development Conference in Las Vegas, Nev. (author's transcription).

89   **The "Alternative Plan":** "Alternative Plan for U.S. National Space Program," by Gerard O'Neill, as posted on the Space Studies Institute web site, www.ssi.org.

Escape Velocity

91   **According to statistics:** "U.S. ELV Launch Record (1988–2000)," Space Consulting Resources Group, www.spacecrg.com.

94   **Through mergers both friendly and hostile:** Information on the consolidated U.S. aerospace industry was taken from a graph titled "U.S. Industry Consolidation: A Work in Progress," *Aviation Week & Space Technology,* December 3, 2001.

96   **In his 1986 state of the union address:** The short, expensive history of the "Orient Express" National Aerospace Plane is from "NASP: Hypervelocity Hyperbole?" Leonard David, *Ad Astra,* July–August 1993; "NASA's Little 'a' Spearheads Future of Orient Express," an uncredited Associated Press story as reported in the *Washington Times,* May 12, 1988; and "Plane of the Future Would Fly into Orbit, Cross U.S. in an Hour," Bob Davis and Andy Pasztor, *Wall Street Journal,* June 20, 1989.

97   **In 1998:** "Airline Ops Offer Paradigm for Reducing Spacelift Costs," William B. Scott, *Aviation Week & Space Technology,* June 15, 1998.

98   **If you were driving:** X-33 History Project, NASA History Office, http://www.hq.nasa.gov/office/pao/History/x-33.

101   **Mitchell Burnside Clapp:** Author interview with Mitchell Burnside Clapp in January 2002 and various e-mails, 2002 and 2003.

103   **It wasn't a rocket plane:** Much of the DC-X chronology was taken from *Halfway to Anywhere,* G. Harry Stine, M. Evans and Company, Inc., New York, 1996; other aspects of DC-X history were from interviews with Rick Tumlinson, Gary Hudson, Walter Kistler, Mitchell Burnside Clapp, and Robert Citron.

103   **The DC-X emerged:** "The Spaceship That Came in from the Cold War: The Untold Story of the DC-X," Andrew J. Butrica, *Ad Astra,* March–April 2001.

105   **A NASA FAQ page:** "DC-X Frequently Asked Questions," NASA History Office, http://www.hq.nasa.gov/office/pao/History/x033/dcx-faq.htm.

106   **In July 1993:** Information on this subject is primarily from two sources: *Halfway to Anywhere,* G. Harry Stine, M. Evans and Company, Inc., New York, 1996; and "The X-33 History Project," NASA History Office, http://www.hq.nasa.gov/office/pao/History/x-33.

110   **The X Prize was launched:** Details about the X Prize are from www

.x-prize.org, as are details of the Orteig Prize and its impact on early commercial aviation.

112   **Rutan's company:** The latest information on Scaled Composites can be found at www.scaled.com; details of the spaceship rollout are from "Affordable Spaceship," Michael Dornheim, *Aviation Week & Space Technology,* April 21, 2003.

Back to the Future

115   **Kistler himself:** Author interviews with Walter Kistler in February 2002 and August 2003.

116   **Things got even tougher:** Information on Kistler Aerospace's Chapter 11 filing can be found at the company website, www.kistleraerospace.com.

119   **Bob Citron:** Author interview with Robert Citron in February 2002 and various e-mails, 2001–2003.

121   **In 1984, Citron founded:** Spacehab information is from their web site, www.spacehab.com.

125   **"If the K-1 vehicle succeeds":** Technical and market information on the K-1 is from various promotional and investor materials provided by Kistler Aerospace Corporation and from the company web site, www.kistleraerospace.com.

127   **That lunch eventually led:** Program information on the Space Launch Initiative, Alternative Access to Space Station and the Orbital Space Plane is from various NASA web sites and press releases.

129   **The Teal Group:** "Teal Survey Counts 600–610 Active Satellites Currently in Orbit," Teal Group news release, October 2, 2001.

129   **Serious money:** Teledesic FAQ and Milestones web pages, http://www.teledesic.com.

131   **Yet even at $17 million:** "Kistler Seeks to Create 'UPS of Space,'" Paul Proctor, *Aviation Week & Space Technology,* June 30, 1997.

131   **Iridium was first:** "McDonnell-Douglas Launches First Iridium Satellites," McDonnell-Douglas News Release, Florida Today Online, May 7, 1997.

132   **Iridium's satellite infrastructure:** "Iridium's Shot at a Second Life," Joanna Glasner, *Wired News,* March 29, 2001.

132   **Teledesic hung on:** "Teledesic Files to Surrender Spectrum," www.spacetoday.net, July 15, 2003.

132   **The Teal Group predicted:** "Industry Sees Launch Slide End in 2002, Has Hopes for Rebound," Jason Bates, *Space News,* January 6, 2003.

133   **"I don't think":** Author interview with Jason Andrews in February 2002.

135   **As for SLI's original goal:** Sean O'Keefe's "bumper sticker" comment is from the news brief "O'Keefe Says SLI's 90% Launch Cost Reduction Goal Was Unrealistic," *Aerospace Daily,* November 18, 2002.

137   **In his biography:** *Powering Apollo: James E. Webb of NASA,* W. Henry

Lambright, Johns Hopkins University Press, Baltimore, Md., and London, 1995.

139   **As we sat:** Author interview with George Mueller in January 2002.

The Belly of the Beast

142   **In September 1999:** Dan Goldin's remarks are from the 1999 Space Frontier Foundation conference, September 24, 1999, as accessed at www .space-frontier.org.

144   **In 1970:** Excerpts from the Thomas Paine meeting are from *The Space Shuttle Decision: NASA's Search for a Reusable Space Vehicle,* T. A. Heppenheimer, NASA History Series, National Aeronautics and Space Administration, 1999.

145   **"NASA should not":** Appeal letter from Spacecause signed by James Beggs, July 1997.

145   **In a February 1996 speech:** Dan Goldin's remarks to the American Association for the Advancement of Science are from the "What's New" online newsletter of February 16, 1996, archived at the American Physical Society web site, www.aps.org.

146   **By October 2000:** Author's transcription of Goldin's remarks at the U.S. Space Foundation Annual Conference, Ronald Reagan International Trade Center, Washington, D.C., October 24, 2000.

146   **In his final days at the agency:** "NASA Fans Should Face Facts—It's Not a Jobs Program, Goldin Says," John Anderson, *Huntsville* (Alabama) *Times,* November 11, 2001.

149   **In an April 2001 report:** Testimony of Russell A. Rau, Assistant Inspector General for Auditing, NASA, before the House Committee on Science, April 4, 2001, as accessed at www.house.gov/science.

151   **"You really can't understand":** Author conversation with Roger A. Launius, March 2002.

151   **Published in 1893:** The version of Frederick Jackson Turner's *The Significance of the Frontier in American History* referenced is from text prepared by Rutgers University professor Jack Lynch, as accessed at http://newark .rutgers.edu/~jlynch/Texts/frontier.html. It was the clearest and most concise presentation of Turner's thesis I found, out of dozens of versions available. Turner's thesis was first offered as a speech at the American Historical Society in 1893 and was first published the following year as *The Significance of the Frontier in American History,* State Historical Society of Wisconsin, Madison, 1894.

153   **The extent of Kennedy's reluctance:** Audiotape of Kennedy-Webb exchange dated November 21, 1962, released in August 2001 by the John F. Kennedy Library and archived at www.space.com.

153   **Webb was NASA's second administrator:** *Powering Apollo: James E. Webb of NASA,* W. Henry Lambright, Johns Hopkins University Press, Baltimore, Md., and London, 1995.

157 **Yet all the while:** The NASA document known as Appendix F, Richard Feynman's personal appendix to the Rogers Commission's *Report on the Space Shuttle Accident,* was published in 1988 and can be found at http://science.ksc.nasa.gov/shuttle/missions/51-1/docs/rogers-commission/Appendix-F.txt.

158 **Much of this cost:** "Space Shuttle Officials Search High and Low for Vintage Spare Parts," Brian Berger, *Space News,* April 29, 2002.

160 **Engineers replaced:** Details of *Columbia's* 1999 repair work are from "Columbia's Final Overhaul Draws NASA's Attention," Andrew Pollack, *New York Times,* February 10, 2003.

160 **The breadth of *Columbia's* needed work:** "NASA Considers Taking Columbia out of Service," Warren Ferster, *Space News,* July 16, 2001.

163 **But Sean O'Keefe:** "Amid Quest for a Safer Shuttle, Budget Fights and Policy Shifts," David Barstow and Michael Moss, *New York Times,* February 9, 2003.

164 **One of the more hair-raising analyses:** Commercial airliner fault and crash analysis, as applied to the shuttle program, was found at www.airsafe.com.

164 **Amazingly, less than a month:** "NASA Will Launch Plan to Keep Shuttles Flying," Shelby Spires, *Huntsville* (Alabama) *Times,* January 4, 2003.

165 **The shuttle was announced:** *Countdown: A History of Space Flight,* T. A. Heppenheimer, John Wiley & Sons, Inc., New York, 1997.

165 **NASA's annual budget:** NASA budget and employment figures are from NASA Pocket Statistics, NASA History Office, as posted at http://history.nasa.gov/pocketstats.

165 **The resulting decline:** Overall loss of NASA workforce from 1966 to 1969 is from "The Post-Apollo Space Program: Directions for the Future, Space Task Group Report to the President," NASA, September 1969.

166 **To get his "all-up" job done:** The promotional zeal of George Mueller and Wernher von Braun for the shuttle is detailed in *The Space Shuttle Decision: NASA's Search for a Reusable Space Vehicle,* T. A. Heppenheimer, NASA History Series, National Aeronautics and Space Administration, 1999.

167 **The rise of the contractor culture:** The shifting balance of bureaucrats and science and technical personnel within NASA is from *Inside NASA: High Technology and Organizational Change in the U.S. Space Program,* Howard McCurdy, Johns Hopkins University Press, Baltimore, Md., 1993.

170 **Military procurements:** The loss of military procurements in the late 1960s and early 1970s is from *NASA and the Space Industry,* Joan Lisa Bromberg, Johns Hopkins University Press, Baltimore, Md., and New York, 1999.

170 **In 1971:** Information on Nixon's New Technology Opportunities Program, or NTOP, is from *Where Do You Go After You've Been to the Moon,* Francis T. Hoban, Krieger Publishing Company, Malabar, Fla., 1997.

Advance Reservations Required

174 **Our longest talk:** Author interview with Alan Ladwig in March 2002 and subsequent e-mails and conversations in 2003.

176 **When Senator Jake Garn:** Garn's comments were made on the *Today* show, January 31, 1985; the *Doonesbury* cartoon ran in the *Washington Post,* among other publications, on February 15, 1985.

177 **Ladwig was one of the few:** "NASA's Midlife Crisis," Beth Dickey, ABC-NEWS.com, September 29, 1998.

180 **"Universal Resurrection":** *N. F. Fedorov, A Study in Russian Eupsychian and Utopian Thought,* Stephen Lukashevich, University of Delaware Press, Newark, Del., 1977.

181 **But Federov's most lasting influence:** *K. E. Tsiolkovsky: Selected Works,* compiled by V. N. Sokolsky, edited by A. A. Blagonravov, translated from the Russian by G. Yankovsky, Mir Publishers, Moscow, 1968.

184 **By 2001:** The budget figure for Russia's space program is from "Russian Debt May Deplete Space Funds," SPACE.com, March 2001.

185 **William Stavros:** Author conversations with William Stavros, April 2001.

186 **Wilshire Associates introduced:** Historic and current information about Wilshire Associates can be found at the company's web site, www.wilshire .com.

186 **Tito himself went:** "Wall St.'s Rocket Scientist; for Master of Investment Formulas Dennis Tito, Life Enters Its 3rd Stage," Debora Vrana, *Los Angeles Times,* June 20, 1999.

186 **Let's do some simple math:** Figures for calculating the rough cost per astronaut on a shuttle flight are taken from *Statistical Abstract of the United States for 1999,* www.census.gov/statab.

189 **But on November 16, 2000:** "Dennis Tito Says It's 'Highly Likely' He Will Go to the ISS in April 2001," Anthony Duignan-Cabrera, SPACE.com, November 16, 2000.

190 **In a February 2001 press conference:** "Europe's Station Chief Opposes Millionaire's Visit to ISS," Peter de Selding and Brian Berger, SPACE.com, February 2, 2001.

190 **Cabana's words:** "One Real Space Cowboy," David France, *Newsweek,* April 2, 2001.

191 **"I had this dream":** "Californian Hopes to Be First Tourist in Orbit," Warren E. Leary, *New York Times,* June 20, 2000.

191 **Speaking to CNN:** "Space Tourist Flap Disrupts Cosmonaut Training," CNN.com, March 20, 2001.

191 **Speaking to CBS Radio:** "ISS Crew Says They'd Welcome Tito Aboard," spacetoday.net, March 31, 2001.

191 **Dan Tam:** "Alpha: Mixing Science and Commerce," Frank Sietzen, Jr., *Aerospace America,* April 2001.

192 **Tito's hottest public comment:** "NASA Chief Raps Tito, Praises Filmmaker Cameron," Deborah Zabarenko, Reuters, May 2, 2001.

192 **Writing in the *National Review*:** "Cosmic Capitalist," Andrew Stutta-ford, *National Review,* May 1, 2001.

192 **In an Associated Press article:** "Dreams of Space Trips Become Reality," Associated Press, April 27, 2000.

193 **On April 24:** "International Space Station Partnership Grants Flight Exemption for Dennis Tito," NASA Headquarters Press Release 01-83, April 24, 2001.

194 **The history of the Baikonur Cosmodrome:** *The Soviet Space Programme,* Ronald D. Humble, Routledge, London and New York, 1988, and *Star-Crossed Orbits: Inside the U.S.-Russian Space Alliance,* James Oberg, McGraw-Hill, New York, 2002.

197 **Baikonur today:** Two articles provided primary source material for Baikonur's recent rejuvenation: "Baikonur: It Ain't Pretty, but It Sure Does Work," John Sotham, *Air & Space Magazine,* February–March 2001, and "Baikonur Cosmodrome Eyeing the Future," Frederic Castel, SPACE.com, July 10, 2000.

197 **In 1991:** "Angry Kazakhstan Again Slams Baikonur Shut," Simon Saradzhyan, *Moscow Times,* October 29, 1999.

199 **"We actually had to":** "Globalstar Sees Need for Staff Rotations at Baikonur," *Space News,* January 10, 2000.

204 **In interviews afterward:** An excellent account of the life and times of the first human space traveler is *Starman: The Truth Behind the Legend of Yuri Gagarin,* Piers Bizony and Jamie Doran, Bloomsbury, London, 1998.

206 **In November 2001:** "Principles Regarding Processes and Criteria for Selection, Assignment, Training and Certification of ISS (Expedition and Visiting) Crewmembers, Multilateral Crew Operations Panel, Revision A, November 2001," a joint document produced by NASA, ESA, NASDA, CSA and Rosaviacosmos.

206 **In May 2001:** The MSNBC space tourism survey was published as part of a news roundup titled "Space Shorts," MSNBC.com, May 7, 2001.

207 **"I think it's real":** Author conversation with Lori Garver in July 2001.

Business in a Vacuum

209 **Tito also warned:** Dennis Tito's postflight perspectives on space tourism and space commerce were taken from his congressional testimony on July 18, 2001, and from the article "Dennis Tito Cautious About Space Tourism Future," Jeff Foust, *Space Flight Now,* January 28, 2002.

212 **Yet NASA's original 1958 charter:** The full text of the Space Act of 1958, which created NASA, can be found at the NASA History Office web site, http://history.nasa.gov.

213 **"When I was in policy and plans":** Author interview with Alan Ladwig in March 2002 and subsequent e-mails and conversations in 2003.

213 **The most cogent rationale:** Hans Mark's comments are excerpted from *Space Commerce Bulletin,* vol. 2, no. 8, April 26, 1985.

214   **Crazy or not:** "Future Privatization, More Competition Urged for Shuttle Program," Gwyneth K. Shaw and Michael Cabbage, *Orlando Sentinel,* October 18, 2002.

214   **A random sampling:** "Annual Audits Released Today: Five Major Agencies Show Improved Results; Two Deteriorate," press release, Office of Management and Budget, February 28, 2002.

214   **This blunt assessment:** *Summary of NASA Office of Inspector General: Audit of Restructuring of the International Space Station Contract,* Report IG-02-002, November 14, 2001.

215   **In a withering report:** *Government at the Brink: An Agency by Agency Examination of Federal Government Management Problems Facing the Bush Administration,* Committee on Governmental Affairs, United States Senate, June 2001.

215   **The NASA Inspector General:** *Consolidated Space Operations Contract: Evaluating and Reporting Cost Savings,* IG-01-29, NASA Office of the Inspector General, August 31, 2001.

215   **As if all this wasn't bad enough:** *Second Annual Performance Report Scorecard: Which Federal Agencies Inform the Public?* Mercatus Center, George Mason University, May 16, 2001.

216   **In the early months:** "Closing a Center Could Be Viable Option for NASA," Shelby G. Spires, *Huntsville* (Alabama) *Times,* December 5, 2001.

217   **The influential *Space News:*** "Time to Close NASA Centers," editorial, *Space News,* September 3, 2001.

217   **"Oh, he'd be":** Author interview with Courtney Stadd in March 2002.

221   **As for any actual competition:** "Further Privatization, More Competition Urged for Shuttle Program," Gwyneth K. Shaw and Michael Cabbage, *Orlando Sentinel,* October 18, 2002.

221   **"If you were to truly":** Author interview with Marcia Smith in March 2002.

222   **Shortly after the September 11, 2001:** "Military Buys Exclusive Rights to Space Imaging's Pictures of Afghanistan War Zone," Associated Press, October 15, 2001.

223   **The popular human:** *General Public Space Travel and Tourism,* a joint report of NASA and the Space Transportation Association, NP-1998-3-11-MSFC, March 1998.

223   **In 2001:** *Future Space Transportation Study, Phase I,* Andrews Space & Technology, January 2001.

223   **In 1994:** *Commercial Space Transportation Study,* May 1994, a joint report of the Boeing Company, General Dynamics, Martin Marietta, McDonnell-Douglas, Lockheed Corporation and Rockwell International.

226   **Thus, far from being received:** The latest facts and figures about the International Space Station can be found at http://spaceflight.nasa.gov/station/isstodate.

226   **In that capacity:** Testimony of Sean O'Keefe, OMB, before the House

Appropriations Subcommittee on VA/HUD and Independent Agencies, press release of the House Appropriations Committee, May 3, 2001.

228 **In his book:** *Voodoo Science: The Road from Foolishness to Fraud,* Robert Park, Oxford University Press, New York, 2000. Park's online newsletter, What's New, can be found at www.aps.org/WN.

228 **In 1991:** Statement on the Manned Space Station, American Physical Society web site, www.aps.org.

229 **As a *Florida Today* article:** "Broken or Not, Scientists Unswayed by Promise of Station," Steven Siceloff, *Florida Today,* as posted on SPACE. com, June 20, 2001.

229 **NASA's "Enhanced Strategy":** *NASA: Enhanced Strategy for the Development of Space Commerce (Draft),* NASA headquarters document, September 24, 2001.

229 **"What they're talking about":** "NASA's New Frontier: Selling Space," Associated Press, October 7, 2001.

230 **In 1998:** Information on ISS commercial opportunities and pricing was taken from the ISS Commercialization web site, http://commercial.hq.nasa .gov, in March 2003 and from "Selling the Station," *Space News,* June 11, 2001.

231 **To date, only one:** Information on ISS experiments flown with StelSys was taken from the NASA Office of Biological & Physical Research web site, http://spaceresearch.nasa.gov/research_projects/ros/all.html, in November 2002.

232 **Many entrepreneurs:** Information on the Kelly Air Mail Act of 1925 is from *Turbulent Skies: The History of Commercial Aviation,* T. A. Heppenheimer, John Wiley & Sons, New York, 1995.

234 **One might reasonably conclude:** "NASA Seeks Money from Space Station Partners," Tamara Lytle, *Orlando Sentinel,* April 5, 2001.

235 **Zero Gravity Corporation:** Information on Zero Gravity Corporation can be found at www.zerogcorp.com.

Welcome to the Revolution

239 **The act's language:** Quoted text from the 1998 Commercial Space Act was found at the ProSpace web site, www.prospace.org.

240 **Dahlstrom laughs:** Author interview with Eric Dahlstrom in July 2001 and various e-mails and conversations, 2001–2003.

241 **By the time this report:** "Lessons Learned from *Challenger,*" a February 1988 report of the NASA Office of Safety, Reliability, Maintainability and Quality Assurance, can be found at www.internationalspace.com.

243 **In May 2001:** "Engineer Fired After Criticizing NASA Cutbacks," John Mangels, *Cleveland Plain Dealer,* May 6, 2001.

243 **A year earlier:** "Safety Officer Alleges Flaws at Marshall," Brett Davis, *Huntsville* (Alabama) *Times,* August 10, 2000.

243 **Only weeks after:** Text of the *Columbia* e-mail exchanges between NASA

and United Space Alliance personnel can be found at www.spaceref.com; "NASA Chief Blasted over Shuttle Memos," Richard Stenger, CNN.com, February 28, 2003.

244 **The riskiest option:** Details on scenarios for rescuing the *Columbia* crew are from the *Columbia* Accident Investigation Board report, August 2003, accessible at http://www.caib.us.

245 **On the heels:** "NASA Was Asked to 'Beg' for Help on Shuttle Photos," Matthew L. Wald and Edward Wong, *New York Times,* March 14, 2003.

245 **For all of O'Keefe's pronouncements:** "Testimony: NASA Unable to Assess Safety," Scott Gold, *Los Angeles Times,* as reported in the *San Jose Mercury News,* March 7, 2003.

249 **"A thousand years":** Author interview with Charles Miller in November 2001 and various conversations, e-mails and telephone calls, 2001–2003.

251 **What Miller called:** "Will American Flag of the Future Display 50 Stars and a Half-Moon?" Sarah Lubman, *Wall Street Journal,* September 3, 1992.

253 **Prospace president:** Author interview with Marc Schlather in June 2001 and various e-mails and conversations, 2001–2003.

254 **Boeing and Lockheed:** Aerospace industry lobbying information was found at the Center for Responsive Politics web site, www.opensecrets.org.

254 **In addition:** "Space Contractors Gave Millions to Lawmakers," Larry Wheeler, *Florida Today,* as reported on SPACE.com, June 20, 2001.

254 **Even those members of Congress:** *NASA Annual Procurement Report: Fiscal Year 2000,* NASA headquarters, Washington, D.C., as accessed at www.nasa.gov.

254 **NASA funding:** "Moving NASA Work to Brevard Draws Ire," Paige St. John, *Florida Today,* February 28, 2002.

255 **NASA money:** "NASA 2002 Budget Contains Record Earmarks," Brian Berger, *Space News,* November 26, 2001.

255 **Since 1998:** "Pork-Barrel Projects Threatening NASA's Core Programs," Michael Cabbage, *Orlando Sentinel,* September 28, 2002.

257 **By early 2002:** "ProSpace Presses Congress to End Funding for SLI," Brian Berger, *Space News,* March 18, 2002.

262 **The week after:** "Revival of NASA Alternate Access to Station Plan Urged," Brian Berger, *Space News,* March 17, 2003.

264 **But one look:** NSS Position Paper: U.S. Space Transportation Policy, February 11, 2003, found at the National Space Society web site, www.nss.org.

265 **"The loss of *Columbia*":** Testimony of Brian Chase before the Science, Technology, and Space Subcommittee of the U.S. Senate Committee on Commerce, Science, and Transportation, April 2, 2003, *The Congressional Record.*

266 **The revamping:** Details about NASA's revamping of the Space Launch Initiative were found in a two-part piece titled "The Big Change" by Keith Cowing, November 11, 2002, as accessed at SpaceRef.com, and in two *Space News* articles by Brian Berger: "NASA 'Refining' Its Plan for Space Station Crew Rescue," October 28, 2002, and "SLI Overhaul Marks Major Shift in NASA Spending," November 18, 2002.

266 **"Are NASA officials"**: "What Will It Really Cost?" editorial, *Space News,* November 18, 2002.

266 **Almost immediately:** "Reps Hall and Gordon Comment on NASA's New Transportation Plan," press release, House Science Committee Democratic Membership, November 14, 2002.

266 **Senator John McCain:** "Replacements for Shuttles Still Far Off," Gwyneth K. Shaw, *Orlando Sentinel,* February 4, 2003.

267 **What NASA calls:** NASA Statement on the Integrated Space Transportation Plan, Release 02-216, November 8, 2002.

268 **As a reality-free exercise:** NeXT information is from NASA's Human Exploration and Development web site, http://hedsadvprograms.nasa.gov/ spaceexploration.html, and the article "NASA Reveals New Plan for the Moon, Mars & Outward," Leonard David, SPACE.com, September 26, 2002.

268 **In this spirit:** NASA's vision statement is taken from the Goddard Space Flight Center web site, www.gsfc.nasa.gov/nasavision/nasa_vision.pdf.

The Emperor of Mars

272 **NASA's very first:** Much of the history of planning for human missions to Mars was extracted from the excellent and very readable *Humans to Mars: Fifty Years of Mission Planning, 1950–2000,* David S. F. Portree, NASA Monograph Series, NASA History Division, 2001.

273 **"It came":** *The Martian Chronicles,* Ray Bradbury, Bantam, New York, 1974 (first published in 1950 by Doubleday, New York).

278 **Like the Americans:** Detailed history on Soviet plans for Mars can be found at Encyclopedia Astronautica, www.astronautix.com.

285 **"In the twenty-first century":** *The Case for Mars: The Plan to Settle the Red Planet and Why We Must,* Robert Zubrin and Richard Wagner, Simon & Schuster, New York, 1997.

285 **One way or another:** Author interview with Robert Zubrin in August 2001 and subsequent e-mails in 2002.

287 **Before that:** *Humans to Mars: Fifty Years of Mission Planning, 1950–2000,* David S. F. Portree, NASA Monograph Series, NASA History Division, 2001.

288 **The term "terraforming":** *Wonder's Child: My Life in Science Fiction,* Jack Williamson, Bluejay Books, New York, 1984.

289 **Two weeks after:** Author interview with Christopher McKay, August 2001.

290 **The result was:** Remarks on the 20th Anniversary of the Apollo 11 Moon Landing, George H. W. Bush, July 20, 1989, George Bush Presidential Library and Museum, http://bushlibrary.tamu.edu.

291 **The report:** *Humans to Mars: Fifty Years of Mission Planning, 1950–2000,* David S. F. Portree, NASA Monograph Series, NASA History Division, 2001; "Politics Not Science: The U.S. Space Program in the Reagan and Bush Years," Lyn Ragsdale, as published in *Spaceflight and the Myth of Presiden-*

tial Leadership, edited by Roger D. Launius and Howard E. McCurdy, University of Illinois Press, Urbana and Chicago, 1997.

292 **From her vantage point:** Remarks at the Space Frontier Foundation's Third Annual Lunar Development Symposium, Las Vegas, July 2001, author's transcription.

292 **American University professor:** *Inside NASA: High Technology and Organizational Change in the U.S. Space Program,* Howard McCurdy, Johns Hopkins University Press, Baltimore, Md., 1993.

294 **Shortly thereafter:** "Mars Timetable Can Be Shorter, Scientists Say," David L. Chandler, *Boston Globe,* June 6, 1990.

295 **The chief innovation:** *American at the Threshold: America's Space Exploration Initiative,* report of the Space Exploration Initiative (SEI) Synthesis Group, 1991, as accessed at http://history.nasa.gov/staffordrep.

297 **In his gripping:** *Red Mars,* Kim Stanley Robinson, Bantam Books, New York, 1993; *Green Mars,* Kim Stanley Robinson, Bantam Books, New York, 1994; *Blue Mars,* Kim Stanley Robinson, Bantam Books, New York, 1996.

298 **"The existence":** *The Significance of the Frontier in American History,* Frederick Jackson Turner. The version referenced here can be accessed at http://www.newark.rutgers.edu/~jlynch/Texts/frontier.html.

298 **"The true value":** *The Case for Mars: The Plan to Settle the Red Planet and Why We Must,* Robert Zubrin and Richard Wagner, Simon & Schuster, New York, 1997.

299 **Zubrin showed:** *Entering Space: Creating a Spacefaring Civilization,* Robert Zubrin, Jeremy P. Tarcher/Putnam, New York, 1999.

299 **In a May 2001 speech:** Testimony of Dr. Robert Zubrin to the Senate Veterans, Housing and Urban Development Appropriations Subcommittee Regarding the National Aeronautics and Space Administration Budget for FY 2002, May 31, 2001, as accessed at www.marssociety.org.

300 **With Gingrich's fall:** *The Martian Race,* Gregory Benford, Aspect/Warner Books, New York, 1999.

Mars on Earth

303 **Since Lee's first:** Author interview with Pascal Lee, August 2001, and subsequent e-mails and conversations in 2001 and 2003.

303 **Following the devastating:** Information on Mars Odyssey can be found at http://mars.jpl.nasa.gov/odyssey.

304 **In October 2000:** "Human Exploration of Mars Is Addressed by NASA," Leonard David and Andrew Bridges, SPACE.com, November 6, 2000.

306 **Cornwallis Island:** *The Last of the Arctic Voyages: Being a Narrative of the Expedition in HMS Assistance . . . in Search of Sir John Franklin, During the Years 1852–53–54,* Edward Belcher, Reeve, London, 1855.

308 **Pascal Lee:** Author interviews for this chapter took place in August and September 2001 with Steven Braham, Charles Cockell, Kelly Snook, Marco Lee.

311 **By 1996:** Author interview with Christopher McKay, August 2001.

313 **"It was just the thing"**: Author interview with Robert Zubrin in August 2001 and subsequent e-mails in 2002.

331 **On the contrary**: Address by Sean O'Keefe at Syracuse University, April 12, 2002, as posted at www.nasa.gov.

332 **When the company**: "PayPal Brings Investors Back to the Net," Associated Press, February 16, 2002.

332 **On the last topic**: Musk Foundation web site, www.muskfoundation.org.

334 **"Haughton crater"**: "From Utah to Mars," editorial, *Boston Globe,* April 6, 2002; "Astrobiologists Try to 'Follow the Water to Life,' " Robert Irion, *Science,* April 26, 2002.

Lighting Out for the Territory

336 **As the work**: "Report on Shuttles Likely to Leave Important Questions Unanswered," Matthew L. Wald, *New York Times,* June 15, 2003.

336 **He had dealt**: H.R. 2450, the Human Space Flight Independent Investigation Commission Act, was introduced on June 12, 2003, by Representative Bart Gordon and announced in a news release from his office titled "Gordon Bill Ensures Independence of Future Space Accident Investigations," June 16, 2003.

337 **O'Keefe told**: The exchange between Harold Gehman, Sean O'Keefe and John McCain was reported by *Government Executive* magazine on its web site, GovExec.com, in an article titled "Investigator Rips NASA Managers for Rejecting Shuttle Images," May 14, 2003.

337 **In his extensive history**: *This New Ocean: The Story of the First Space Age,* William E. Burrows, Modern Library, New York, 1999.

338 **Sanguinely, the *Columbia***: *Columbia* Accident Investigation Board report, August 2003, accessible at http://www.caib.us.

338 **The foundation**: "Lead, Follow, or Get Out of the Way: The Case for Privatizing NASA," Jennifer DeButts, Policy Paper 121, National Taxpayers Union Foundation, September 29, 1999.

339 **One of the brightest**: Information on Project Prometheus is from http://spacescience.nasa.gov/missions/prometheus.htm; "NASA's Nuclear Prometheus Project Viewed as Major Paradigm Shift," Leonard David, SPACE.com, February 7, 2003.

340 **"Because of the risks"**: *Columbia* Accident Investigation Board report, August 2003, accessible at http://www.caib.us.

341 **Some longtime critics**: "Understanding *Columbia*—and Fixing NASA," Keith Cowing, www.SpaceRef.com, August 26, 2003; Roger Launius's comments are from "Echoes of *Challenger* in *Columbia* Blast," Todd Ackerman, *Houston Chronicle,* August 26, 2003.

344 **Speaking at a space tourism conference**: Rich Clifford's prescription for astronaut training was transcribed by the author from Clifford's remarks at "Going Public 2001," conference of the Space Transportation Association, Washington, D.C., June 2001.

345 **The foundation for such**: E-mail exchanges with Rick Tumlinson, March

and April 2003, and "The Space Settlement Summit," John Carter Mc-
Knight, Spacefaring Web (www.spacedaily.com), March 20, 2003.

346   **As for the ISS:** Author interview with Rick Tumlinson in August 2001 and
various conversations, e-mails and telephone calls, 2001–2003; testimony
to the House Subcommittee on Space and Aeronautics, "Space Tourism
and Private Space Travel (a Few Thoughts)," Rick N. Tumlinson, President, Space Frontier Foundation, and Executive Director, FINDS, June 25,
2001.

347   **Just such a conclusion:** The full report of the Commission on the Future
of the United States Aerospace Industry can be found at www.aero
spacecommission.gov.

347   **Anyone with even a passing familiarity:** Information on the *Star Trek*
franchise can be found at www.startrek.com, www.trekweb.com and www
.treknation.com.

348   **According to Roddenberry's:** The various draft missions of the Starship
Enterprise were excerpted from http://www.roddenberry.com/creations/
bio/groddenberry.bio.html, by David Alexander from his book, *Star Trek
Creator: The Authorized Biography of Gene Roddenberry,* Pan Macmillan,
1995.

348   **The Soviets:** A fine account of the U.S.-Soviet military Space Race is *Star-
Crossed Orbits: Inside the U.S.-Russian Space Alliance,* James Oberg,
McGraw-Hill, New York, 2002.

348   **This is true:** "The US Rules in Space, at Risk of Arms Race: SIPRI," Agence
France-Presse, June 13, 2002.

349   **In the summer:** "Army's First Space Operations Course Ends," Daniel J.
Montoya, U.S. Army press release, August 8, 2001.

349   **"I think it's imperative":** "New NASA Chief Sees Closer Ties to Pentagon," Deborah Zabarenko, Reuters, January 9, 2002.

350   **If military needs:** Information on FY 2003 U.S. Department of Defense
budget is from "FYI: The A/P Bulletin of *Science Policy News,*" published
by the American Institute of Physics at www.aip.org/enews/fyi.

350   **"The whole area":** "Military Space Battle Looms for U.S. Giants," Frank
Morring, Jr., *Aviation Week & Space Technology,* December 10, 2001.

350   **According to the Stockholm Institute:** Data on worldwide military
spending is from the Stockholm International Peace Research Institute,
http://projects.sipri.se/milex/mex_data_index.html.

352   **There is a nonfictional place:** International Space University web site,
www.isunet.net.

352   **To date, Hawley:** More information on Celestis can be found at http://
www.celestis.com.

Acknowledgments

I could not have written *Lost in Space* without the input of those individuals I interviewed at length or talked with casually. As a list of such individuals would run to many pages, to each of you I offer my profound gratitude for your time and openness. That said, the following people gracefully dealt with a greater than average share of pestering on my part: Keith Cowing, Alan Ladwig, Jeffrey Manber, Charles Miller, Courtney Stadd and Rick Tumlinson. Thanks again: *ad astra per aspera*.

Individual thanks also are due to Steve Garber and Jane Odom in the NASA History Office, who allowed me to sit in their midst and pepper them with questions and requests for days on end. To Mark Wade, whom I have never met, thank you for creating and maintaining the *Encyclopedia Galactica* (www.astronautix .com), the planet's best online space history resource.

My colleagues at the SETI Institute offered uniform support for this book even as it drew me away from my work with them; particular appreciation goes to Frank Drake, Tom Pierson, Diane Richards, Brenda Simmons and Jill Tarter. The encouragement and skill of my literary agent, Joe Spieler, made this book possible in the first place; Dan Frank's gentle guidance and editorial wizardry made it far better than I could have made it alone. Marina Benjamin, Barry Bone and Iris Yokoi read through rough cuts of the manuscript and provided essential insight for its improvement. Throughout the development of this book, David King and Barbara Steinmetz sent me helpful bits and pieces of "space information," which were much appreciated.

I have never valued the support of my family more than on the long road leading to these final pages. This holds especially true for my father, Barry Bone, my brother, David Klerkx (always a willing copilot of the makeshift spaceships we "flew" in our basement during those cold Michigan winters) and most of all my mother, Mary Ellen Sands King: you always said I should aim for the stars, although perhaps you didn't mean for me to take that advice quite so literally.

Index

Access to Space study, 106–7
Aerospace America, 191
Aerospace Corp., 97
aerospace industry:
  economic influence of, 16, 27, 44,
    62, 64, 65, 79, 94–95, 98, 165–66,
    219–20
  entrepreneurship in, 18, 22, 30–53,
    61–63, 73, 74–75, 79–88, 91,
    95–96, 113–14, 116–18, 121–22,
    124, 136–37, 169, 191, 209–39,
    267, 345, 350
  government sponsorship of, 44, 98,
    150–51, 342, 346–47
  investment in, 116–19, 121, 124,
    132–33, 139
  military contracts of, 349–50
  monopoly of, 16, 211, 214–15,
    218–20, 223–25, 233, 239, 263
  political influence of, 16, 44–45, 62,
    64, 65, 79, 98, 166–67, 254, 265,
    292
  see also specific companies
Afanasyev, Viktor, 39
AHL 84001 meteorite, 331
Air Force, U.S., 65, 96, 102–3, 105,
  137, 166, 169, 348
airline industry, 17–18, 35–36, 97,
  164, 232–33, 346–47, 351
AirSafe, 164
Akiyama, Toyohiro, 209
À la recherche d'une vie sur Mars
  (Ducrocq), 309
Albaugh, James, 350

Aldrin, Edwin "Buzz," 208, 290,
  345
Alexander, David, 348
Allen, Joseph, 83, 88
"all-up" testing, 137–39, 166
Alternate Access to Space Station
  program, 134–36, 249, 257–63,
  265, 340
"Alternative Plan for U.S. National
  Space Program" (O'Neill), 89
alternative space community, 106,
  127, 142–43, 146–67, 169–70,
  238–39, 247–50, 253–64, 270–73,
  343–45
America at the Threshold, 295
American Association for the
  Advancement of Science, 145
American Physical Society (APS),
  228–29
Ames Research Center, 311, 312–13
Anderman, David, 41
Anderson, Walter, 37, 38, 41, 43–44,
  51, 189, 202, 210–11, 240, 352
Andrews, Jason, 133, 134
Andrews Space & Technology, 135,
  223, 249
Apollo 1 disaster, 54–56, 138
Apollo 8 mission, 60
Apollo 11 mission, 14, 153, 165, 166,
  208, 280, 281, 283, 290, 338
Apollo 13 mission, 244–46
Apollo 17 mission, 282
Apollo Applications program, 139–40,
  281

Apollo space program, 164–71
  astronauts in, 206, 208
  contractors for, 26, 166
  cost of, 77–78, 154, 282, 294, 305
  historic importance of, 10, 24–25,
      36, 56, 89–90, 144, 147, 208,
      249–50, 338
  mission objectives of, 26, 68–69,
      137–39, 152–55, 165, 166, 268,
      279, 280–81
  public support for, 10, 11–12, 13,
      54–55, 342
  research programs of, 81, 278
  schedule for, 13, 14–15, 55–56,
      137–39, 152–53, 290
  spacecraft of, 61, 266–67
  technology for, 28, 268, 281
  termination of, 164–71, 185, 268,
      279, 281–83, 337
  training for, 206, 313
  U.S.-Soviet rivalry and, 9, 55, 138,
      152–55, 166, 202, 278
Arc Technologies, 218–19
Ares rocket, 293–94, 301
Ariane rocket, 132, 171, 175, 257, 259
Armageddon, 45–46
Armstrong, Neil, 10, 13, 60, 93, 208,
      290, 338
Army, U.S., 151, 348, 349
Arrow spacecraft (Canadian), 111
Arthur Andersen, 215
Arthur C. Clarke Mars Greenhouse,
      333
artificial gravity, 71, 279, 293
Asimov, Isaac, 73, 249
asteroids, 69, 256, 306–7
astronauts:
  civilians as, 68, 98, 167, 172–73,
      176–211, 286, 343–44
  popular image of, 5–6, 8, 33,
      46
  training of, 37, 38, 39, 68, 122, 125,
      167, 180, 181–85, 190, 206, 235,
      309–10, 313, 344
Atlantis space shuttle, 161, 162, 244

Atlas rockets, 91, 95, 101, 106, 132,
      133, 134, 135
Atomic Energy Commission (AEC),
      150, 212
Aurora program, 342–43
Aviation Week, 131, 350

Baikonur Cosmodrome, 44, 52, 182,
      193–205, 325
Baker, David, 293–94, 296, 301
Baker, James, 86
Baturin, Yuri, 190
Baxter, Stephen, 73
Beagle 2 spacecraft, 325
Becker, Karl, 93
Beggs, James, 124, 144–45
Beim, Howard, 260–61
Bekey, Ivan, 38
Benford, Gregory, 300
Bernal, J. D., 70–71
Beyond the Planet Earth (Tsiolkovsky),
      70
Bezos, Jeff, 219
Bigelow, Robert, 235
Binder, Alan, 88
Blitch, John, 322, 323, 325
Boeing 777 aircraft, 97, 160, 344
Boeing Company, 16, 34, 45, 83, 91,
      94, 95, 97, 100, 108, 109, 113,
      126, 130, 135, 149, 160, 166,
      214–15, 221, 223, 233, 241, 254,
      323, 344, 348, 350
Bonestell, Chesley, 64
Borg (*Star Trek*), 75
Boston Globe, 294, 332, 334
Bowersox, Ken, 191
Bradbury, Ray, 175, 273–74
Braham, Steven, 314, 315, 320, 324,
      334
Brand, Stewart, 72
Brilliant Pebbles program, 103, 105
Brin, David, 73
Bristol Corp., 219
British Interplanetary Society, 167
Brower, David, 66

Brown, Jerry, 72
Buran space shuttle, 182, 200–201
Burnside Clapp, Mitchell, 101–3, 104,
    109, 110, 130, 232, 296
Burrows, William, 337
Bush, George H. W., 251, 252, 272,
    283, 290–93, 294
Bush, George W., 155, 163, 217,
    225–26, 227, 254, 259, 266
Bush, Jeb, 254

Cabana, Robert, 50, 190
Cameron, James, 192
"Can We Get to Mars?" (von Braun),
    277
Carmack, John, 219
Case for Mars, The (Zubrin and Wag-
    ner), 285, 287, 296–99
Case for Mars conference series,
    287–89, 311
Chaffee, Roger, 54–55
Challenger space shuttle disaster:
    cause of, 25, 86, 161, 241–43
    Columbia disaster compared with,
        10, 25, 54, 157, 158–59, 161, 243,
        246–67, 341
    congressional investigation of, 75,
        157–58, 162–63, 164, 242
    crew in, 12, 25, 55, 173, 176–77,
        353
    independent investigation of,
        157–58, 241–42, 246–47, 341
    public reaction to, 24–25, 54–55
    space program affected by, 10, 12,
        16, 27, 33, 54–55, 56, 79, 84, 86,
        87, 91, 94, 96, 161, 163, 171, 172,
        178–80, 212, 218, 222, 250–51,
        261, 289–90, 291, 292, 342
Chandra X-ray observatory, 159
Chase, Brian, 265
Cheney, Dick, 96, 155
Childhood's End (Clarke), 73
China, 9, 92, 95, 98, 342, 346
cities, space, 69–72, 77–78, 79, 81,
    88–89, 93, 98

Citron, Bob, 119–25, 127, 135, 136–37,
    175, 234, 235
Citron, Rick, 119, 120
Clark, Benton, 287–88
Clarke, Arthur C., 3, 24, 60, 71, 73,
    249, 333, 353
Clifford, Rich, 344
Cline, Lynn, 52
Clinton, Bill, 28, 29–30, 47, 105, 107,
    177, 238, 252, 292
CNN, 191, 229, 332, 333
Cockell, Charles, 317, 319, 334
Cohen, Aaron, 126
Collier's, 64, 65, 277, 279
Collins, Michael, 290
Collision Orbit (Williamson), 288
Colonies in Space (Heppenheimer), 68
"Colonization of Space, The" (O'Neill),
    70
Columbia space shuttle:
    as first shuttle, 31, 161, 240, 288
    Hubble repair mission of, 155–56,
        160, 161
    launch of (2002), 155–57
    overhaul of, 159–60, 163–64, 245
Columbia space shuttle disaster:
    Apollo 13 emergency compared
        with, 244–46
    causes of, 160, 162, 243–47, 336–37
    Challenger disaster compared with,
        10, 25, 54, 157, 158–59, 161, 243,
        246–47, 341
    congressional investigation of, 75,
        158, 159, 243–44, 247
    crew in, 12, 54, 55, 136, 164, 244,
        353
    independent investigation of, 336,
        340–41, 347
    public reaction to, 9–10, 54–55
    space program affected by, 9–10,
        12, 23, 31, 52–55, 56, 98, 101,
        113, 117, 122, 134, 163–64, 180,
        210, 228, 231, 239, 249, 258, 259,
        261–62, 265, 336, 340–44, 345
comets, 256, 297, 306–7

Commerce Department, U.S., 32, 33, 85, 111
Commercial Space Act (1998), 232, 238–39, 250, 265
Commercial Space Launch Act (1984), 222
"Commercial Space Transportation Study," 223–35
Commission on the Future of the United States Aerospace Industry, 347
Committee for the Future, 175–76
Computerized Launch and Control System (shuttle), 216
Comsat Corp., 222
Conrad, Pete, 104, 107, 130
Consolidated Space Operations, 215
Constellation Services International, 249
Contact (Sagan), 66
Cosmic Connection, The (Sagan), 309
Cosmos (Sagan), 65, 67, 72–73, 79
Cosumano, Joseph, 349
Cousteau, Jacques, 286
Cowing, Keith, 214, 217, 341
Cruise, Tom, 57

Dahlstrom, Emeline, 240
Dahlstrom, Eric, 239–42, 246–47
Darwin, Charles, 325, 348
Dasch, Pat, 229, 230
Davidson, Keay, 72
Da Vinci Project, 111
DC-X spacecraft, 101–10, 112, 113, 121, 143, 348, 351
DC-XA spacecraft, 109, 110
DC-Y spacecraft, 109
Defense Advanced Research Projects Agency (DARPA), 105, 350
Defense Department, U.S., 42, 84, 105, 106–7, 165, 169, 170, 350
Delta Clipper Experimental program, 101–10
Delta rockets, 91, 95, 106, 126, 130, 131, 132, 133, 134, 135, 171, 292

Devon Island, 301–8, 311–34, 346, 352
Diamandis, Peter, 110, 235–36, 240, 352
Discovery Channel, 321, 327
Discovery space shuttle, 161, 178
Disney, Walt, 64, 277
dot-com bubble, 50–51, 118–19, 132, 314, 332
Ducrocq, Albert, 309
Dyna-Soar, 96
Dyson, Freeman, 76, 88, 101, 265

Early Manned Planetary-Interplanetary Roundtrip Expeditions (EMPIRE), 278–80
Earth:
    atmosphere of, 277, 284
    environmental degradation of, 66–68, 72, 120, 297, 298
    magnetic field of, 38–39, 310
    Mars compared with, 275–76, 277, 284
    population of, 66–68, 72, 285
Edwards Air Force Base, 150
Eisenhower, Dwight D., 150–51, 337
Electric Power Research Institute, 256
electrodynamic tethers, 38–43, 53
electrophoresis, 81, 84
Endeavour space shuttle, 161, 172, 177
Energia (company), 33–34, 39–43, 47–49, 51–52, 188, 189, 202, 204, 234–35, 346
Energiya rocket, 61, 200–201, 301
"Enhanced Strategy for the Development of Space Commerce," 229–30
Entering Space (Zubrin), 299
Enterprise space module, 234, 346
Esprit Telecom, 37, 41
European Space Agency (ESA), 30, 50, 124, 190, 206, 257, 259, 309, 325, 342–43
Expedition One crew, 49–50, 51
Explorer 1 satellite, 151
extraterrestrial life, 6–7, 66, 319, 331

Faget, Maxime, 82, 112
Farnum's Freehold (Heinlein), 74
Federal Aviation Administration
    (FAA), 168, 218, 236
Federal Emergency Management
    Agency (FEMA), 214
Federal Energy Research and Develop-
    ment Administration, 72
Federov, Nikolai Federovich, 180–82
Feustel-Buechl, Jorg, 190
Feynman, Richard, 157–58, 162–63,
    164, 242
Fletcher, James, 86–87, 144, 165, 171
Foale, Michael, 37
Forum for the Advancement of Stu-
    dents in Technology (FAST),
    176
Foster-Miller Inc., 322, 323
Foundation and Empire (Asimov), 73
Foundation for International Non-
    governmental Development of
    Space (FINDS), 37, 39–42, 53
Foundation for the Future, 119, 175
Franklin, John, 306
Futron, 15–16
"Future Space Transportation Study,"
    223

Gagarin, Yuri, 33, 178, 180, 181, 182,
    196, 203, 204, 278
Gardellini, Gus, 41
Garn, Jake, 176–77, 178
Garver, Lori, 46, 207, 292
Garvin, Jim, 334
Gates, Bill, 119, 130
Gaubatz, William, 110
Gehman, Harold W., Jr., 336–37
Gemini space program, 8, 137, 166,
    175, 206
General Dynamics Corp., 94, 223
geology, 306, 310, 316–20
geostationary orbit, 129
Gillin, Joe, 260–61
Gingrich, Newt, 74, 251–52, 299–300
Glaser, Peter, 71

Glenn, John, 178, 179, 221
Glenn Research Center, 150, 243, 278
global positioning system (GPS), 104,
    222
Globalstar, 129–31, 132, 199
Goddard, Robert, 115–16
Goddard Space Flight Center, 150,
    217, 240–41
Goldin, Daniel Saul, 34, 35, 36, 45, 46,
    48, 105, 106–8, 110, 113, 127,
    142–44, 145–46, 147, 156, 163,
    174, 177–79, 190, 191, 192, 206,
    207–8, 216, 217, 218, 223, 243,
    293
Gorbachev, Mikhail, 32
Gore, Al, 108–9, 254
Greene, Graham, 10
Grissom, Virgil "Gus," 54–55

Hall, Ralph, 266
Hanuska, Karl, 203
Haughton Crater, 307–8, 311, 315–16,
    318, 334
Haughton-Mars Project (HMP),
    301–35
Hawes, Mike, 207–8
Hawley, Todd, 240, 352–53
Heinlein, Robert, 60, 73–74, 91, 249
Helms, Susan, 191
Henrich, Joan, 187
Heppenheimer, Thomas, 68
Hickam, Homer, 285–86
High Frontier, The (O'Neill), 68, 70–72,
    76, 80, 83, 89, 110, 119, 217, 240,
    252, 296
Hines, Donna, 122
HMS *Assistance*, 306
Holmes, D. Brainerd, 137
Hubbard, Barbara Marx, 175
Hubble Space Telescope, 7, 155–56,
    160, 161
Hudson, Gary, 130
Humans to Mars (Portree), 279
Hunter, Maxwell, 110, 121
hybrid fuel, 219

Ikonos satellite, 222

IL Aerospace Technologies, 111

Illuminatus! Trilogy (Wilson), 43

Industrial Space Facility (ISF), 80–88,
90, 144, 211, 228

In-Situ Resource Utilization (ISRU),
287–88, 289, 292, 293, 294, 295,
303

intercontinental ballistic missiles
(ICBMs), 65, 92, 93–94, 103, 130,
138, 171, 194, 195, 196, 197, 274,
342, 348, 349

International Launch Services (ILS),
199

International Space Station (ISS),
80–88
access to, 134–46, 143, 228–32, 239,
249, 257–63, 265, 339–40
assembly of, 45, 48–49, 83–84
budget for, 39, 51, 108, 121, 140,
143, 146, 163, 225–27, 229, 258,
295–96, 303–4
commercial use of, 81–82, 144,
213–14, 223, 225, 228–34,
238–39, 258–59, 346–47
congressional oversight of, 28–29,
36, 45
contractors for, 44, 149, 226
costs of, 23, 26, 45, 51, 78, 99, 108,
226–27, 230–32, 303–4
crew for, 30, 49–50, 51, 83, 134–36,
172, 189, 191, 193, 226, 227, 231,
266, 339–41
design of, 28–29, 83, 140, 183, 276,
290, 296
energy sources for, 243, 269
European participation in, 30, 50,
190, 228, 262
experiments conducted on, 83,
87–88, 228–34, 266
human exploration based from, 12,
30, 50, 145–46, 190, 228, 240,
262, 342, 346
international partners in, 189, 206,
228, 234

media coverage of, 44, 45, 49–50,
57, 87, 190, 191, 206–7, 227, 229
Mir space station compared with,
22, 27, 39, 42, 43, 44–45, 48,
49–52, 88, 120, 125, 183, 225,
227
modules of, 42, 45, 48, 51–52,
83–84, 122, 183, 227, 233–34,
346
NASA control of, 22, 113, 121, 140,
143, 149–50, 225, 269, 295–96,
339, 346
political considerations for, 26,
85–87, 148–49, 225–27, 240–41,
267
private space station as alternative
to, 80–88
resupply of, 48, 52–53, 83, 125,
134–36, 172, 189, 257, 258, 265
Russian participation in, 22, 29–30,
47–52, 172, 201, 206, 228,
233–34, 262
safety of, 192–93
space shuttle and, 26, 52–53, 83, 85,
86, 87, 144, 145–46, 155, 161,
163, 172, 226, 231, 244, 246, 249,
266, 269
Tito's visit to, 189–93, 200–11, 233,
234, 240
training program for, 37, 38, 39,
122, 125
International Space University (ISU),
240, 352–55
International Traffic in Arms Regulations (ITAR), 42–43, 53
Iridium, 129–32

Jarvis, Greg, 177
Jemison, Mae, 351–52
Jet Propulsion Laboratory (JPL), 150,
151, 185, 186, 217
Johnson, Lyndon B., 55–56, 153, 169,
281
Johnson Space Center, 108, 126,
149–50, 153, 175, 180, 190, 210,

214–15, 235, 236, 295, 303, 314, 331–32

Jones, Jeffrey, 303, 304, 334

K-1 spacecraft, 116, 121, 125–27, 130–33, 134, 135, 141, 224

Kalery, Alexander, 44

Karash, Yuri, 39–40

Kazakhstan, 194–95, 197–98, 203, 240

Kelly Air Mail Act (1925), 232–33, 346–47

Kennedy, John F., 55, 56, 137, 138, 152–55, 169, 174, 178, 180, 278, 281, 291

Kennedy Space Center, 138, 150, 172, 216, 254–55, 299

Khrushchev, Nikita, 56, 182, 196

King, Martin Luther, Jr., 139

Kirsch, Steve, 314

Kistler, Walter, 115–19, 132–33, 135, 175

Kistler Aerospace, 114, 115–21, 122, 125–27, 128, 130–33, 134, 135, 139, 140, 224

Klebanov, Ilya, 51

Kohrs, Richard, 126

Koptev, Yuri, 352

Korolev, Sergei, 33, 55, 125, 180, 182, 196, 203, 278

Kraft, Chris, 175

Krikalev, Sergei, 24, 49

Kubrick, Stanley, 15, 24, 60

Kursk submarine disaster, 51

L5 Society, 78, 79–80, 217, 250–51

Ladwig, Alan, 174–80, 213, 235, 236

Lagrangian Point Five (L5), 69–70, 72, 78, 79–80, 91, 93, 268

Lambright, W. Henry, 137, 155

Langley Research Center, 150, 241

Launch Services Purchase Act (1990), 250–51

Launius, Roger, 151, 341

Lawton, Graham, 317, 319, 320, 327, 331

Lee, Marco, 302–3, 308–9

Lee, Pascal, 300, 301–5, 308–14, 318, 319, 320, 321–24, 326, 327, 328, 329, 332, 333, 334–35

Lem, Stanislaw, 60

Lemke, Larry, 326, 327, 329

Lenorovitz, Jeff, 49

"Lessons Learned from Challenger," 241–42, 246–47

Lewis Research Center, 150, 277–78

Ley, Willy, 277

libertarianism, 74–75, 80–81, 252

Lichtenberg, Byron, 235–36

Limits to Growth, 66

Lindbergh, Charles, 17–18, 110–13, 338

Llewellyn, Peter, 40

Lockheed Martin Corp., 16, 34, 45, 91, 94, 95, 98–101, 108–9, 113, 130, 133, 135, 166, 214, 215, 221, 223, 241, 247, 254, 340, 348, 350

Long March launch vehicle, 342

low-Earth orbits, 16, 39, 91, 96, 129, 196, 344

Lowell, Percival, 316

Lukashevich, Stephen, 181

Luna 2010 proposal, 251–52

Lunar Prospector, 88–89

McAuliffe, Christa, 55, 173, 176, 178, 179, 180

McCain, John, 266, 337

McCaw, Craig, 130

McCaw, John, 128

McCurdy, Howard, 292–93

McDonnell-Douglas, 81–82, 84, 95, 103–5, 108, 166, 171, 223

McKay, Christopher, 289, 311, 326, 333

Magellan spacecraft, 280

Manber, Jeffrey, 30–34, 36, 40–44, 48–49, 50, 51, 53, 57, 88, 188–89, 201, 202, 235, 352

Mangolds, Arnis, 322–23, 325

Manhattan Project, 166, 212

Manned Orbital Laboratory, 65
"Man Will Conquer Space Soon"
    (von Braun), 64
MarchStorm, 238, 239, 247–50,
    253–54, 259–64, 343
Mariner space probes, 280
Mark, Hans, 213–14
Mars:
    atmosphere of, 276, 277, 283–84,
        297, 326
    colonization of, 62, 69, 144, 145,
        167, 269, 270–73, 283–335
    commercial possibilities of,
        272–73
    Earth compared with, 275–76, 277,
        284
    as frontier, 273–74, 278, 298–99
    human space mission to, 12, 13,
        14–15, 27, 69, 83, 91, 139, 165,
        169, 186, 240, 251, 269, 279,
        280–83, 287–88, 290–335, 342–43
    life on, 319, 331
    media coverage of, 294, 295, 296,
        304–5, 312, 314, 315, 321, 326,
        327–31, 332, 333, 334
    NASA planning for, 272, 277–83,
        289–96, 301, 303–5, 309, 331–34
    robotic probes for, 8, 145, 146, 149,
        279–80, 287–88, 293, 303, 309,
        310, 311–12, 325, 331, 333
    rovers designed for, 289, 294,
        327–28
    science fiction on, 273–77, 285
    shuttle system and space station
        for, 27, 277, 279, 281, 282, 283,
        288, 291, 294, 295, 301
    simulated environment of, 294,
        301–35, 346, 352
    terraforming of, 288–89, 297, 298
    von Braunian program for, 276–78,
        279, 280, 281–83, 291, 294, 296
    water on, 289, 297–98, 331
Mars Climate Orbiter, 145, 146, 303
Mars Design Reference Mission, 295,
    301, 304

Mars Direct program, 293–97, 299,
    301
Mars Express spacecraft, 325
Mars Global Surveyor probe, 280
Marshall Space Flight Center, 106–7,
    108, 135, 149–50, 215, 217, 243,
    279
Mars Institute, 334, 345
Mars Oasis mission, 333
Mars Observer probe, 310
Mars Odyssey probe, 303, 331
Mars Pathfinder, 8, 150, 325
Mars Polar Lander, 145, 146, 303, 315,
    325
Mars Prize, 299–300
Marsprojekt, Das (von Braun), 276–78,
    280, 294, 296
Mars Settlement Design Competition,
    332
Mars Society, 272, 313, 314, 322,
    332–33, 345
Mars Underground, 311, 312–13
Martian Chronicles, The (Bradbury),
    273–74
Martian Piloted Complex (MPK),
    278
Martian Race, The (Benford), 300
Martian Radiation Environment
    Experiment (MARIE), 303
Martin Marietta Corp., 95, 166, 223,
    287, 289–90, 293, 294, 296,
    300
Marzec, Ed, 192, 198
mass drivers, 69, 71, 89
Mercury space program, 8, 16, 26, 82,
    93, 166, 175, 177, 206
meteorites, 310, 311–12
Michaud, Michael, 70
microgravity, 33, 62, 71, 86, 224, 228,
    231, 235–36, 293
Mikulski, Barbara, 190
Miller, Charles, 249–53, 259–60, 264,
    299, 343
Mini Station 1 space station, 235–36,
    237, 346

Minuteman missile, 138

MirCorp, 43–51, 62, 80, 120, 187–89, 211, 224, 234–35, 237, 346

Mir Electrodynamic Tether System (METS), 42–43

Mir space station, 30–54
  accidents on, 37, 38, 46, 47
  corporation established for, 43–51, 62, 80, 120, 187–89, 211, 224, 234–35, 237, 346
  crews for, 29, 37, 39, 44, 46, 47
  design of, 183, 276
  electrodynamic tethers for, 38–43, 53
  experiments conducted on, 32–33
  final descent of, 20–23, 37, 41, 42, 47, 49, 51, 53–54, 61, 120, 189
  ISS compared with, 22, 27, 39, 42, 43, 44–45, 48, 49–52, 88, 120, 125, 183, 225, 227
  launch of, 24, 27, 61
  maintenance of, 22–23, 24, 29–30, 37–38, 47–52
  media coverage of, 33, 39, 40, 44, 45–46, 50, 51
  NASA opposition to, 23, 30, 33, 38–40, 41, 43–53, 57, 193
  orbit of, 38–39, 42
  political considerations for, 37, 39–40, 47–48, 49
  private refurbishment of, 22–23, 30–53, 62, 120, 211, 262
  public attitudes toward, 38, 39, 40–41, 45–46
  resupply of, 43, 44, 48, 51, 125
  safety of, 33, 37–38, 45–47
  Skylab compared with, 27, 28–30, 33, 34
  Spektr module of, 37, 38
  termination of, 30, 88, 90, 120, 146, 182, 190, 193, 225, 353
  Tito's proposed visit to, 44, 49, 50, 143–44, 187–89

Mitrofanov, Alexei, 204

Moon:
  base on, 279, 280–83, 291, 294, 295, 348
  colonization of, 69, 79, 139, 144, 167, 169, 217, 240, 251–52, 268
  geochemical mapping of, 88–89
  landings on, see Apollo space program
  mining operations on, 69, 89
  nuclear waste dump on, 224
  terrain of, 275

Moon Is a Harsh Mistress, The (Heinlein), 74

Morgan, Barbara, 178–80, 343–44

Mueller, George, 126–27, 137–41, 165, 166, 167, 168, 281, 282

Musabayev, Talgat, 190

Musk, Elon, 114, 332–33, 345

Myers, Dale, 126, 127

N-1 rocket, 125, 202, 278

NASA Exploration Team (NeXT), 267–68, 269

National Advisory Committee for Aeronautics (NACA), 36, 150–51, 166, 233, 277–78, 337, 339

National Aeronautics and Space Administration (NASA):
  accounting methods of, 214–15, 236–37
  administrative codes used by, 148–50, 258
  administrators of, 142–46, 147, 148, 150, 153–55, 165, 167, 174, 177–79, 216–18, 226–27, 237, 239, 281, 292–93, 338
  budget of, 7, 14, 16, 26, 27, 28–29, 30, 76, 77, 85–86, 94, 95, 108, 134, 135, 137, 140, 144, 154, 163, 165, 169–71, 177, 184, 185, 216, 225–27, 236–37, 253, 254–55, 259, 266, 268, 281–83, 299, 305, 349–50

National Aeronautics and Space
   Administration (NASA)
   (continued):
 bureaucracy of, 7, 23, 56, 75–80,
    136–37, 143, 144, 145, 147–50,
    151, 165, 166–67, 213–14,
    216–17, 220, 246–47, 261–62,
    311, 337
 charter of, 212, 213
 Commercial Programs Office of, 85
 congressional oversight of, 28–29,
    74, 75, 76, 77, 86–87, 165, 167,
    208, 226, 237, 238–39, 253–55,
    257, 259, 292, 336
 contractors for, 16, 17, 26, 27, 34,
    44–45, 62, 64, 79, 87–88, 94–95,
    98, 99–100, 108–9, 113–14,
    121–22, 133, 135, 136, 143,
    166–67, 219, 220–21, 236–37,
    242, 254–55, 292, 293, 296, 339,
    343, 346, 349–50
 culture of, 36, 46, 54–57, 75, 113,
    134, 216, 219–20, 242–43, 246,
    268–69, 336–38, 341
 economic influence of, 10, 44–45,
    62, 77, 139–40, 216–17, 338–39
 employees of, 10, 26, 114, 124,
    126–27, 147–48, 165–66, 185,
    217, 239, 242–43, 246, 290–91,
    345
 engineers of, 164, 166–67, 240–41,
    243–45, 258
 Exploration Office of, 295
 formation of, 23–24, 36, 150–55,
    212, 213
 future viability of, 15–19, 141,
    142–47, 165–66, 216–18, 268–69,
    336–47
 as government agency, 5, 31–32,
    75–80, 89–90, 99–100, 133, 144,
    197, 212, 214, 216, 217, 255,
    338–39
 headquarters of, 61, 108, 182, 183,
    225, 242, 247
 Human Resources Office of, 149

 Inspector General of, 149, 214–15
 management of, 148–50, 158, 164,
    165, 166–67, 226, 240–44, 245,
    258, 265, 337, 341
 media coverage of, 18, 49–50, 57,
    76, 87, 107, 146, 243, 255, 266
 military agenda of, 63, 64, 143, 222,
    292, 342, 349–50
 "paper" vs. "real," 76–77, 85–86,
    101, 211, 265, 268, 269, 279, 293
 Planetary Projects Office of, 295
 political influence on, 5, 7, 13–14,
    16, 26, 46, 62, 76–77, 86–87,
    144–50, 167, 169–71, 212–13,
    216–18, 240–41, 254–55, 292–93
 post-Apollo survival of, 13–14,
    60–63, 139–40, 144, 165–71, 185,
    268, 279, 281–83, 337
 private enterprise opposed by,
    31–32, 44–45, 48, 79–88, 90,
    108–9, 113–14, 127–28, 136–37,
    141, 142–43, 146, 147, 209–37,
    250–51, 252, 255–56, 265, 267,
    269, 338–39
 Procurement Office of, 149
 Public Affairs Office of, 150
 public relations campaigns of, 18,
    45–46, 56–57, 64, 76–77, 83,
    105–6, 147, 157, 167, 171–73,
    177–78, 215–16, 337
 Space Flight Office of, 137, 148–49,
    258
 Space Science Office of, 149
 Tito's space trip opposed by, 187,
    189–93, 202, 206–8, 209
 Turnerian ethos of, 151–52, 164,
    180
 vision statement of, 268–69
 see also Apollo space program;
    International Space Station (ISS);
    shuttle, space
NASA Watch, 214
National Aerospace Initiative, 349
National Aerospace Plane (NASP), 96,
    340

National Air and Space Museum, 172, 209
National Commission on Space, 78–79, 289
National Geographic Society, 312
National Imagery and Mapping Agency, 337
National Research Council (NRC), 311, 312
National Science Foundation, 150, 220, 256
National Space Council, 290
National Space Institute (NSI), 78, 250–51
National Space Society (NSS), 78, 145, 250–51, 263–66, 292, 294, 343, 345
National Taxpayers Union Foundation, 338–39, 347
Native Americans, 152, 194, 252–53, 273
Negev-5 spacecraft, 111
Nelson, Bill, 177, 178
NERVA propulsion system, 282–83
New Technology Opportunities Program (NTOP), 170
New York Times, 33, 65, 87, 163, 191, 332
Nichols, Nichelle, 351
90-Day Study, 291–93, 294, 295
Niven, Larry, 73
Nixon, Richard M., 144, 169–71, 281, 282
Northrop Grumman, 94, 133
nuclear power, 65, 67, 170, 212, 224, 269, 274–75, 278, 282–83, 287, 294, 307, 310, 339, 348

Oberg, James, 192–93
Oberth, Hermann, 115–16
Office of Management and Budget (OMB), 85, 135, 169–70, 214, 226, 258
O'Keefe, Sean, 135, 144, 147, 154–55, 163, 179, 216–18, 226–27, 236,

243–44, 245, 258, 263, 268–69, 331, 337, 339, 352
O'Neill, Gerard K., 63, 67–81, 82–83, 84, 88–90, 91, 93, 98, 101, 103, 110, 117, 119, 123, 167, 217, 225, 240, 249, 250, 252, 269, 272, 289, 296, 332–33, 352, 353
Orbital Sciences, 91, 123
Orbital Space Plane (OSP), 134–36, 228, 257, 258, 259, 265, 266–67, 339–40
Orteig Prize, 110–12
Ortho Pharmaceuticals, 82, 84, 211
Osinski, Gerald "Oz," 316–20
Outer Space Treaty (1967), 62, 251

Pace, Scott, 262
Pacific Rocket Society, 120
Paine, Thomas, 78, 144, 145, 169, 281, 282, 289
Pan Am, 15, 17
parabolic flights, 235–36, 313
Park, Robert, 228
Pathfinder rocket plane, 102
PayPal, 332
pharmaceutical experiments in space, 32, 62, 81–82, 84
Philosophy of the Common Task, The (Federov), 181
Pioneer Astronautics, 296
Pioneer Rocketplane, 101–3, 109, 130, 296, 350
Pizza Hut, 233–34
Planetary Society, 250
Planning Research Corp., 241–42
Portree, David, 278, 279
Powell, Colin, 299
Powers, Gary Francis, 195–96
Pratt & Whitney, 104, 233
probes, space, 88, 145, 146, 275, 279–80
Progress and Freedom Foundation, 299
Progress spacecraft, 43, 48, 51, 52–53, 136, 172, 257
Project Harvest Moon, 175–76

Project Prometheus, 339
ProSpace, 238–89, 247–50, 253–64,
    265, 267, 343
Proton rocket, 171, 198, 228, 233–34
Putin, Vladimir, 44, 51, 53, 189

Qualcomm Inc., 130
Quayle, Dan, 290, 291, 292

Ramon, Ilan, 55
RAND Corp., 221
Ranger probes, 280
Rascal rocket, 118
"Reactive Vehicle Will Save Us from
    Calamities That Await the Earth,
    The" (Tsiolkovsky), 181
Readdy, William, 164, 191, 243–44
Reagan, Ronald, 15, 27, 33, 78, 80, 81,
    82, 83, 84, 96, 103, 172, 218, 222,
    226, 240, 289, 290, 291, 340
Red Mars trilogy (Robinson), 297–98,
    334
Red Planet, 46
Redstone rocket, 203
Rendezvous with Rama (Clarke), 73
Richards, Robert, 352
Right Stuff, The (Wolfe), 33
Road Ahead, The (Gates), 119
Robinson, Kim Stanley, 297–98, 334
robots, 322–23
Rocketdyne, 162
rocket planes, 92–93, 98–114, 167,
    168, 296, 340, 350
rockets:
    amateur, 115–16, 119–20, 285–86
    development of, 91–92, 104, 111,
        118, 137–39, 151, 171, 274–75,
        276, 278
    heavy-lift, 279, 281, 291, 293–94
    as launch vehicles, 91–95, 96, 98,
        99, 101, 104, 114, 115–16, 118,
        120, 125, 130
    nuclear, 278, 282–83
    Russian, 93, 94, 95, 125, 194, 195,
        196, 199, 200–201, 202, 278

"sounding," 219
U.S., 91, 95, 101, 106, 126, 130–35,
    138, 171, 279, 281, 292
Rocketship X-M, 275
Rockwell International, 100, 108, 109,
    123, 223
Roddenberry, Gene, 6, 175, 348,
    353
Rogers Commission, 157–58, 241–42,
    246–47, 341
Rohrbacher, Dana, 74, 252
Rossi, David, 123
Rotary Rocket, 130
Russian Space Agency (Rosaviacos-
    mos), 47–48, 49, 189, 202, 203,
    204, 206
Russian space program:
    achievements of, 9, 13, 17, 23–24,
        33, 39–40, 120, 178, 196, 204,
        209, 285
    bureaucracy of, 47–48, 49
    commercialization of, 30–53,
        143–44, 212–13, 233–35, 346
    economic factors in, 51, 61, 183–84,
        197–98, 212–13, 343
    Federovian ethos of, 180–82
    funding of, 183–84
    Mars exploration planning by, 278,
        325, 333
    military control of, 197, 348
    political factors in, 13, 52, 197–98
    schedule for, 13, 29, 39, 55
    tourism and, 44, 49, 50, 143–44,
        180–93, 200–213, 234, 240
    see also Mir space station
Rutan, Burt, 112
Ryan Airlines, 113
Ryumin, Valeri, 41

Sagan, Carl, 65–67, 72–73, 79, 81, 250,
    289, 309, 325
Salyut space stations, 24, 53, 71, 348
satellites:
    assembly of, 83, 354
    commercial, 172, 221–22

communications, 11, 128–32, 133, 221–22
data collected by, 9–10
launch of, 27, 94, 126, 128–32, 133, 160, 176, 196, 198
market for, 114, 131–33
military, 198, 222, 348–49
orbits of, 38, 91, 129
photographic imaging by, 222, 245
Russian, 23–24, 33, 93, 94, 120, 150, 169, 196, 198, 277–78, 285
spy, 11, 142, 169
U.S., 11, 142, 151, 169, 349
Saturn rockets, 14, 26, 28, 61, 63, 84, 125, 138–39, 140, 156, 167, 175, 201, 202, 279, 282, 301
Scaled Composites, 112
Schlather, Marc, 253–54, 259, 262–63, 267
Schutt, John, 310–11, 321–22
Schweikart, Russell "Rusty," 76
science fiction, 5, 6, 7, 31, 43, 45–46, 59–60, 66, 70–71, 73–74, 119, 273–77, 285, 309, 325, 327, 351–52
Science Policy Research Division (SPRD), 29
SeaLaunch, 34, 91
Seamans, Robert, 169
SEI Outreach Program, 293, 294–95
SEI Synthesis Group, 294–95
Sellars, Ron, 183
Semenov, Yuri, 33, 39, 40, 41–44
September 11th attacks, 227, 308, 349
SETI Institute, 6–7, 302, 334
Shapiro, Robert, 45–46
Sharman, Helen, 209
Shaw, George Bernard, 343
Shenzhou spacecraft, 98, 342
Shepard, Alan, 16, 93, 203
Shepherd, William, 49, 191
shuttle, space, 155–73
accident rate of, 157–59, 161, 162, 164

alternatives to, 96–114, 125–27, 134–36, 223–25, 253, 256–59, 264–67, 339–40
average number of missions flown by, 160–61
budget for, 27, 32, 77, 94, 106, 108, 140, 163, 165, 169–71, 177, 178, 253, 254–55, 264–65, 266, 295–96
commercial use of, 167, 171, 172, 176, 211–14, 223–25, 238–39, 250, 265
contractors for, 27, 45, 99–100, 162, 170–71, 176, 214, 220–21, 292
costs of, 94, 97–100, 106, 158, 160–61, 169–71, 186, 209, 211–12, 223–24, 255
crews of, 12, 25, 54, 55, 136, 164, 168, 172–73, 176–77, 244, 353
delta wings of, 168–69, 243, 245
design of, 82, 96–98, 100–101, 105–6, 140, 156, 160, 162–72, 201, 290, 340
development of, 69, 71, 82–83, 86, 96–97, 269, 276, 281
economic influence of, 10, 99–100, 165–66, 170–71, 178, 211–14, 220–21, 223–25, 238–39
engines of, 97, 162, 294
experiments conducted on, 26–27, 123–25, 176
external tank of, 156, 157, 162, 243, 294
grounding of, 33, 52–53, 171, 221
human spaceflight in, 12, 13–14, 21, 25–26, 69, 77, 140, 160, 164–65, 166, 167–68, 172–73, 265–66
inauguration of, 165–73
International Space Station (ISS) and, 26, 52–53, 83, 84, 85, 86, 87, 144, 145–46, 155, 161, 163, 172, 226, 231, 244, 246, 249, 266, 269

shuttle, space *(continued)*:
   landings of, 168, 172
   launches of, 46, 52, 155–57, 168,
      171, 172, 205, 216, 223–24
   maintenance of, 8, 97, 100, 158,
      159–64, 186, 254–55, 340–41
   media coverage of, 77, 163, 177,
      178, 255
   military use of, 94, 169, 170, 172,
      292
   NASA control of, 100–101, 104–14,
      121–28, 133–36, 140, 149–50,
      164–73, 241–47, 258, 264–67,
      295–96, 336–37, 339, 346
   as obsolete, 100–101, 105–6,
      159–63, 292
   orbiter of, 83, 100, 108, 121, 156,
      157, 160–64, 168, 243–44, 245
   o-rings of, 25, 161
   passengers for, 123, 172–73,
      176–77, 209–10, 212–13
   payloads of, 94, 97, 121–25, 159,
      168, 170, 171, 176, 201, 239,
      250
   political considerations for, 32, 86,
      98, 148–49, 158, 162–63, 166,
      176–77, 178, 212–13
   public relations campaigns for, 157,
      164, 171–73, 177–78
   reusability of, 96–98, 106, 107, 124,
      140, 167, 168, 169, 170, 171, 172,
      239, 253
   rocket boosters of, 156, 157, 168,
      294
   safety of, 57, 157–60, 161, 162–63,
      164, 168, 245–47, 340–41
   satellites released by, 27, 94, 160,
      176
   schedule for, 15, 25–28, 30, 85, 98,
      179–80, 212, 226, 231
   technology of, 157–64, 221, 265
   thermal tiles of, 156, 160, 162,
      243–44, 337
   see also specific shuttle orbiters
Shuttleworth, Mark, 206, 210, 343

"Significance of the Frontier in Ameri-
   can History, The" (Turner),
   151–52
Single Stage to Orbit (SSTO) program,
   103, 105, 107
Skylab space station, 21, 24, 27,
   28–30, 33, 34, 61, 65, 71, 84, 140,
   227, 281, 282
Smith, Albert, 350
Smith, Marcia, 29, 221
Smith, Peter, 325–26, 327, 328–29, 334
Smithsonian Astrophysical Observatory, 120
Snook, Kelly, 317
Solaris (Lem), 60
solar power, 69, 70, 72, 76, 77, 89, 256
Songs of Distant Earth (Clarke), 73
Soyuz spacecraft, 34, 44, 48, 52–53,
   61, 98, 134, 136, 172, 182, 184,
   188, 190, 199, 200, 203–5, 210,
   228, 234, 325, 342
space:
   access to, 91–92, 95–96, 98, 101,
      106–7, 160, 171, 223–25, 346–47
   colonization of, 68–78, 79, 81, 83,
      89, 119, 167, 217–18, 225,
      251–52, 270–73
   commercialization of, 32, 80–87,
      209–17, 261
   Federovian concept of, 180–82
   as frontier, 59–60, 64, 65, 73–74,
      75, 151–52, 164, 180, 238, 251,
      252, 261, 269, 273–74, 278, 288,
      291, 298–99
   government control of, 62–63, 64,
      74–75, 79–81
   international treaties on, 62, 72, 81,
      251
   military use of, 63, 64, 81, 347–51
   O'Neillian concept of, 63, 67–81,
      82–83, 88–90, 101, 103, 110, 117,
      119, 167, 217, 225, 250, 252, 269,
      272, 332–33, 352
   Saganian concept of, 65–67, 72–73,
      81

spiritual aspect of, 58, 60–61
von Braunian concept of, 63–65, 66,
    67, 78, 89, 167, 213, 250, 265–66,
    269, 272, 283, 291
Space: 1999, 224
Space Adventures, 110, 240
Space Age, 5–19, 23–24, 25, 56–57,
    98, 197, 269, 296–97, 341
Space Age Management (Webb), 155
Spacecause, 145, 264
SPACE.com, 47
Space Consulting Resources Group, 91
spacecraft, reusable, 91–114
    costs of, 95–96, 97, 103, 106, 108–9,
        126, 128, 131, 137, 220–21
    design of, 125–28, 133–34
    engines of, 104, 125, 126, 133, 269
    escape velocity for, 91–92, 95–96,
        99
    external shell of, 107, 112, 133
    government contracts for, 99–103,
        108–9
    landing apparatus of, 104, 107, 109,
        125–26
    maintenance of, 107–8
    military use of, 102–3, 105
    NASA proposals for, 125–28,
        134–36
    payloads of, 91, 94, 95, 106
    private development of, 17–18,
        95–96, 101–14, 115–19, 125–27,
        133–34
    reusability of, 96–114, 116, 121,
        125–28, 133–34, 253, 344,
        350–51
    satellites launched by, 128–32, 133
    suborbital, 92–93, 109–14, 128
    technology for, 103–4, 110, 112–13,
        125
Space Exploration Initiative (SEI),
    290–95
Space Flight Participant Program,
    176–80
Space Frontier Foundation, 36, 38, 41,
    89, 142–43, 146, 230, 345

Spacehab, 121–25, 234, 346
Space Imaging, 222
Space Industries, 83, 84–88
Spacelab, 124
Space Launch Corp., 350
Space Launch Initiative (SLI), 127–28,
    133–34, 135, 143, 209, 223,
    257–58, 265, 266, 267
Space Media, Inc., 234
Space News, 217, 257, 262, 266
Space Operations Officers Qualifica-
    tion Course, 349
Spaceport Equality Act, 256
spaceports, 78, 256
space program, U.S.:
    author's interest in, 5–19
    aviation industry compared with,
        17–18, 35–36, 61–62, 96, 97,
        110–13, 150–51, 164, 232–33,
        346–47
    "brother-in-law problem" in, 124,
        127, 136, 175, 219, 235–36,
        323
    civilian participation in, 68, 98,
        123, 167, 172–213, 286, 343–44
    Cold War and, 5, 9, 13, 26, 27, 30,
        33, 76, 94, 152, 183–84, 194, 196,
        275, 337–38, 351
    economics of, 9, 15–18, 30–53,
        93–98, 99, 147, 160–61, 163,
        169–71, 210–11, 214, 290–91,
        338–39
    established interests in, 57–58,
        79–80, 142–46
    goals of, 11–19, 25–26, 32, 54,
        55–57, 60–61, 70, 78–79, 98, 140,
        290–93, 296
    human exploration in, 8–19, 25–26,
        59–61, 65, 67, 89–90, 119, 143,
        149–50, 167–68, 172–73, 210,
        212–13, 222–25, 261, 263–64,
        267–73, 339–42
    Kennedy's mandate for, 55, 56, 137,
        138, 152–55, 169, 174, 178, 278,
        281, 291

space program, U.S. *(continued)*:
NASA monopoly on, 7, 14–15, 16,
31–32, 55–58, 61–63, 75–80,
88–90, 99–100, 113–14, 152–55,
169–70, 172–73, 209–10, 212,
225, 239, 261–62, 292
political considerations in, 52, 55,
56, 74–75, 95, 137, 138, 152–55,
169, 174, 178, 232, 238–39,
247–50, 278, 281, 290–91
post-Apollo era of, 13–14, 60–63,
89–90, 139–40, 144, 165–71, 185,
268, 279, 281–83, 337
privatization of, 15–18, 22, 30–34,
61–63, 73, 74–75, 80–88, 91,
95–96, 113–14, 116–18, 121–22,
124, 136–37, 169, 191, 209–39,
267, 345, 350
public interest in, 5–19, 45–46,
56–57, 64, 76–77, 83, 105–6, 119,
147, 157, 167, 171–73, 177–78,
215–16, 266, 345–46, 353–54
research and development in, 24,
35, 42–43, 53, 61–63, 69, 73,
80–88, 102–4, 145, 233, 257
robotic exploration in, 8–9, 11, 12,
67, 88, 149, 150, 342
taxpayer funding of, 97, 99, 108,
133, 147, 161, 214, 338–39
timeline for, 14–15, 55–56, 137–39
tourism and, 15–18, 44, 49, 50, 62,
110–14, 206–13, 223–25, 234,
240, 344
two-part promise of, 13–16, 18, 25,
56–57, 98
U.S.-Soviet Space Race and, 9, 13,
29–30, 33–34, 42, 47, 55–56,
60–61, 62, 64, 80, 94, 95, 138,
150–55, 166, 178, 184, 191, 197,
198, 202, 274, 278, 337–38, 342,
348
see also National Aeronautics and
Space Administration (NASA);
Russian space program
SpaceRef.com, 214, 333

Space Resources and Space Settlements,
76
SpaceShipOne rocket plane, 112
space shuttle, *see* shuttle, space
Space Station Freedom, 27–28, 29, 65,
83–84, 85, 240–41, 247, 295
space stations:
commercial possibilities of, 82, 144,
228–35
design of, 14, 240–41, 246, 277
Russian, *see* Mir space station
unmanned, 9–10, 80–88, 144, 211,
228
see also International Space Station
(ISS)
Space Studies Institute, 88, 89, 345
Space Systems, 350
Space Tourism Society, 187
Space Transportation Association, 223
SpaceX, 114, 333
Sputnik satellite, 23–24, 33, 93, 94,
120, 150, 196, 277–78, 285
Sputnik Two satellite, 285
SS-6 missile, 93–94, 194, 196
Stadd, Courtney, 217–20, 230, 232,
236–37
Star City, 180, 190, 206
Starsem, 199, 200
Starstruck, 218–19
Star Trek, 6, 59, 75, 76, 79, 276, 288,
332, 347–48, 351–52
Star Wars, 256, 349
Stavros, William, 185, 187
Steinbrenner, George, 45
StelSys, 231–32
Stevens, Tom, 185, 192
Stockholm International Peace
Research Institute, 349, 350
Stoker, Carol, 289, 326, 327, 329, 334
Stranger in a Strange Land (Heinlein),
60, 73
Strategic Defense Initiative (SDI), 103,
105
Stuttaford, Andrew, 192
Syromiatnikov, Vladimir, 278

Takei, George, 351
Tam, Dan, 191
Teacher in Space Program, 176, 178–80, 343–44
Teal Group, 129, 132
Team Encounter, 175, 235
telecommunications industry, 128–32, 171
Teledesic, 129–30, 132
terraforming, 288–89, 297, 298
Thagard, Norman, 47
This New Ocean (Burrows), 337
Thornton, Billy Bob, 46
Titan rockets, 91, 95, 138, 292
Tito, Dennis, 44, 49, 50, 110, 112, 143–44, 180–93, 200–211, 233, 234, 240, 283, 343, 345
Tito, Suzanne, 185, 186
Toffler, Alvin, 119, 252
Toffler, Heidi, 252
Transportation Department, U.S., 85, 218, 225, 256, 346–47
Truly, Richard, 165, 178–79, 290, 292–93, 295
Tsibilev, Vasily, 37
Tsiolkovsky, Konstantin, 70, 115, 137–38, 180, 181, 353
Tumlinson, Rick, 34–37, 38, 39, 41–43, 48, 53, 57, 63, 73, 79–80, 89, 109, 127, 142–43, 144, 187–88, 207, 240, 345, 346, 352
Turner, Frederick Jackson, 151–52, 164, 180, 252, 273–74, 278, 298–99
2001: A Space Odyssey, 15, 24, 80, 99

U-2 spy planes, 195–96
Udall, Morris, 72
United Aircraft and Transport, 233
United Nations, 68, 72
United Space Alliance (USA), 99–100, 216, 220–21, 244, 254
Universal Space Lines, 130
Usachev, Yuri, 193

U.S. Space Transportation Policy (1994), 107

V-2 rockets, 104, 111, 118, 151, 276, 278
Van Allen belts, 280
Vandenberg Air Force Base, 102–3, 256
VentureStar space vehicle, 99, 109
Venus, 59, 67, 69, 275, 280
Verne, Jules, 59, 73, 325
Viking probes, 287–88, 293, 309
"Viking Results—the Case for Man on Mars, The" (Clark), 287–88
von Braun, Wernher, 63–65, 66, 67, 78, 82, 90, 93, 104, 118, 138, 151, 166, 167, 168, 213, 250, 265–66, 269, 272, 276–78, 279, 280, 281–83, 288, 291, 294, 296
Voodoo Science (Park), 228
Voyager aircraft, 112
Voyager space probes, 8

Wagner, Richard, 296
Walker, Joseph, 92–93, 111
Walker, Robert, 252
Wall Street Journal, 251–52
Wang, Robert, 128
Webb, James, 86, 137, 153–55, 281, 282
Weiler, Ed, 304
Wells, H. G., 73
When Worlds Collide, 275
"Which Agencies Inform the Public?" survey, 216
White, Edward, 54–55
White Knight aircraft, 112
White Sands Missile Range, 104
Whitten, Jamie, 86
Williams, Robert, 236
Williamson, Jack, 288
Wilshire Associates, 185–87, 210
Wilson, Robert Anton, 43
Wolfe, Tom, 33
Worden, Simon "Pete," 110

World, the Flesh and the Devil, The (Bernal), 70–71
World Space Congress, 267–68, 334
Wright, Wilbur and Orville, 17

X-1 aircraft, 92, 118
X-15 rocket plane, 92–93, 111, 112
X-33 spacecraft, 98–101, 102, 107, 108–9, 113, 117, 127, 146, 253, 257, 340, 351
X-38 escape vehicle, 48, 228, 266
X-2000 spacecraft, 106–7, 108

X Prize, 110–14, 235, 240, 344, 345, 352

Yeager, Chuck, 92, 118
Yuri Gagarin Cosmonaut Training Center, 180, 181–85

Zaletin, Sergei, 44
Zarya module, 51–52
Zero Gravity Corp., 235–36, 237, 240, 352
Zubrin, Robert, 270–73, 283, 284–300, 301, 311, 320–21, 323–24, 326, 332–33, 334, 352
Zvezda module, 51, 233–34